MATLAB® FOR CONTROL ENGINEERS

KATSUHIKO OGATA

Upper Saddle River, New Jersey 07458

Library of Congress Cataloging-in-Publication Data on File

Vice President and Editorial Director, ECS: *Marcia J. Horton*
Executive Editor: *Michael McDonald*
Associate Editor: *Alice Dworkin*
Editorial Assistant: *William Opaluch*
Managing Editor: *Scott Disanno*
Production Editor: *Rose Kernan*
Director of Creative Services: *Paul Belfanti*
Creative Director: *Paul Belfanti*
Cover Designer: *Bruce Kenselaar*
Managing Editor, AV Management and Production: *Patricia Burns*
Art Editor: *Gregory Dulles*
Manufacturing Manager, ESM: *Alexis Heydt-Long*
Manufacturing Buyer: *Lisa McDowell*
Marketing Manager: *Tim Galligar*

 © 2008 Pearson Education, Inc.
Pearson Prentice Hall
Pearson Education, Inc.
Upper Saddle River, NJ 07458

All rights reserved. No part of this book may be reproduced in any
form or by any means, without permission in writing from the publisher.

Pearson Prentice Hall™ is a trademark of Pearson Education, Inc.
All other trademarks or product names are the property of their respective owners.

MATLAB is a registered trademark of The MathWorks, Inc.,
The MathWorks, Inc., 3 Apple Hill Drive, Natick, MA 01760-2098
WWW: http://www.mathworks.com

The author and publisher of this book have used their best efforts in preparing this book. These
efforts include the development, research, and testing of the theories and programs to determine
their effectiveness. The author and publisher make no warranty of any kind, expressed or implied,
with regard to these programs or the documentation contained in this book. The author and
publisher shall not be liable in any event for incidental or consequential damages in connection
with, or arising out of, the furnishing, performance, or use of these programs.

Printed in the United States of America

10 9 8 7 6 5 4 3 2 1

ISBN: 0-13-615077-2
 978-0-13-615077-0

Pearson Education Ltd., *London*
Pearson Education Australia Pty. Ltd., *Sydney*
Pearson Education Singapore, Pte. Ltd.
Pearson Education North Asia Ltd., *Hong Kong*
Pearson Education Canada, Inc., *Toronto*
Pearson Educación de Mexico, S.A. de C.V.
Pearson Education—Japan, *Tokyo*
Pearson Education Malaysia, Pte. Ltd.
Pearson Education, Inc., *Upper Saddle River,
New Jersey*

Contents

Preface vii

Chapter 1 Introduction to MATLAB 1

1–1 Introduction   1
1–2 Addition, Subtraction, Multiplication, and Division with MATLAB   19
1–3 Computing Matrix Functions   26
1–4 Plotting Response Curves   34
1–5 Three-Dimensional Plots   42
1–6 Drawing Geometrical Figures with MATLAB   46

Chapter 2 Preliminary Study of MATLAB Analysis of Dynamic Systems 55

2–1 Partial-Fraction Expansion with MATLAB   55
2–2 Transformation of Mathematical Models of Dynamic Systems   66
2–3 MATLAB Representation of Systems in Block Diagram Form   73

Chapter 3 Transient-Response Analysis   85

3–1   Introduction   85
3–2   Step Response   85
3–3   Impulse Response   118
3–4   Ramp Response   122
3–5   Response to Arbitrary Input   128
3–6   Response to Arbitrary Initial Condition   136
3–7   Three-Dimensional Plots   144

Chapter 4 Root-Locus Analysis   151

4–1   Introduction   151
4–2   Root Locus Plots with Polar Grids   165
4–3   Finding the Gain Value K at an Arbitrary Point on the Root Locus   169
4–4   Root-Locus Plots of Non–Minimum-Phase Systems   173
4–5   Root-Locus Plots of Conditionally Stable Systems   176
4–6   Root Loci for Systems with Transport Lag   180
4–7   Root-Locus Approach to Control Systems Compensation   186

Chapter 5 Frequency-Response Analysis   221

5–1   Plotting Bode Diagrams with MATLAB   221
5–2   Plotting Nyquist Diagrams with MATLAB   234
5–3   Log-Magnitude-Versus-Phase Plots   250
5–4   Phase Margin and Gain Margin   262
5–5   Frequency-Response Approach to Control Systems Compensation   271

Chapter 6 MATLAB Approach to the State-Space Design of Control Systems   305

6–1   Introduction   305
6–2   Controllability and Observability   305
6–3   Pole Placement   313
6–4   Solving Pole-Placement Problems with MATLAB   321
6–5   Design of State Observers with MATLAB   325
6–6   Minimum-Order Observers   335
6–7   Observer Controllers   349

Chapter 7 Some Optimization Problems Solved with MATLAB   375

7–1 Computational Approach to Obtaining Optimal Sets of Parameter Values   375

7–2 Solving Quadratic Optimal Control Problems with MATLAB   395

Appendix   409

References   425

Index   427

Preface

This book is written to assist those students and practicing engineers who wish to study MATLAB to solve control engineering problems. It is written at the level of the senior engineering student.

The book is organized into seven chapters. Chapter 1 presents an introduction to MATLAB. Chapter 2 deals with preliminary materials that the reader must know prior to applying MATLAB to the analysis and design of control systems. Chapter 3 is a detailed discussion of how to apply MATLAB to obtain transient response outputs of dynamic systems to time-domain inputs. Chapter 4 treats root-locus analysis and design with MATLAB. Detailed frequency-response analysis and design with MATLAB are given in Chapter 5. Chapter 6 discusses state-space design problems, such as pole placement and state observers, solved with MATLAB. Finally, Chapter 7 presents a computational approach to obtaining optimal sets of parameter values in connection with control systems design. The book concludes with a discussion of MATLAB's approach to solving quadratic optimal control problems.

The book includes some of the MATLAB materials presented in my previous publications *Modern Control Engineering* (4th ed.) and *System Dynamics* (4th ed.).

All sample problems discussed in the book are given detailed explanations so that the reader will acquire a good understanding of MATLAB's approach to solving the analysis and design problems presented.

The book assumes that the reader has a relatively new version of MATLAB in his or her computer. In plotting root-locus diagrams or Nyquist diagrams with MATLAB, the reader's grid command may produce grid lines or curves different from those presented here; it all depends on the version of MATLAB. (For problems that may arise with regard to grid lines or curves, see the appendix.)

It is hoped that the reader will find this book useful in applying MATLAB to solve many control engineering problems.

KATSUHIKO OGATA

Introduction to MATLAB

1-1 INTRODUCTION

MATLAB® is a matrix-based system for performing mathematical and engineering calculations. We may think of MATLAB as a language of technical computing. All variables handled in MATLAB are matrices. That is, MATLAB has only one data type: a matrix, or rectangular array, of numbers. MATLAB has an extensive set of routines for obtaining graphical outputs.

This section presents background material necessary for the effective use of MATLAB in solving control engineering problems. First, we introduce MATLAB commands and mathematical functions. Then we present matrix operators, relational and logical operators, and special characters used in MATLAB. Finally, we introduce the semicolon operator, MATLAB ways to enter vectors and matrices into the computer, the colon operator, and other important material that we must become familiar with before writing MATLAB programs to solve control engineering problems.

MATLAB is used with a variety of toolboxes. (A toolbox is a collection of special files called M-files.) For control systems analysis and design, MATLAB is used with the control system toolbox. When we refer to MATLAB in this book, we include the basic programs of MATLAB and the control system toolbox.

MATLAB is basically command driven. Therefore, the user must know the commands that are used in solving computational problems. Table 1–1 lists various types of MATLAB commands and predefined functions that are frequently utilized in solving control engineering problems.

Table 1–1 MATLAB Commands and Matrix Functions

Commands and Matrix Functions Commonly Used in Solving Control Engineering Problems	Explanations of What Commands Do and Matrix Functions Mean
abs	Absolute value, complex magnitude.
acker	Compute a state-feedback gain matrix for pole placement, using Ackermann's formula.
angle	Phase angle.
ans	Answer when expression is not assigned.
atan	Arctangent.
axis	Manual axis scaling.
bode	Plot Bode diagram.
clear	Clear workspace.
clf	Clear current figure.
computer	Type of computer.
conj	Complex conjugate.
conv	Convolution, multiplication.
corrcoef	Correlation coefficients.
cos	Cosine.
cosh	Hyperbolic cosine.
cov	Covariance.
ctrb	Compute the controllability matrix.
c2d	Conversion of continuous-time models to discrete-time models.
deconv	Deconvolution, division.
det	Determinant.
diag	Diagonal matrix.
eig	Eigenvalues and eigenvectors.
end	Terminate scope of for, while, switch, try, and if statements.
exit	Terminate program.
exp	Exponential base e.
expm	Matrix exponential.
eye	Identity matrix.
feedback	Feedback connection of two LTI models.
filter	Direct filter implementation.
for	Repeat statement(s) a specified number of times.
format long	Fifteen-digit scaled fixed point. (Example: 1.33333333333333)
format long e	Fifteen-digit floating point. (Example: 1.33333333333333e + 000)
format short	Five-digit scaled fixed point. (Example: 1.3333)

Table 1–1 (*continued*)

Commands and Matrix Functions Commonly Used in Solving Control Engineering Problems	Explanations of What Commands Do and Matrix Functions Mean
format short e	Five-digit floating point. (Example: 1.3333e + 000)
freqs	Laplace transform frequency response.
freqz	z-Transform frequency response.
gram	Controllability and observability gramians.
grid	Toggles the major lines of the current axes.
grid off	Removes major and minor grid lines from the current axes.
grid on	Adds major grid lines to the current axes.
help	Lists all primary help topics.
hold	Toggles the hold state.
hold off	Returns to the default mode whereby plot commands erase the previous plots and reset all axis properties before drawing new plots.
hold on	Holds the current plot and all axis properties so that subsequent graphing commands add to the existing graph.
i	$\sqrt{-1}$
imag	Imaginary part.
impulse	Impulse response of LTI models.
inf	Infinity (∞)
inv	Inverse
j	$\sqrt{-1}$
legend	Graph legend.
length	Length of vector.
linspace	Linearly spaced vector.
load	Load workspace variables from disk.
log	Natural logarithm.
loglog	Loglog x–y plot.
logm	Matrix logarithm.
logspace	Logarithmically spaced vector.
log10	Log base 10.
lqe	Linear quadratic estimator design.
lqr	Linear quadratic regulator design.
lsim	Simulate time response of LTI models to arbitrary inputs.
lyap	Solve continuous-time Lyapunov equations.

Table 1-1 (*continued*)

Commands and Matrix Functions Commonly Used in Solving Control Engineering Problems	Explanations of What Commands Do and Matrix Functions Mean
margin	Gain and phase margins and crossover frequencies.
max	Maximum value.
mean	Mean value.
median	Median value.
mesh	Three-dimensional mesh surface.
min	Minimum value.
minreal	Minimal realization and pole–zero cancellation.
NaN	Not a number.
ngrid	Generate grid lines for a Nichols plot.
nichols	Draw the Nichols plot of the LTI model.
num2str	Convert number to string.
nyquist	Plot Nyquist frequency response.
obsv	Compute the observability matrix.
ode45	Solve nonstiff differential equations, medium-order method. (If time constants involved do not vary by several orders of magnitude or more, the differential equations are called nonstiff differential equations.)
ode23	Solve nonstiff differential equations, low-order method.
ode113	Solve nonstiff differential equations, variable-order method.
ones	Constant.
ord2	Generate continuous-time second-order system.
pade	Pade approximation of time delays.
parallel	Parallel interconnection of two LTI models.
pi	Pi(π).
place	Compute a state-feedback gain matrix for pole placement.
plot	Linear *x–y* plot.
polar	Polar plot.
pole	Compute the poles of LTI models.
poly	Convert roots to polynomial.
polyfit	Polynomial curve fitting.
polyval	Polynomial evaluation.
polyvalm	Matrix polynomial evaluation.
printsys	Print system in pretty format.
prod	Product of elements.
pzmap	Pole–zero map of LTI models.

Table 1–1 (*continued*)

Commands and Matrix Functions Commonly Used in Solving Control Engineering Problems	Explanations of What Commands Do and Matrix Functions Mean
quit	Terminate program
rand	Generate random numbers and matrices.
rank	Calculate the rank of a matrix.
real	Real part.
rem	Remainder after division.
residue	Partial-fraction expansion.
rlocfind	Find root-locus gains for a given set of roots.
rlocus	Plot root loci.
rmodel	Generate random stable continuous-time nth-order test models.
roots	Polynomial roots.
semilogx	Semilog x–y plot (x-axis logarithmic).
semilogy	Semilog x–y plot (y-axis logarithmic).
series	Interconnect two LTI models in series.
shg	Show graph window.
sign	Signum function.
sin	Sine.
sinh	Hyperbolic sine.
size	Size of matrix.
sqrt	Square root.
sqrtm	Matrix square root.
ss	Create state-space model or convert LTI model to state-space model.
ss2tf	Convert state-space model to transfer-function model.
std	Standard deviation.
step	Plot unit-step response.
subplot	Create axes in tiled positions.
sum	Sum of elements.
switch	Switch among several cases, based on expression.
tan	Tangent.
tanh	Hyperbolic tangent.
text	Arbitrarily positioned text.
tf	Create transfer-function model or convert LTI model to transfer-function model.
tf2ss	Convert transfer-function model to state-space model.
tf2zp	Convert transfer-function model to zero–pole model.
title	Plot title.
trace	Trace of a matrix.

Table 1–1 (*continued*)

Commands and Matrix Functions Commonly Used in Solving Control Engineering Problems	Explanations of What Commands Do and Matrix Functions Mean
who whos	Lists all the variables currently in memory. List all the variables in the current workspace, together with information about their size, bytes, class, etc.
xlabel	x-axis label.
ylabel	y-axis label.
zero zeros zlabel zpk zp2tf	Transmission zeros of LTI systems. Zeros array. z-axis label Create zero–pole–gain models or convert to zero–pole–gain format. Convert zero–pole model to transfer-function model.

Accessing and exiting MATLAB. On most systems, once MATLAB has been installed, execute the command MATLAB to invoke MATLAB. To exit MATLAB, execute the command exit or quit.

MATLAB has an online help facility that may be invoked whenever the need arises. The command help will display a list of predefined functions and operators for which online help is available. The command

help 'function name'

will give information on the purpose and use of the specific function named. The command

help help

will give information on how to use the online help.

Matrix operators. The following notation is used in matrix operations (if multiple operations are involved, the order of the arithmetic operations can be altered with the use of parentheses):

+	Addition
−	Subtraction
*	Multiplication
^	Power
'	Conjugate transpose
/ or \	Matrix division
./ or .\	Array division

(We shall not discuss matrix division, such as **A/B** or **A\B**, in this book. Scalar division (e.g., **A**/c or c**A**) and array division (e.g., **A.**/**B** or **A.\B**, where **A** and **B** may be vectors) are presented.

Relational and logical operators. The following relational operators are used in MATLAB:

<	Less than
<=	Less than or equal to
>	Greater than
>=	Greater than or equal to
==	Equal
~=	Not equal

The first four operators compare the real parts only. The last two (==, ~=) compare both the real and imaginary parts. Note that "=" is used in an assignment statement while "==" is used in a relation.

The logical operators are

&	AND
|	OR
~	NOT

Special characters. The following special characters are used in MATLAB:

[ ]	Used to form vectors and matrices
( )	Arithmetic expression precedence
,	Separate subscripts and function arguments
;	End rows, suppress printing
:	Subscripting, vector generation
!	Execute operating system command
%	Comment

Use of semicolon. The semicolon is used to suppress printing. If the last character of a statement is a semicolon, printing is suppressed; the command is still executed, but the result is not displayed. This is a useful feature, since one may not need to print intermediate results. Also, in entering a matrix, a semicolon is used to indicate the end of a row, except for the last row.

Use of colon. The colon plays an important role in MATLAB, being involved in creating vectors, subscripting matrices, and specifying iterations. For example, the statement

$$t = 1:5$$

generates a row vector containing the numbers from 1 to 5 with unit increment — that is,

$$t = $$
$$1 \quad 2 \quad 3 \quad 4 \quad 5$$

An increment other than unity can be used. For example,

$$t = 1:0.5:3$$

will result in

t =
1.0000 1.5000 2.0000 2.5000 3.0000

Negative increments also may be used. For example, the statement

$$t = 5:-1:2$$

gives

t =
5 4 3 2

Other MATLAB commands that generate sequential data, such as linspace and logspace, are presented later in this section.

The colon is frequently used to subscript matrices. A(:,j) is the jth column of **A** and A(i,:) is the ith row of **A**. For example, if matrix **A** is given by

$$\mathbf{A} = \begin{bmatrix} 1 & 2 & 3 \\ 4 & 5 & 6 \\ 7 & 8 & 9 \end{bmatrix}$$

then A(:,3) gives the third element in all of the rows (i.e., the third column), as follows:

3
6
9

A(2,:) gives the second row of **A**, namely,

4 5 6

A(:) returns a long column vector consisting of elements of the first column, second column, and third column:

1
4
7
2
5
8
3
6
9

Note that A(i,j) denotes the entry in the ith row, jth column, of matrix **A**. For example, A(2,3) is 6.

An individual vector can be referenced with indexes inside parentheses. For example, if a vector **x** is given by

$$x = [2 \quad 4 \quad 6 \quad 8 \quad 10]$$

then x(3) is the third element of **x** and x([1 2 3]) gives the first three elements of **x** (that is, 2, 4, 6).

Entering vectors into MATLAB programs. In entering vectors and matrices into MATLAB programs, no dimension statements or type statements are needed. Vectors, which are $1 \times n$ or $n \times 1$ matrices, are used to hold ordinary one-dimensional sampled data signals, or sequences. One way to introduce a sequence into MATLAB is to enter it as an explicit list of elements separated by blank spaces or commas, as in

$$x = [1 \quad 2 \quad 3 \quad -4 \quad -5]$$

or

$$x = [1, 2, 3, -4, -5]$$

For readability, it is better to provide spaces between the elements. The values must always be entered within brackets.

The statement

$$x = [1 \quad 2 \quad 3 \quad -4 \quad -5]$$

creates a simple five-element real sequence in a row vector. The sequence can be turned into a column vector by transposition. That is,

$$y = x'$$

results in

$$y = \begin{matrix} 1 \\ 2 \\ 3 \\ -4 \\ -5 \end{matrix}$$

How to enter matrices into MATLAB programs. A matrix

$$\mathbf{A} = \begin{bmatrix} 1.2 & 10 & 15 \\ 3 & 5.5 & 2 \\ 4 & 6.8 & 7 \end{bmatrix}$$

may be entered into a MATLAB program by a row vector as follows:

$$A = [1.2 \quad 10 \quad 15; 3 \quad 5.5 \quad 2; 4 \quad 6.8 \quad 7]$$

Again, the values must be entered within brackets. As with vectors, the elements of any row must be separated by blanks (or by commas). The end of each row, except the last, is indicated by a semicolon.

Note that the elements of the matrix **A** are automatically displayed after the statement is executed following the carriage return:

$$A =$$
$$\begin{array}{ccc} 1.2000 & 10.0000 & 15.0000 \\ 3.0000 & 5.5000 & 2.0000 \\ 4.0000 & 6.8000 & 7.0000 \end{array}$$

If we add a semicolon at the end of the matrix statement such that

$$A = [1.2 \quad 10 \quad 15; 3 \quad 5.5 \quad 2; 4 \quad 6.8 \quad 7];$$

the output is suppressed and no output will be seen on the screen.

A large matrix may be spread across several input lines. For example, consider the matrix

$$\mathbf{B} = \begin{bmatrix} 1.5630 & 2.4572 & 3.1113 & 4.1051 \\ 3.2211 & 1.0000 & 2.5000 & 3.2501 \\ 1.0000 & 2.0000 & 0.6667 & 0.0555 \\ 0.2345 & 0.9090 & 1.0000 & 0.3333 \end{bmatrix}$$

This matrix may be spread across four input lines as follows:

$$\begin{array}{rcccc} B = [1.5630 & 2.4572 & 3.1113 & 4.1051 \\ 3.2211 & 1.0000 & 2.5000 & 3.2501 \\ 1.0000 & 2.0000 & 0.6667 & 0.0555 \\ 0.2345 & 0.9090 & 1.0000 & 0.3333] \end{array}$$

Note that carriage returns replace the semicolons.

As another example, a matrix

$$\mathbf{C} = \begin{bmatrix} 1 & e^{-0.02} \\ \sqrt{2} & 3 \end{bmatrix}$$

may be entered as

$$C = [1 \quad \exp(-0.02); \text{sqrt}(2) \quad 3]$$

After the carriage return, the following matrix will be seen on the screen:

$$C =$$
$$\begin{array}{cc} 1.0000 & 0.9802 \\ 1.4142 & 3.0000 \end{array}$$

Generating vectors. Besides the colon operator, the linspace and logspace commands generate sequential data:

$$x = \text{linspace}(n1, n2, n)$$
$$w = \text{logspace}(d1, d2, n)$$

The linspace command generates a vector from n1 to n2 with n points (including both endpoints):

```
>> x = linspace (-10, 10, 5)
x =
    -10  -5  0  5  10
```

The logspace command generates a logarithmically spaced vector from 10^{d1} to 10^{d2} with n points (again, including both endpoints). Frequently n is chosen to be 50 or 100, but it can be any number. For example,

$$w = \text{logspace}(-1, 1, 10)$$

generates 10 points from 0.1 to 10 (note that the 10 points include both endpoints):

```
>> w = logspace (-1,1,10)
w =
  Columns 1 through 8
    0.1000  0.1668  0.2783  0.4642  0.7743  1.2915  2.1544  3.5938
  Columns 9 through 10
    5.9948  10.0000
```

Transpose and conjugate transpose. The apostrophe or prime denotes the conjugate transpose of a matrix. If the matrix is real, the conjugate transpose is simply the transpose. An entry such as

$$A = [1 \quad 2 \quad 3; 4 \quad 5 \quad 6; 7 \quad 8 \quad 9]$$

will produce the following matrix on the screen:

$$A =$$
$$\begin{matrix} 1 & 2 & 3 \\ 4 & 5 & 6 \\ 7 & 8 & 9 \end{matrix}$$

Also, if

$$B = A'$$

is entered, then

$$B = \begin{matrix} 1 & 4 & 7 \\ 2 & 5 & 8 \\ 3 & 6 & 9 \end{matrix}$$

appears on the screen.

Entering complex numbers. Complex numbers may be entered with the function i or j. For example, a number $1 + j\sqrt{3}$ may be entered as

$$x = 1+\text{sqrt}(3)*i$$

or

$$x = 1+\text{sqrt}(3)*j$$

This complex number, $1 + j\sqrt{3} = 2 \exp[(\pi/3)j]$, may also be entered as

$$x = 2*\exp((pi/3)*j)$$

It is important to note that, when complex numbers are entered as matrix elements within brackets, we avoid blank spaces. For example, $1 + j5$ should be entered as

$$x = 1+5*j$$

If spaces are provided around the $+$ sign, as in

$$x = 1 + 5*j$$

the representation is of two separate numbers.

If i and j are used as variables, a new complex unit may be generated as follows:

$$ii = \text{sqrt}(-1)$$

or

$$jj = \text{sqrt}(-1)$$

Then $-1 + j\sqrt{3}$ may be entered as

$$x = -1+\text{sqrt}(3)*ii$$

or

$$x = -1+\text{sqrt}(3)*jj$$

If we defined ii $= \sqrt{-1}$ and want to change ii to the predefined i $= \sqrt{-1}$, enter clear ii in the computer. Then the predefined variable i can be reset.

Entering complex matrices. For the complex matrix

$$\mathbf{X} = \begin{bmatrix} 1 & j \\ -j5 & 2 \end{bmatrix}$$

an entry such as

$$X = [1 \quad j; -j*5 \quad 2]$$

will produce the following matrix on the screen:

```
X =
   1.0000                0 + 1.0000i
   0 - 5.0000i           2.0000
```

Note that

$$Y = X'$$

will yield

```
Y =
   1.0000                0 + 5.0000i
   0 - 1.0000i           2.0000
```

which is

$$\mathbf{Y} = \begin{bmatrix} 1 & j5 \\ -j & 2 \end{bmatrix}$$

Since the prime denotes the complex conjugate transpose, use one of the following two entries for an unconjugated transpose:

$$Y.' \quad \text{or} \quad \text{conj}(Y')$$

If we enter

$$Y.'$$

then the screen shows

```
ans =
   1.0000                0 - 1.0000i
   0 + 5.0000i           2.0000
```

Also, if

$$\text{conj}(Y')$$

is entered, then the screen shows

```
ans =
   1.0000                0 - 1.0000i
   0 + 5.0000i           2.0000
```

Entering a long statement that will not fit on one line. A statement is normally terminated with a carriage return or the enter key. If the statement being entered is too long for one line, an ellipsis consisting of three or more periods (...), followed by the carriage return, can be used to indicate that the statement continues on the next line. An example is

$$x = 1.234 + 2.345 + 3.456 + 4.567 + 5.678 + 6.789 \ldots$$
$$+ 7.890 + 8.901 - 9.012;$$

Note that the blank spaces around the =, +, and − signs are optional. Such spaces are often provided to improve readability.

Entering several statements on one line. Several statements may be placed on one line if they are separated by commas or semicolons. Examples are

$$x1 = [1 \quad 2 \quad 3], x2 = [4 \quad 5 \quad 6], x3 = [7 \quad 8 \quad 9]$$

and

$$x1 = [1 \quad 2 \quad 3]; x2 = [4 \quad 5 \quad 6]; x3 = [7 \quad 8 \quad 9]$$

Selecting the output format. All computations in MATLAB are performed in double precision. However, the displayed output may have a fixed point with four decimal places. For example, for the vector

$$\mathbf{x} = [1/3 \quad 0.00002]$$

MATLAB exhibits the following output:

$$x =$$
$$0.3333 \quad 0.0000$$

If at least one element of a matrix is not an exact integer, there are four possible output formats. The displayed output can be controlled with the use of the following commands:

> format short
> format long
> format short e
> format long e

Once invoked, the chosen format remains in effect until changed.

For systems analysis, format short and format long are commonly used. Whenever MATLAB is invoked and no format command is entered, MATLAB shows the numerical results in format short as follows:

```
>> x = [1/3   0.00002];
>> x

x =

    0.3333    0.0000
```

For x = [1/3   0.00002], the commands format short; x and format long; x yield the following output:

```
>> x = [1/3   0.00002];
>> format short; x

x =

    0.3333    0.0000

>> format long; x

x =

    0.33333333333333    0.00002000000000
```

If all elements of a matrix or vector are exact integers, then format short and format long yield the same result:

```
>> y = [2   5   40];
>> y

y =

    2   5   40

>> format short; y

y =

    2   5   40

>> format long; y

y =

    2   5   40
```

Utility matrices.   In MATLAB, the functions

$$\text{ones}(n)$$
$$\text{ones}(m,n)$$
$$\text{zeros}(n)$$
$$\text{zeros}(m,n)$$

generate special matrices. The function ones(n) produces an $n \times n$ matrix of ones, while ones(m,n) produces an $m \times n$ matrix of ones. Similarly, zeros(n) produces an $n \times n$ matrix of zeros, while zeros(m,n) produces an $m \times n$ matrix of zeros.

Identity matrix.   We often need to enter an identity matrix **I** in MATLAB programs. The statement eye(n) gives an $n \times n$ identity matrix. For example,

```
>> eye(5)
ans =
     1     0     0     0     0
     0     1     0     0     0
     0     0     1     0     0
     0     0     0     1     0
     0     0     0     0     1
```

Diagonal matrix.   If x is a vector, the statement diag(x) produces a diagonal matrix with x on the diagonal line. For example, for a vector

$$x = [ones(1,n)]$$

diag([ones(1,n)]) gives the following $n \times n$ identity matrix:

```
>> diag([ones(1,5)])
ans =
     1     0     0     0     0
     0     1     0     0     0
     0     0     1     0     0
     0     0     0     1     0
     0     0     0     0     1
```

If **A** is a square matrix, then diag(A) is a vector consisting of the diagonal of **A** and diag(diag(A)) is a diagonal matrix with elements of diag(A) appearing on the diagonal line:

```
>> A = [1 2 3;4 5 6;7 8 9];
>> diag(A)

ans =

     1
     5
     9

>> diag(diag(A))

ans =
     1     0     0
     0     5     0
     0     0     9
```

Note that diag(1:5) gives

```
>> diag(1:5)

ans =

    1    0    0    0    0
    0    2    0    0    0
    0    0    3    0    0
    0    0    0    4    0
    0    0    0    0    5
```

and diag(0:4) gives

```
>> diag(0:4)

ans =

    0    0    0    0    0
    0    1    0    0    0
    0    0    2    0    0
    0    0    0    3    0
    0    0    0    0    4
```

Hence, diag(1:5) − diag(0:4) is an identity matrix.

It is important to note that diag(0,n) is quite different from diag(0:n). The command diag(0,n) gives an $(n + 1) \times (n + 1)$ matrix consisting of all zero elements:

```
>> diag(0,4)

ans =

    0    0    0    0    0
    0    0    0    0    0
    0    0    0    0    0
    0    0    0    0    0
    0    0    0    0    0
```

Variables in MATLAB. A convenient feature of MATLAB is that variables need not be dimensioned before they are used. This is because a variable's dimensions are generated automatically upon the first use of the variable. (The dimensions of the variables can be altered later if necessary.) Such variables (and their dimensions) remain in memory until the command exit or quit is entered.

Section 1–1 / Introduction

Suppose that we enter the following statements in MATLAB workspace:

A = [1   2   3;4   5   6;7   8   9];
x = [3   4   5];

To obtain a list of the variables in the workspace, simply type the command who. Then all of the variables currently in the workspace appear on the screen:

```
>> who
Your variables are:

A     x
```

If, instead of entering the command who, we enter the command whos, the screen will show a list of all the variables in the current workspace, together with information about their size, number of bytes, and class:

```
>> A = [1  2  3;4  5  6;7  8  9];
>> x = [3  4  5];
>> whos
   Name    Size    Bytes    Class

   A       3x3     72       double array
   x       1x3     24       double array

Grand total is 12 elements using 96 bytes
```

The command clear will clear all nonpermanent variables from the workspace. If it is desired to clear only a particular variable, say, 'x', from the workspace, enter the command clear x.

If you need to know the time and date. Statement clock gives the year, month, day, hour, minute, and second. That is, clock returns a six-element row vector containing the current time and date in decimal form:

clock
ans =
   [year   month   day   hour   minute   second]

or

```
>> clock

ans =

   1.0e+003*

   2.0070   0.0010   0.0090   0.0150   0.0070   0.0150
```

Statement date gives the current date:

```
>> date
ans =
09-Jan-2007
```

Correcting mistyped characters. Use the arrow keys on the keypad to edit mistyped commands or to recall previous command lines. For example, if we typed

$$x = (1 \quad 1 \quad 2]$$

then the parenthesis must be corrected. Instead of retyping the entire line, hit the *up-arrow* key. The incorrect line will be displayed again. Using the *left-arrow* key, move the cursor over the parenthesis, type [, and then hit the *delete* key.

How MATLAB is used. MATLAB is usually used in a command-driven mode. When single-line commands are entered, MATLAB processes them immediately and displays the results. MATLAB is also capable of executing sequences of commands that are stored in files.

The commands that are typed may be accessed later with the *up-arrow* key. It is possible to scroll through some of the latest commands that are entered and recall a particular command line.

How to enter comments in a MATLAB program. If you desire to enter comments that are not to be executed, use the % symbol at the start of the line. That is, the % symbol indicates that the rest of the line is a comment and should be ignored. If comments or remarks require more than one program line, each line must begin with %.

1-2 ADDITION, SUBTRACTION, MULTIPLICATION, AND DIVISION WITH MATLAB

Addition and subtraction. Matrices of the same dimension may be added or subtracted. Consider the following matrices:

$$\mathbf{A} = \begin{bmatrix} 2 & 3 \\ 4 & 5 \\ 6 & 7 \end{bmatrix}, \quad \mathbf{B} = \begin{bmatrix} 1 & 0 \\ 2 & 3 \\ 0 & 4 \end{bmatrix}$$

If we enter

$$A = [2 \quad 3;4 \quad 5;6 \quad 7]$$

then the screen shows

$$A = \begin{matrix} 2 & 3 \\ 4 & 5 \\ 6 & 7 \end{matrix}$$

If matrix **B** is entered as

$$B = [1 \quad 0; 2 \quad 3; 0 \quad 4]$$

then the screen shows

$$B = \begin{matrix} 1 & 0 \\ 2 & 3 \\ 0 & 4 \end{matrix}$$

For the addition of two matrices, such as **A** + **B**, we enter

$$C = A + B$$

Then matrix **C** appears on the screen as

$$C = \begin{matrix} 3 & 3 \\ 6 & 8 \\ 6 & 11 \end{matrix}$$

If a vector **x** is given by

$$\mathbf{x} = \begin{bmatrix} 5 \\ 4 \\ 6 \end{bmatrix}$$

then we enter this vector as

$$x = [5;4;6]$$

The screen shows the column vector as follows:

$$x = \begin{matrix} 5 \\ 4 \\ 6 \end{matrix}$$

The following entry will subtract 1 from each element of vector **x**:

$$y = x - 1$$

The screen will show

$$y = \begin{matrix} 4 \\ 3 \\ 5 \end{matrix}$$

Matrix multiplication. Consider the matrices

$$\mathbf{x} = \begin{bmatrix} 1 \\ 2 \\ 3 \end{bmatrix}, \quad \mathbf{y} = \begin{bmatrix} 4 \\ 5 \\ 6 \end{bmatrix}, \quad \mathbf{A} = \begin{bmatrix} 1 & 1 & 2 \\ 3 & 4 & 0 \\ 1 & 2 & 5 \end{bmatrix}$$

or, in MATLAB,

x = [1;2;3];   y = [4;5;6];   A = [1 1 2;3 4 0;1 2 5]

If the expression

$$z = x'*y$$

is entered into the computer, the result is

$$z = 32$$

(Multiplication of matrices is denoted by *.) Note that when the variable name and = are not included in the expression, as in

$$x'*y$$

the result is assigned to the generic variable ans:

```
>> x = [1;2;3];
>> y = [4;5;6];
>> x'*y

ans =

    32
```

Also, the entry

$$x*y'$$

will yield a 3 × 3 matrix as follows:

```
>> x = [1;2;3];
>> y = [4;5;6];
>> x*y'

ans =

    4    5    6
    8   10   12
   12   15   18
```

Similarly, if we enter

$$y*x'$$

then the screen shows

```
ans =
     4    8   12
     5   10   15
     6   12   18
```

Matrix–vector products are a special case of general matrix–matrix products. For example, the entry

$$b = A*x$$

will produce

```
b =
     9
    11
    20
```

Note that a scalar can multiply, or be multiplied by, any matrix. For example, entering

$$5*A$$

gives

```
ans =
     5    5   10
    15   20    0
     5   10   25
```

and entering

$$A*5$$

will also give

```
ans =
     5    5   10
    15   20    0
     5   10   25
```

Magnitude and phase angle of a complex number. The magnitude and phase angle of the complex number $z = x + iy = re^{i\theta}$ are respectively given by

$$r = abs(z)$$
$$theta = angle(z)$$

and the statement

$$z = r*\exp(i*\text{theta})$$

converts them back to the original complex number z.

Array multiplication. Array, or element-by-element, multiplication is denoted by .*. If **x** and **y** have the same dimension, then

$$x.*y$$

denotes the array whose elements are simply the products of the individual elements of x and y. For example, if

$$\mathbf{x} = \begin{bmatrix} 1 & 2 & 3 \end{bmatrix}, \quad \mathbf{y} = \begin{bmatrix} 4 & 5 & 6 \end{bmatrix}$$

then

$$z = x.*y$$

results in

$$\mathbf{z} = \begin{bmatrix} 4 & 10 & 18 \end{bmatrix}$$

Obtaining squares of entries of vector x. For a vector **x**, x.^2 gives the vector of the square of each element. For example, for

$$x = \begin{bmatrix} 1 & 2 & 3 \end{bmatrix}$$

x.^2 is given as shown in the following MATLAB output:

```
>> x = [1 2 3];
>> x.^2

ans =

     1     4     9
```

Also, for the vector

$$\mathbf{y} = \begin{bmatrix} 2 + 5j & 3 + 4j & 1 - j \end{bmatrix}$$

y.^2 is given as follows:

```
>> y = [2+5*i  3+4*i  1-i];
>> y.^2

ans =

  -21.0000 + 20.0000i   -7.0000 + 24.0000i        0 - 2.0000i
```

Similarly, if matrices **A** and **B** have the same dimensions, then **A.*B** denotes the array whose elements are simply the products of the corresponding elements of **A** and **B**. For example, if

$$\mathbf{A} = \begin{bmatrix} 1 & 2 & 3 \\ 0 & 9 & 8 \end{bmatrix}, \quad \mathbf{B} = \begin{bmatrix} 4 & 5 & 6 \\ 7 & 6 & 5 \end{bmatrix}$$

then

$$C = A.*B$$

results in

$$\mathbf{C} = \begin{bmatrix} 4 & 10 & 18 \\ 0 & 54 & 40 \end{bmatrix}$$

Obtaining squares of entries of matrix A. For a matrix **A**, A.^2 gives the matrix consisting of the square of each element. For example, for matrices

$$\mathbf{A} = \begin{bmatrix} 1 & 2 \\ 3 & 4 \end{bmatrix}, \quad \mathbf{B} = \begin{bmatrix} 1+j & 2-2j \\ 3+4j & 5-j \end{bmatrix}$$

A.^2 and B.^2 are given as follows:

```
>> A = [1   2;3   4];
>> A.^2

ans =

    1    4
    9   16

>> B = [1+i   2−2*i;3+4*i   5−i];
>> B.^2

ans =
           0 + 2.0000i              0 − 8.0000i
     −7.0000 + 24.0000i       24.0000 − 10.0000i
```

Absolute values. The command abs(A) gives the matrix consisting of the absolute value of each element of **A**. If **A** is complex, abs(A) returns the complex modulus (magnitude):

$$abs(A) = sqrt(real(A).^2 + imag(A).^2)$$

The command angle(A) returns the phase angles, in radians, of the elements of the complex matrix **A**. The angles lie between $-\pi$ and π. The following example is illustrative:

```
>> A = [2+2*i   1+3*i;4+5*i   6−i];
>> abs(A)

ans =

    2.8284    3.1623
    6.4031    6.0828

>> angle(A)

ans =

    0.7854    1.2490
    0.8961   −0.1651
```

Array division. The expressions x./y, x.\y, A./B, and A.\B give the quotients of the individual elements. Thus, for

$$\mathbf{x} = \begin{bmatrix} 1 & 2 & 3 \end{bmatrix}, \quad \mathbf{y} = \begin{bmatrix} 4 & 5 & 6 \end{bmatrix}$$

the statement

$$u = x./y$$

gives

$$\mathbf{u} = \begin{bmatrix} 0.25 & 0.4 & 0.5 \end{bmatrix}$$

and the statement

$$v = x.\backslash y$$

results in

$$\mathbf{v} = \begin{bmatrix} 4 & 2.5 & 2 \end{bmatrix}$$

Similarly, for matrices **A** and **B**, where

$$\mathbf{A} = \begin{bmatrix} 1 & 2 & 3 \\ 1 & 9 & 8 \end{bmatrix}, \quad \mathbf{B} = \begin{bmatrix} 4 & 5 & 6 \\ 7 & 6 & 5 \end{bmatrix}$$

the statement

$$C = A./B$$

gives

$$C = \begin{bmatrix} 0.2500 & 0.4000 & 0.5000 \\ 0.1429 & 1.5000 & 1.6000 \end{bmatrix}$$

and the statement

$$D = A.\backslash B$$

yields

$$D = \begin{bmatrix} 4.0000 & 2.5000 & 2.0000 \\ 7.0000 & 0.6667 & 0.6250 \end{bmatrix}$$

Note that whenever a division of a number by zero occurs, MATLAB gives a warning, as in the following outputs:

```
>> 5/0
Warning: Divide by zero.

ans =

   Inf
```

```
>> 0/0
Warning: Divide by zero.

ans =

   NaN
```

(Inf denotes infinity and NaN means "not a number.")

1-3 COMPUTING MATRIX FUNCTIONS

In this section, we discuss computations of norms, eigenvalues, and eigenvectors; polynomial evaluation; and finding the inverse of a square matrix, among other topics.

Norms. The norm of a matrix is a scalar that gives some measure of the size of the matrix. Several different definitions are commonly used. One is

$$\text{norm}(A) = \text{largest singular value of } A$$

Similarly, several definitions are available for the norm of a vector. One commonly used definition is

$$\text{norm}(x) = \text{sum}(\text{abs}(x).^2)^{0.5}$$

The following example illustrates the use of the norm command:

```
>> x = [2  3  6];
>> norm(x)

ans =

     7
```

Characteristic equation. The roots of the characteristic equation of a square matrix **A** are the same as the eigenvalues of **A**. The characteristic equation of **A** is computed with

$$p = \text{poly}(A)$$

For example, if

$$\mathbf{A} = \begin{bmatrix} 0 & 1 & 0 \\ 0 & 0 & 1 \\ -6 & -11 & -6 \end{bmatrix}$$

then the command poly(A) will yield

```
>> A = [0  1  0;0  0  1;-6  -11  -6];
>> p = poly(A)

p =

    1.0000    6.0000   11.0000    6.0000
```

This is the MATLAB representation of the characteristic equation

$$s^3 + 6s^2 + 11s + 6 = 0$$

Note that polynomials are represented as row vectors containing the polynomial coefficients in descending order; that is, in the present example,

$$p = [1 \quad 6 \quad 11 \quad 6]$$

The roots of the characteristic equation $p = 0$ can be obtained by entering the command r = roots(p):

```
>> r = roots(p)

r =

   -3.0000
   -2.0000
   -1.0000
```

The roots are $s = -3$, $s = -2$, and $s = -1$. Note that the commands poly and roots can be combined into a single expression, such as

$$\text{roots(poly(A))}$$

The roots of the characteristic equation may be reassembled back into the original polynomial with the command q = poly(r). For r = [−3 −2 −1], poly(r) will produce the polynomial equation

$$s^3 + 6s^2 + 11s + 6 = 0$$

The following MATLAB output is illustrative:

```
>> r = [−3  −2  −1];
>> q = poly(r)

q =

    1    6    11    6
```

Addition or subtraction of polynomials. If the two polynomials are of the same order, add the arrays that describe their coefficients. If the polynomials are of different order (n and m, where $m < n$), then add $n - m$ zeros to the left-hand side of the coefficient array of the lower order polynomial. The following MATLAB output is illustrative:

```
>> a = [3  10  25  36  50];
>> b = [0  0  1  2  10];
>> a+b

ans =

    3    10    26    38    60
```

For the subtraction of b from a, consider the subtraction as an addition of a and $-b$.

Eigenvalues and eigenvectors. If **A** is an $n \times n$ matrix, then the n numbers λ that satisfy

$$\mathbf{Ax} = \lambda \mathbf{x}$$

are the eigenvalues of **A**. They are obtained with the command

$$\text{eig}(A)$$

which returns the eigenvalues in a column vector.

If **A** is real and symmetric, the eigenvalues will be real. But if **A** is not symmetric, the eigenvalues will frequently be complex numbers.

For example, with

$$\mathbf{A} = \begin{bmatrix} 0 & 1 & 0 \\ -1 & 0 & 2 \\ 3 & 0 & 5 \end{bmatrix}$$

the command eig(A) produces the eigenvalues shown in the following output:

```
>> A = [0   1   0;-1   0   2;3   0   5];
>> eig(A)

ans =

   5.2130
  -0.1065 + 1.4487i
  -0.1065 - 1.4487i
```

MATLAB functions may have single- or multiple-output arguments. For example, as seen previously, eig(A) produces a column vector consisting of the eigenvalues of **A**, while the double-assignment statement

$$[X,D] = \text{eig}(A)$$

produces eigenvalues and eigenvectors. The diagonal elements of the diagonal matrix **D** are the eigenvalues, and the columns of **X** are the corresponding eigenvectors such that

$$\mathbf{AX} = \mathbf{XD}$$

For example, if

$$\mathbf{A} = \begin{bmatrix} 0 & 1 & 0 \\ 0 & 0 & 1 \\ -6 & -11 & -6 \end{bmatrix}$$

then the statement

$$[X,D] = \text{eig}(A)$$

gives the following result:

```
>> A = [0   1   0;0   0   1;–6   –11   –6];
>> [X, D] = eig(A)

X =

    –0.5774      0.2182     –0.1048
     0.5774     –0.4364      0.3145
    –0.5774      0.8729     –0.9435

D =

    –1.0000           0           0
          0     –2.0000           0
          0           0     –3.0000
```

The eigenvectors are scaled so that the norm of each is unity.

If the eigenvalues of a matrix are distinct, the eigenvectors are always independent and the eigenvector matrix **X** will diagonalize the original matrix **A** if **X** is applied as a similarity transformation matrix. However, if a matrix has repeated eigenvalues (multiple eigenvalues), it is not diagonalizable unless it has a full (independent) set of eigenvectors. If the eigenvectors are not independent, the original matrix is said to be defective. Even if a matrix is defective, the solution from eig satisfies the relationship **AX** = **XD**.

Convolution (product of polynomials). Consider the polynomials

$$a(s) = 3s^4 + 10s^3 + 25s^2 + 36s + 50$$
$$b(s) = s^2 + 2s + 10$$

The product of the polynomials is the convolution of the coefficients. The product of polynomials $a(s)$ and $b(s)$ can be obtained by entering the command c = conv(a,b).

```
>> a = [3   10   25   36   50];
>> b = [1    2   10];
>> % Define the product of a and b as c.
>> c = conv(a,b)

c =
     3    16    75    186    372    460    500
```

The foregoing is the MATLAB representation of the polynomial

$$c(s) = 3s^6 + 16s^5 + 75s^4 + 186s^3 + 372s^2 + 460s + 500$$

Deconvolution (division of polynomials). To divide the polynomial $a(s)$ by $b(s)$, use the deconvolution command

$$[q,r] = \text{deconv}(a,b)$$

The following MATLAB output is illustrative:

```
>> a = [3   10   25   36   50];
>> b = [1   2   10];
>> % Define the quotient and remainder of a/b as q and r, respectively.
>> [q, r] = deconv (a,b)

q =

    3    4    -13

r =

    0    0    0    22    180
```

This MATLAB output means that

$$3s^4 + 10s^3 + 25s^2 + 36s + 50$$
$$= (s^2 + 2s + 10)(3s^2 + 4s - 13) + 22s + 180$$

Polynomial evaluation. If p is a vector whose elements are the coefficients of a polynomial in descending powers, then polyval(p,s) is the value of the polynomial, evaluated at s. For example, to evaluate the polynomial

$$p(s) = 3s^2 + 2s + 1$$

at $s = 5$, enter the command

$$p = [3 \quad 2 \quad 1];$$
$$\text{polyval}(p,5)$$

Then we get

$$\text{ans} =$$
$$86$$

Next, consider the matrix

$$\mathbf{J} = \begin{bmatrix} -2 + j2\sqrt{3} & 0 & 0 \\ 0 & -2 - j2\sqrt{3} & 0 \\ 0 & 0 & -10 \end{bmatrix}$$

The command poly(J) gives the characteristic polynomial for **J**.

```
>> J = [ -2+i*2*sqrt(3)   0   0;0   -2-i*2*sqrt(3)   0;0   0   -10];
>> p = poly (J)

p =

       1.0000   14.0000   56.0000   160.0000
```

This is the MATLAB expression for the characteristic polynomial for **J**:

$$\text{poly}(\mathbf{J}) = \phi(\mathbf{J}) = \mathbf{J}^3 + 14\mathbf{J}^2 + 56\mathbf{J} + 160\mathbf{I}$$

Here, **I** is the identity matrix. For the matrix

$$\mathbf{A} = \begin{bmatrix} 0 & 1 & 0 \\ 0 & 0 & 1 \\ -6 & -11 & -6 \end{bmatrix}$$

the command polyvalm(poly(J),A) evaluates the following $\phi(\mathbf{A})$:

$$\phi(\mathbf{A}) = \mathbf{A}^3 + 14\mathbf{A}^2 + 56\mathbf{A} + 160\mathbf{I} = \begin{bmatrix} 154 & 45 & 8 \\ -48 & 66 & -3 \\ 18 & -15 & 84 \end{bmatrix}$$

The MATLAB output is as follows:

```
>> J = [-2+i*2*sqrt(3)   0   0;0   -2-i*2*sqrt(3)   0;0   0   -10];
>> A = [0   1   0;0   0   1;-6   -11   -6];
>> polyvalm (poly (J), A)

ans =

     154.0000      45.0000       8.0000
     -48.0000      66.0000      -3.0000
      18.0000     -15.0000      84.0000
```

Matrix exponential. The command expm(A) gives the matrix exponential of an $n \times n$ matrix **A**. That is,

$$\text{expm}(\mathbf{A}) = \mathbf{I} + \mathbf{A} + \frac{\mathbf{A}^2}{2!} + \frac{\mathbf{A}^3}{3!} + \cdots$$

Note that a transcendental function is interpreted as a matrix function if an "m" is appended to the function name, as in expm(A) or sqrtm(A).

As an example, consider the matrix

$$\mathbf{A} = \begin{bmatrix} 0 & 1 & 0 \\ 0 & 0 & 1 \\ -6 & -11 & -6 \end{bmatrix}$$

Then the matrix exponential $e^{\mathbf{A}}$ can be obtained as follows:

```
>> A = [0  1  0;0  0  1;-6  -11  -6];
>> expm(A)

ans =

    0.7474    0.4530    0.0735
   -0.4410   -0.0611    0.0121
   -0.0723   -0.5735   -0.1334
```

The following MATLAB output affords another example:

```
>> expm(eye (3))

ans =

    2.7183         0         0
         0    2.7183         0
         0         0    2.7183
```

Inverse of a square matrix. The inverse of a square matrix **A** can be obtained with the command

$$\text{inv}(A)$$

For example, if matrix **A** is given by

$$\mathbf{A} = \begin{bmatrix} 1 & 1 & 2 \\ 3 & 4 & 0 \\ 1 & 2 & 5 \end{bmatrix}$$

then the inverse of matrix **A** is obtained as follows:

```
>> A = [1   1   2;3   4   0;1   2   5];
>> inv(A)

ans =

    2.2222   -0.1111   -0.8889
   -1.6667    0.3333    0.6667
    0.2222   -0.1111    0.1111
```

That is,

$$\mathbf{A}^{-1} = \begin{bmatrix} 2.2222 & -0.1111 & -0.8889 \\ -1.6667 & 0.3333 & 0.6667 \\ 0.2222 & -0.1111 & 0.1111 \end{bmatrix}$$

MATLAB is case sensitive. It is important to note that MATLAB is case sensitive. That is, MATLAB distinguishes between upper- and lowercase letters. Thus, x and X are not the same variable. All function names must be in lowercase, such as inv(A), eig(A), and poly(A).

1–4 PLOTTING RESPONSE CURVES

MATLAB has an extensive set of routines for obtaining graphical output. The plot command creates linear x–y plots. (Logarithmic or polar plots are created by substituting the word loglog, semilogx, semilogy, or polar for plot.) All such commands are used the same way: They affect only how the axis is scaled and how the data are displayed. This section treats two-dimensional plots only. (Three-dimensional plots are discussed briefly in Section 1-5.)

If **x** and **y** are vectors of the same length, the command

$$\text{plot(x,y)}$$

plots the values in **y** against the values in **x**.

Plotting multiple curves. To plot multiple curves on a single graph, use the plot command with multiple arguments:

$$\text{plot(X1, Y1, X2, Y2, \ldots , Xn, Yn)}$$

The variables X1, Y1, X2, Y2, and so on, are pairs of vectors. Each x–y pair is graphed, generating multiple curves on the plot. Multiple arguments have the benefit of allowing vectors of different lengths to be displayed on the same graph. Each pair uses a different type of line.

Plotting more than one curve on a single graph may also be accomplished by using the command hold on or the command hold, either of which freezes the current plot and inhibits erasure and rescaling. Hence, subsequent curves will be plotted over the original curve. Entering the command hold off or the command hold again releases the current plot.

Adding grid lines, title of the graph, *x*-axis label, and *y*-axis label. Once a graph is on the screen, grid lines may be drawn, the graph may be titled, and *x*- and *y*-axes may be labeled. MATLAB commands to produce a grid, title, *x*-axis label, and *y*-axis label are as follows:

> grid   (grid lines)
> title   (graph title)
> xlabel   (x-axis label)
> ylabel   (y-axis label)

Note that, once the command display has been brought back, grid lines, the graph title, and *x* and *y* labels can be put on the plot by successively entering the preceding commands.

Writing text on the graph. To write text on the plot, use the command text. The command text (X, Y, 'string') adds the text in quotes to location (X, Y) on the current axes, where (X, Y) is expressed in units of the current plot. For example, the statement

> text(3,0.45,'sin t')

will write sin *t* horizontally, beginning at point (3,0.45).

Another command used frequently to add text to a two-dimensional plot is gtext. The command gtext('string') displays the graph window, puts up a cross hair, and waits for a mouse button or keyboard key to be pressed. The cross hair can be positioned with the mouse. Pressing a mouse button or any key writes the text string onto the graph at the selected location.

Imaginary and complex data. If **z** is a complex vector, then plot(z) is equivalent to plot(real(z), imag(z)). That is, plot(z) will plot the imaginary part of **z** against the real part of **z**.

Polar plots. The command polar(theta,rho) will give a plot of the angle theta (in radians) versus the radius rho, in polar coordinates. Subsequent use of the grid command draws polar grid lines.

Logarithmic plots. The following commands produce the indicated plots:

> log log: a plot using $\log_{10}$–$\log_{10}$ scales
> semilogx: a plot using semilog scales; the x–axis is $\log_{10}$, while the y–axis is linear
> semilogy: a plot using semilog scales; the y–axis is $\log_{10}$, while the x–axis is linear

Automatic plotting algorithms. In MATLAB, a plot is automatically scaled. If another plot is requested, the old plot is erased and the axis is automatically rescaled. The automatic plotting algorithms for transient-response curves, root loci, Bode diagrams, Nyquist plots, and the like are designed to work with a wide range of systems, but are not always perfect. Thus, in certain situations, it may become desirable to override the automatic axis scaling feature of the plot command and to select the plotting limits manually.

Manual axis scaling. If it is desired to plot a curve in a region specified by

$$v = [\text{x-min} \quad \text{x-max} \quad \text{y-min} \quad \text{y-max}]$$

enter the command axis(v), where v is a four-element vector. This command sets the axis scaling to the prescribed limits. For logarithmic plots, the elements of v are $\log_{10}$ of the minima and maxima.

Executing axis(v) freezes the current axis scaling for subsequent plots. Typing axis again resumes automatic scaling.

The command axis('square') sets the plot region on the screen to be square. With an equal aspect ratio, a line with unity slope is at a true 45°, not skewed by the irregular shape of the screen. The command axis('normal') sets the aspect ratio back to normal. The command axis('equal') sets the aspect ratio so that equal tick mark increments on each axis are equal in size.

Plot types. The command

$$\text{plot}(x,y,'o')$$

draws a point plot using o-mark symbols. Note that the o-marks will not be connected with lines or curves. To connect them with solid lines or curves, plot the data twice by entering the command

$$\text{plot}(x,y,x,y,'o')$$

The command

$$\text{plot}(X1,Y1,':',X2,Y2,'+')$$

uses a dotted line for the first curve and the plus symbol (+) for the second curve. Available line and point types are as follows:

Line types		Point types	
solid	-	point	.
dashed	--	plus	+
dotted	:	star	*
dash-dot	-.	circle	O
		x-mark	×

Available colors on the screen. The statements

$$\text{plot}(X,Y,'r')$$
$$\text{plot}(X,Y,'+g')$$

indicate the use of a red line on the first graph and green + marks on the second. The available colors are as follows:

red	r
green	g
blue	b
white	w
yellow	y
magenta	m
cyan	c
black	black

(Other colors can be generated with a color map.)

Generating Greek letters, short arrows, etc. To generate Greek letters, such as $\alpha, \beta, \gamma, \delta, \omega$, etc., use '\character', as the following list of some characters from the Greek alphabet illustrates:

α	\alpha	θ	\theta
β	\beta	ζ	\zeta
γ	\gamma	Δ	\Delta
δ	\delta	Θ	\Theta
ω	\omega	Σ	\Sigma
σ	\sigma	Ω	\Omega
ϕ	\phi	Γ	\Gamma
ψ	\psi	π	\Pi

Similarly, to generate the infinity sign, horizontal and vertical short arrows, etc., use the following commands:

∞	\infinity	$\circ$	\circ
$\leftarrow$	\leftarrow	$\rightarrow$	\rightarrow
$\uparrow$	\uparrow	$\downarrow$	\downarrow
$\pm$	\pm		etc.

Plotting and printing curves. Let us enter MATLAB Program 1–1 into the computer and print the resulting plot.

MATLAB Program 1–1

```
>> t = 0 : 0.01*pi:2*pi;
>> alpha = 3;
>> y = sin(alpha*t);
>> plot(t,y)
>> grid
>> title('Plot of sin(\alphat)    (\alpha = 3)')
>> xlabel('t (sec)')
>> ylabel('sin (\alphat)')
```

Note that the vector t is a partition of the domain $0 \leq t \leq 2\pi$ with mesh size 0.01π, while y is a vector giving the values of the sine at the nodes of the partition.

Figure 1–1 shows a plot of the output $y = \sin(\alpha t)$ versus t, reduced to 50% of the actual print size. The letter size of the title, xlabel, and ylabel may be adequate for the plot as seen on the screen. However, if the plot size is reduced to half or less, the letters become too small.

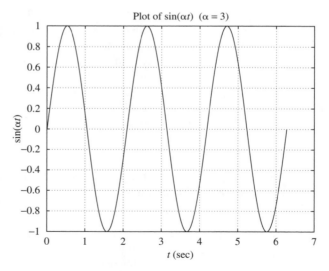

Figure 1–1
Plot of $y = \sin(\alpha t)$ versus t.

To enlarge the letter size, specify a larger font size in the title, xlabel, and ylabel. (See MATLAB Program 1–2.) The size of the plot in Figure 1–2 is half that of the actual printout from the computer. Notice that in Figure 1–2 the letter size of the title, xlabel, and ylabel appears sufficiently large.

MATLAB Program 1–2

```
>> t = 0 : 0.01*pi:2*pi;
>> alpha = 3;
>> y = sin(alpha*t);
>> plot(t,y)
>> grid
>> title('Plot of sin(\alphat) (\alpha = 3)', 'Fontsize', 20)
>> xlabel('t (sec)','Fontsize', 20)
>> ylabel('sin(\alphat)','Fontsize', 20)
```

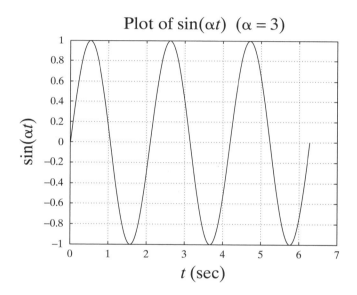

Figure 1–2
Plot of $\sin(\alpha t)$ versus t.

In addition to setting the font size with commands such as

'Fontsize', 15
'Fontsize', 20

it is possible to set the font angle and font name by using commands such as

'Fontangle', 'italic'
'Fontname', 'Times New Roman'

Subscripts can be obtained with "_". For example,

y_1 generates y_1
y_2 generates y_2

Section 1–4 / Plotting Response Curves

Superscripts can be generated by using "^". For example,

$$x\wedge2 \text{ generates } x^2$$
$$x\wedge3 \text{ generates } x^3$$

As another example, let us plot the graph of

$$y = x^2$$

over the interval $0 \le x \le 3$ with increments of 0.1. MATLAB Program 1–3 plots this graph.

MATLAB Program 1–3

>> x = 0:0.1:3;
>> y = x.^2;
>> plot(x,y)
>> grid
>> title('Plot of y = x^2','Fontsize',20,'Fontangle','italic')
>> xlabel('x','Fontsize',20,'Fontangle','italic')
>> ylabel('y','Fontsize',20,'Fontangle','italic')

Note that it is necessary that '^2' be preceded by a period to ensure that it operates entrywise. Figure 1–3 shows the resulting plot.

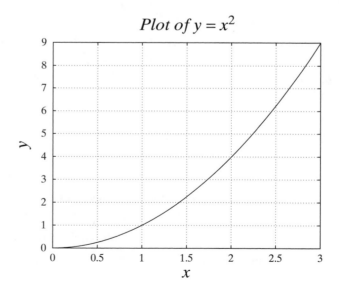

Figure 1–3
Plot of $y = x^2$.

Use of subplot command. Multiple curves on one screen may be split into multiple windows with the use of the command

$$\text{subplot(m,n,p)}$$

The graph display is subdivided into m × n smaller subwindows numbered from left to right and top to bottom. The integer p specifies the window. For example, subplot(2,3,5) or subplot(235) splits the graph display into six subwindows numbered from left to right and top to bottom. The integer p = 5 means the fifth window (that is, the window located in the second row and second column).

Let us next plot the four curves

$$y_1 = \sin t$$
$$y_2 = \sin 2t$$
$$y_3 = \sin t + \sin 2t$$
$$y_4 = (\sin t)(\sin 2t)$$

for $0 \leq t \leq 2\pi$. We shall plot the curves in four subwindows, one curve to a window. MATLAB Program 1–4 produces the plots shown in Figure 1–4.

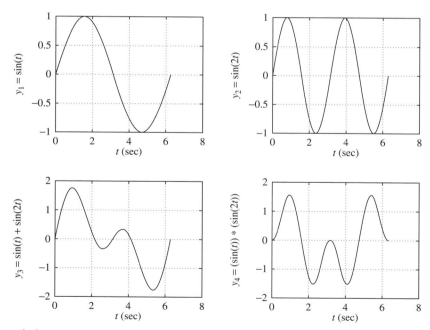

Figure 1–4
Plots of sin t, sin $2t$, sin t + sin $2t$, and (sin t)(sin $2t$).

MATLAB Program 1–4

```
>> t = 0 : 0.01*pi:2*pi;
>> y1 = sin(t);
>> y2 = sin(2*t);
>> y3 = sin(t) + sin(2*t);
>> y4 = (sin(t)).* (sin(2*t));
>> subplot(2,2,1), plot(t,y1), grid
>> xlabel('t (sec)'), ylabel('y_1 = sin(t)')
>> subplot(2,2,2), plot(t,y2), grid
>> xlabel('t (sec)'), ylabel('y_2 = sin(2t)')
>> subplot(2,2,3), plot(t,y3), grid
>> xlabel('t (sec)'), ylabel('y_3 = sin(t) + sin(2t)')
>> subplot(2,2,4), plot(t,y4), grid
>> xlabel('t (sec)'), ylabel('y_4 = (sin(t))*(sin(2t))')
```

1–5 THREE-DIMENSIONAL PLOTS

MATLAB can handle three-dimensional plots. Three-dimensional curves may be plotted with the use of the command plot3, command mesh, or command surf, among many other plotting commands available in MATLAB.

The three-dimensional curve is a curve that is expressed in a three-dimensional coordinate system. Consider a simple three-dimensional coil defined by

$$x = \cos(t), \quad y = \sin(t), \quad z = t$$

The command plot3 produces the three-dimensional curve shown in Figure 1–5 by using MATLAB Program 1–5.

MATLAB Program 1–5

```
>> t = 0:pi/10:6*pi;
>> x = cos (t);
>> y = sin (t);
>> z = t;
>> plot3(x,y,z)
>> grid
>> title ('Three-dimensional coil')
>> axis ('square')
>> xlabel ('x-axis')
>> ylabel ('y-axis')
>> zlabel ('z-axis')
```

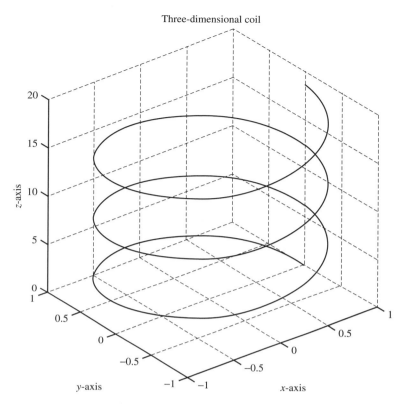

Figure 1–5
Three-dimensional curve.

To write text on the three-dimensional plot, use the command text(X, Y, Z, 'string').

To plot a general three-dimensional surface, use the command mesh or surf. For example, to plot a sphere of radius unity with $n \times n$ facets, we may use MATLAB Program 1–6.

MATLAB Program 1–6
>> n = 10; >> [x, y, z] = sphere (n); >> mesh (x, y, z) >> title ('Mesh plot'); >> axis ('equal') >> xlabel ('x-axis') >> ylabel ('y-axis') >> zlabel ('z-axis')

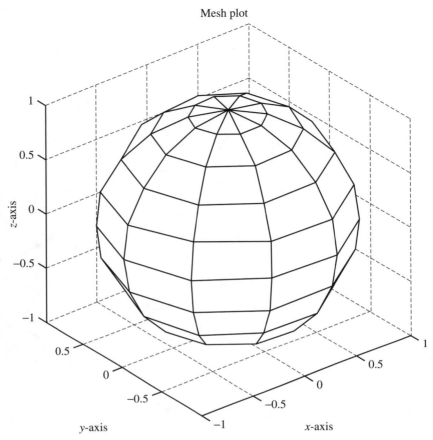

Figure 1–6
10 × 10-faceted sphere in the form of a mesh plot.

Note that

$$[x,y,z] = \text{sphere}(n)$$

generates an $n \times n$-faceted sphere. MATLAB Program 1–6 generates a 100-faceted sphere, as shown in Figure 1–6.

The surf plot (the plot generated by the command surf) looks like a mesh plot. The difference between the mesh plot and surf plot is that in the former only the lines have colors (the color varies along the z-axis, with each line in a facet having constant color) and in the latter the spaces between the lines (facets) are filled in by colors. The boundary lines of the facets are shown in black. MATLAB Program 1–7 generates an $n \times n$-faceted sphere in the form of a surf plot. (See Figure 1–7.)

MATLAB Program 1–7

```
>> n = 10;
>> [x, y, z] = sphere (n);
>> surf (x, y, z)
>> title ('Surf plot');
>> axis ('equal')
>> xlabel ('x-axis')
>> ylabel ('y-axis')
>> zlabel ('z-axis')
```

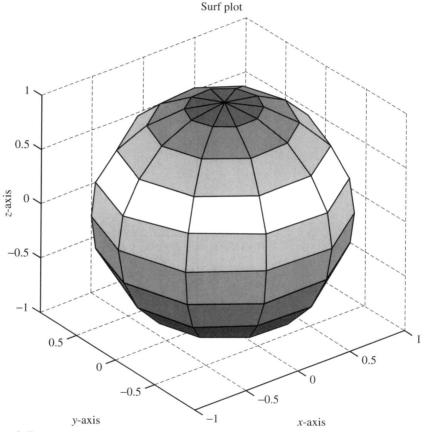

Figure 1–7
10 × 10-faceted sphere in the form of a surf plot.

In Chapter 4, we shall use the command mesh to generate three-dimensional plots related to step responses and impulse responses.

1-6 DRAWING GEOMETRICAL FIGURES WITH MATLAB

In this section, we present commonly used MATLAB programs to draw geometrical figures such as circles, triangular waves, boxes, and spirals.

Circle. Let us write a MATLAB program to draw a circle defined by

$$(x + 2)^2 + (y - 1)^2 = 2^2$$

The center of the circle is

$$(x_c, y_c) = (-2, 1)$$

and the radius of the circle is 2. Assume that the xy region of the plot is given by $-5 \leq x \leq 5$ and $-5 \leq y \leq 5$.

MATLAB Program 1–8 draws this circle, which is shown in Figure 1–8.

MATLAB Program 1–8

```
>> t = 0:0.01:2*pi;
>> x = 2*cos(t);
>> y = 2*sin(t);
>> plot(x−2,y+1)
>> axis([−5  5  −5  5]); axis('square')
>> grid
>> title('A circle of radius 2 located at x = −2 and y = 1')
>> xlabel('x')
>> ylabel('y')
```

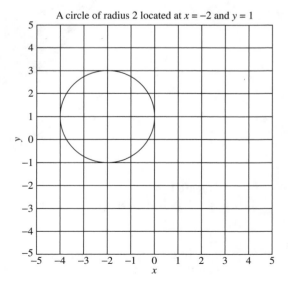

Figure 1–8
Circle of radius 2 located at $x = -2$, $y = 1$.

Triangular wave. Write a MATLAB program to draw the triangular wave shown in Figure 1–9.

MATLAB Program 1–9 draws such a triangular wave, shown in Figure 1–10.

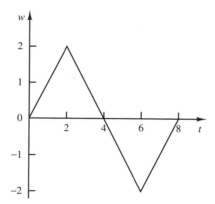

Figure 1–9
Triangular wave.

MATLAB Program 1–9

```
>> u1 = [0:0.5:2]';   u2 = [1.5:-0.5:-2]';   u3 = [-1.5:0.5:0]';
>> w = [u1;u2;u3];
>> t = 0:0.5:8.0;
>> plot(t,w), grid
>> xlabel ('t')
>> ylabel ('w')
```

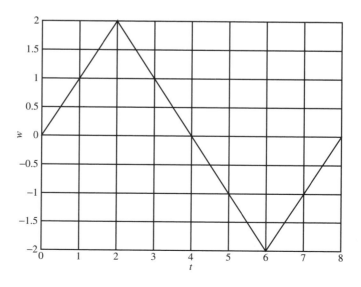

Figure 1–10
Triangular wave.

Rectangles. Write a MATLAB program to draw two rectangles as shown in Figure 1–11. Assume that the region for the plot is given by $-20 \leq x \leq 20$ and $-20 \leq y \leq 20$.

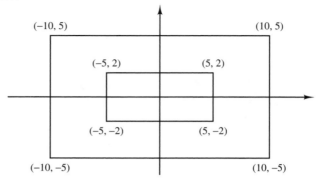

Figure 1–11
Two rectangles.

MATLAB Program 1–10 draws the rectangles, as in Figure 1–12.

MATLAB Program 1–10

```
>> x1 = [−10  10   10  −10  −10];
>> y1 = [−5   −5    5    5   −5];
>> x2 = [−5    5    5   −5   −5];
>> y2 = [−2   −2    2    2   −2];
>> plot(x1,y1,x2,y2);
>> v = [−20  20  −20  20];
>> axis(v); axis('square');
>> grid;
>> xlabel('x')
>> ylabel('y')
```

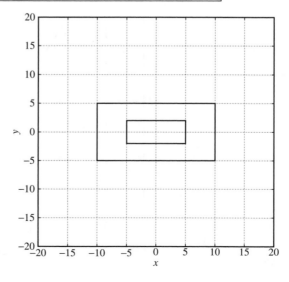

Figure 1–12
Two rectangles produced by MATLAB Program 1–10.

Inverted-pendulum system. Write a MATLAB program to draw the inverted-pendulum system shown in Figure 1–13.

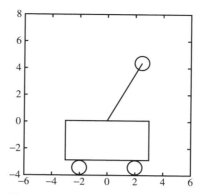

Figure 1–13
Inverted-pendulum system.

MATLAB Program 1–11 draws this inverted-pendulum system, as shown in Figure 1–14.

MATLAB Program 1–11

```
>> xp = [0   5*sin(pi/6)];
>> yp = [0   5*cos(pi/6)];
>> plot(xp,yp)
>> v = [-6   6   -4   8]; axis(v);
>> hold
Current plot held
>> a = 0:0.01:2*pi;
>> x = 5*sin(pi/6)+0.5*cos(a); y = 5*cos(pi/6)+0.5*sin(a);
>> plot(x,y)
>> xb = [-3   3   3   -3   -3];
>> yb = [-3   -3   0   0   -3];
>> plot(xb,yb)
>> plot(2+0.5*cos(a), -3.5+0.5*sin(a))
>> plot(-2+0.5*cos(a), -3.5+0.5*sin(a))
>> axis('square')
```

Section 1–6 / Drawing Geometrical Figures with MATLAB

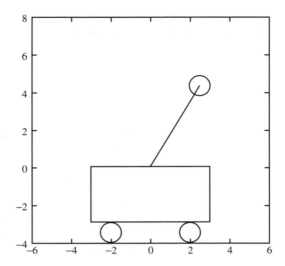

Figure 1–14
Inverted-pendulum system produced by MATLAB Program 1–11.

Three-dimensional coil. MATLAB Program 1–12 produces the three-dimensional (3D) coil shown in Figure 1–15. The program also produces a zx-plane plot for the coil. This plot is shown in Figure 1–16. Yet another thing MATLAB Program 1–12 does is to produce a plot of the coil rotated about the x-axis by 15°. The zx-plane plot of the rotated coil is shown in Figure 1–17.

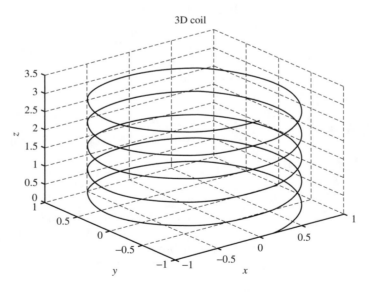

Figure 1–15
3D coil.

Chapter 1 / Introduction to MATLAB

MATLAB Program 1–12

```
>> t = 0:pi/50:10*pi;
>> x = cos(t+3*pi/2);
>> y = sin(t+3*pi/2);
>> z = 0.1*t;
>> plot3(x,y,z)
>> grid
>> title('3D Coil');
>> xlabel('x')
>> ylabel('y')
>> zlabel('z')
>>
>> % zx plane plot
>>
>> plot(z,x)
>> grid
>> title('zx plane plot for coil')
>> xlabel('z')
>> ylabel('x')
>>
>> % Rotate about x axis (zx plane rotated about x axis by 15 degrees)
>>
>> a = 15*pi/180;
>> c = cos(a); s = sin(a);
>> zp = s*y + c*z;    % Rotate about x axis
>> xp = x;
>> yp = c*y – s*z;
>> plot(zp,xp)    % Rotated zx plane plot
>> grid
>> title('zx plane rotated about x axis by 15 degrees')
>> xlabel('zp')
>> ylabel('xp')
```

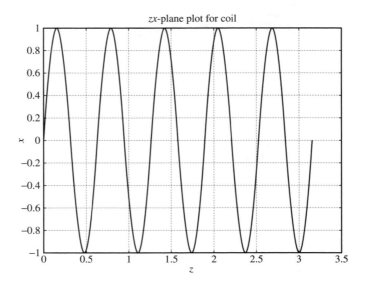

Figure 1–16
zx-plane plot of 3D coil.

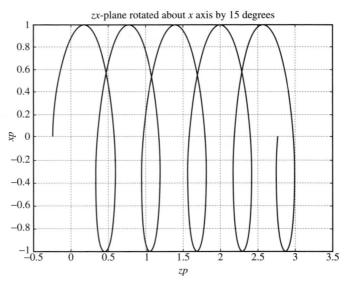

Figure 1–17
zx-plane plot of 3D coil with zx-plane rotated about x-axis by 15 degrees.

Spirals. Many MATLAB programs draw spirals. MATLAB Programs 1–13 through 1–16 show some of them.

MATLAB Program 1–13

```
>> q = [0:0.1:10*pi];
>> r = (0.5).^(q/pi);
>> x = r.*cos(q);
>> y = r.*sin(q);
>> plot(x,y)
>> grid
>> v = [-1 1 -1 1]; axis(v); axis('square')
>> title('Spiral')
>> xlabel('x')
>> ylabel('y')
```

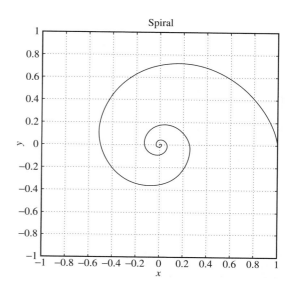

MATLAB Program 1–14

```
>> theta = -10:0.01:0;
>> r = -0.1*theta;
>> x = r.*cos(theta);
>> y = r.*sin(theta);
>> plot(x,y)
>> v = [-1 1 -1 1];axis(v); axis('square');
>> grid
>> title('Spiral')
>> xlabel('x')
>> ylabel('y')
```

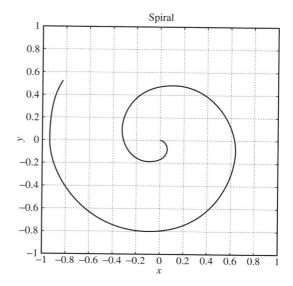

Section 1–6 / Drawing Geometrical Figures with MATLAB

MATLAB Program 1–15

```
>> t = 0:0.01:10;
>> r = 0.08*t;
>> x = −r.*cos(t);
>> y = −r.*sin(t);
>> plot(x,y)
>> v = [ −1 1 −1 1];axis(v); axis('square');
>> grid
>> title('Spiral')
>> xlabel('x')
>> ylabel('y')
```

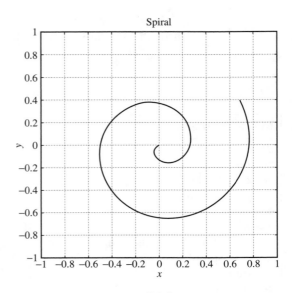

MATLAB Program 1–16

```
>> t = 0:0.01:10*pi;
>> x = 1./pow2(t).*cos(pi*t);
>> y = 1./pow2(t).*sin(pi*t);
>> plot(x,y)
>> grid
>> v = [−1 1 −1 1]; axis(v); axis('square')
>> title('Spiral')
>> xlabel('x')
>> ylabel('y')
>> % The spiral obtained here is the
>> % same as that obtained by
>> % MATLAB Program 1–13.
```

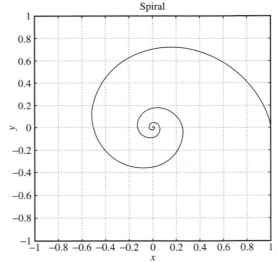

Note: Command pow2(t) produces base 2 power as follows:

For t = 0:1:5

pow2(t) produces

$$2^0 \quad 2^1 \quad 2^2 \quad 2^3 \quad 2^4 \quad 2^5$$

See the following MATLAB output.

```
>> t = 0:1:5;
>> pow2(t)
ans =
     1    2    4    8   16   32
```

Preliminary Study of MATLAB Analysis of Dynamic Systems

2-1 PARTIAL-FRACTION EXPANSION WITH MATLAB

For transfer functions with denominators involving higher-order polynomials, hand computation of partial-fraction expansion may be quite time consuming. In such a case, the use of MATLAB is recommended. MATLAB has a command to obtain the partial-fraction expansion of $B(s)/A(s)$. It also has a command to obtain the zeros and poles of $B(s)/A(s)$.

First we present the MATLAB approach to obtaining the partial-fraction expansion of $B(s)/A(s)$. Then we discuss the MATLAB approach to obtaining the zeros and poles of $B(s)/A(s)$.

Partial-fraction expansion with MATLAB.   Consider the function

$$\frac{B(s)}{A(s)} = \frac{\text{num}}{\text{den}} = \frac{b_0 s^n + b_1 s^{n-1} + \cdots + b_n}{s^n + a_1 s^{n-1} + \cdots + a_n}$$

where some of a_i and b_j may be zero. In MATLAB, row vectors num and den specify the coefficients of the numerator and denominator of the transfer function. That is,

$$\text{num} = [b_0 \ b_1 \ \ldots \ b_n]$$
$$\text{den} = [1 \ a_1 \ \ldots \ a_n]$$

The command

$$[r,p,k] = \text{residue(num,den)}$$

55

finds the residues (r), poles (p), and direct terms (k) of a partial-fraction expansion of the ratio of two polynomials $B(s)$ and $A(s)$.

The partial-fraction expansion of $B(s)/A(s)$ is given by

$$\frac{B(s)}{A(s)} = \frac{r(1)}{s - p(1)} + \frac{r(2)}{s - p(2)} + \cdots + \frac{r(n)}{s - p(n)} + k(s) \qquad (2\text{-}1)$$

EXAMPLE 2-1 Consider the transfer function

$$\frac{B(s)}{A(s)} = \frac{2s^3 + 5s^2 + 3s + 6}{s^3 + 6s^2 + 11s + 6}$$

For this function,

$$\text{num} = [2 \quad 5 \quad 3 \quad 6]$$
$$\text{den} = [1 \quad 6 \quad 11 \quad 6]$$

The command

$$[r,p,k] = \text{residue}(\text{num},\text{den})$$

gives the following result:

```
>> num = [2    5    3    6];
>> den = [1    6    11   6];
>> [r, p, k] = residue (num,den)

r =

    -6.0000
    -4.0000
     3.0000

p =

    -3.0000
    -2.0000
    -1.0000

k =

     2
```

(Note that the residues are returned in column vector r, the pole locations in column vector p, and the direct term in row vector k.) This is the MATLAB representation of the following partial-fraction expansion of $B(s)/A(s)$:

$$\frac{B(s)}{A(s)} = \frac{2s^3 + 5s^2 + 3s + 6}{s^3 + 6s^2 + 11s + 6}$$

$$= \frac{-6}{s + 3} + \frac{-4}{s + 2} + \frac{3}{s + 1} + 2$$

The residue command can also be used to form the polynomials (numerator and denominator) from the partial-fraction expansion. That is, the command

$$[\text{num},\text{den}] = \text{residue}(r,p,k)$$

where r, p, and k are as given in the previous MATLAB output, converts the partial-fraction expansion back to the polynomial ratio $B(s)/A(s)$, as follows:

```
>> r = [-6    -4     3];
>> p = [-3    -2    -1];
>> k = 2;
>> [num,den] = residue(r,p,k)

num =

        2     5     3     6


den =

        1     6    11     6
```

Or we may enter the following MATLAB program to get $B(s)/A(s)$ = num/den:

```
>> [num,den] = residue(r,p,k);
>> printsys(num,den,'s')
num/den =

   2s^3 + 5s^2 + 3s + 6
   ────────────────────
   s^3 + 6s^2 + 11s + 6
```

The command

$$\text{printsys}(\text{num},\text{den},\text{'s'})$$

prints num/den in terms of the ratio of polynomials in s.

Note that if $p(j) = p(j+1) = \cdots = p(j+m-1)$ (that is, $p_j = p_{j+1} = \cdots = p_{j+m-1}$), then the pole $p(j)$ is a pole of multiplicity m. In such a case, the expansion includes terms of the form

$$\frac{r(j)}{s - p(j)} + \frac{r(j+1)}{[s-p(j)]^2} + \cdots + \frac{r(j+m-1)}{[s-p(j)]^m}$$

(For details, see Example 2–2.)

Section 2–1 / Partial-Fraction Expansion with MATLAB

EXAMPLE 2-2 Expand

$$\frac{B(s)}{A(s)} = \frac{s^2 + 2s + 3}{(s+1)^3} = \frac{s^2 + 2s + 3}{s^3 + 3s^2 + 3s + 1}$$

into partial-fractions with MATLAB.

For this function, we have

$$\text{num} = [0 \quad 1 \quad 2 \quad 3]$$
$$\text{den} = [1 \quad 3 \quad 3 \quad 1]$$

Note that for the numerator either [0 1 2 3] or [1 2 3] may be used. The command

$$[r,p,k] = \text{residue}(\text{num},\text{den})$$

gives the following result:

```
>> num = [1   2    3];
>> den = [1   3    3    1];
>> [r,p,k] = residue(num,den)
r =

    1.0000
    0.0000
    2.0000

p =

   -1.0000
   -1.0000
   -1.0000

k =

    []
```

This output is the MATLAB representation of the partial-fraction expansion of

$$\frac{B(s)}{A(s)} = \frac{1}{s+1} + \frac{0}{(s+1)^2} + \frac{2}{(s+1)^3}$$

Note that the direct term k is zero.

To obtain the original function $B(s)/A(s)$ from r, p, and k, enter the following program into the computer:

```
>> [num,den] = residue(r,p,k);
>> printsys(num,den,'s')
```

The computer will then show

$$\text{num/den} = \frac{s^{\wedge}2 + 2s + 3}{s^{\wedge}3 + 3s^{\wedge}2 + 3s + 1}$$

Finding zeros and poles of $B(s)/A(s)$ with MATLAB. The MATLAB command

$$[z,p,K] = \text{tf2zp(num,den)}$$

obtains the zeros, poles, and gain K of $B(s)/A(s)$.

Consider the system defined by

$$\frac{B(s)}{A(s)} = \frac{4s^2 + 16s + 12}{s^4 + 12s^3 + 44s^2 + 48s}$$

To obtain the zeros (z), poles (p), and gain (K), enter the following MATLAB program into the computer:

```
>> num = [4   16   12];
>> den = [1   12   44   48   0];
>> [z,p,K] = tf2zp(num,den)
```

The computer will then produce the following output on the screen:

```
z =

    -3
    -1

p =

     0
   -6.0000
   -4.0000
   -2.0000

K =

    4
```

The zeros are at $s = -3$ and -1. The poles are at $s = 0, -6, -4,$ and -2. The gain K is 4.

If the zeros, poles, and gain K are given, then the following MATLAB program will yield the original num/den:

```
>> z = [-1; -3];
>> p = [0; -2; -4; -6];
>> K = 4;
>> [num,den] = zp2tf(z,p,K);
>> printsys(num,den,'s')
```

num/den =

$$\frac{4s^{\wedge}2 + 16s + 12}{s^{\wedge}4 + 12s^{\wedge}3 + 44s^{\wedge}2 + 48s}$$

Section 2–1 / Partial-Fraction Expansion with MATLAB

EXAMPLE 2-3 Obtain the zeros (z), poles (p), and gain (K) of the system

$$\frac{B(s)}{A(s)} = \frac{5s^3 + 30s^2 + 55s + 30}{s^3 + 9s^2 + 33s + 65}$$

MATLAB Program 2–1 gives the desired result.

MATLAB Program 2–1

```
>> num = [5   30   55   30];
>> den = [1   9   33   65];
>> [z,p,K] = tf2zp(num,den)
```

z =

   −3.0000
   −2.0000
   −1.0000

p =

   −5.0000
   −2.0000 + 3.0000i
   −2.0000 − 3.0000i

K =

   5

Next, given the zeros (z), poles (p), and gain (K), as obtained in MATLAB Program 2–1, determine the numerator and denominator of $B(s)/A(s)$ and also the transfer function $B(s)/A(s)$ = num/den. MATLAB Program 2–2 yields the desired result.

MATLAB Program 2–2

```
>> z = [−1; −2; −3];
>> p = [−2+j*3;−2−j*3;−5];
>> K = 5;
>> [num,den] = zp2tf(z,p,K)
```

num =

   5   30   55   30

den =

   1   9   33   65

```
>> printsys(num,den,'s')
```

num/den =

$$\frac{5\,s^{\wedge}3 + 30\,s^{\wedge}2 + 55\,s + 30}{s^{\wedge}3 + 9\,s^{\wedge}2 + 33\,s + 65}$$

EXAMPLE 2-4 Consider the mechanical system shown in Figure 2–1. The system is at rest initially. The displacements x and y are measured from their respective equilibrium positions. Assuming that $p(t)$ is a step force input and the displacement $x(t)$ is the output, obtain the transfer function of the system. Then, assuming that $m = 0.1$ kg, $b_2 = 0.4$ N-s/m, $k_1 = 6$ N/m, $k_2 = 4$ N/m, and $p(t)$ is a step force of magnitude 10 N, obtain an analytical solution $x(t)$.

The equations of motion for the system are

$$m\ddot{x} + k_1 x + k_2(x - y) = p$$
$$k_2(x - y) = b_2 \dot{y}$$

Laplace transforming these two equations, assuming zero initial conditions, we obtain

$$(ms^2 + k_1 + k_2)X(s) = k_2 Y(s) + P(s) \tag{2-2}$$

$$k_2 X(s) = (k_2 + b_2 s)Y(s) \tag{2-3}$$

Solving Equation (2–3) for $Y(s)$ and substituting the result into Equation (2–2), we get

$$(ms^2 + k_1 + k_2)X(s) = \frac{k_2^2}{k_2 + b_2 s}X(s) + P(s)$$

or

$$[(ms^2 + k_1 + k_2)(k_2 + b_2 s) - k_2^2]X(s) = (k_2 + b_2 s)P(s)$$

from which we obtain the transfer function

$$\frac{X(s)}{P(s)} = \frac{b_2 s + k_2}{mb_2 s^3 + mk_2 s^2 + (k_1 + k_2)b_2 s + k_1 k_2} \tag{2-4}$$

Substituting the given numerical values for m, k_1, k_2, and b_2 into Equation (2–4), we have

$$\frac{X(s)}{P(s)} = \frac{0.4s + 4}{0.04s^3 + 0.4s^2 + 4s + 24}$$

$$= \frac{10s + 100}{s^3 + 10s^2 + 100s + 600} \tag{2-5}$$

Since $P(s)$ is a step force of magnitude 10 N,

$$P(s) = \frac{10}{s}$$

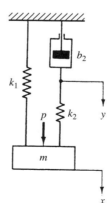

Figure 2–1
Mechanical system.

Then, from Equation (2–5),

$$X(s) = \frac{10s + 100}{s^3 + 10s^2 + 100s + 600} \cdot \frac{10}{s}$$

To find an analytical solution, we need to expand $X(s)$ into partial fractions. For this purpose, we may use the following MATLAB program to find the residues, poles, and direct term:

```
>> num = [100  1000];
>> den = [1  10  100  600  0];
>> [r,p,k] = residue(num, den)

r =
    -0.6845 + 0.2233i
    -0.6845 - 0.2233i
    -0.2977
     1.6667

p =
    -1.2898 + 8.8991i
    -1.2898 - 8.8991i
    -7.4204
     0

k =
     []
```

On the basis of the MATLAB output, $X(s)$ can be written as

$$X(s) = \frac{-0.6845 + j0.2233}{s + 1.2898 - j8.8991} + \frac{-0.6845 - j0.2233}{s + 1.2898 + j8.8991}$$

$$+ \frac{-0.2977}{s + 7.4204} + \frac{1.6667}{s}$$

$$= \frac{-1.3690(s + 1.2898) - 3.9743}{(s + 1.2898)^2 + 8.8991^2} - \frac{0.2977}{s + 7.4204} + \frac{1.6667}{s}$$

The inverse Laplace transform of $X(s)$ gives

$$x(t) = -1.3690 e^{-1.2898t} \cos(8.8991 t)$$

$$-0.4466 e^{-1.2898t} \sin(8.8991 t) - 0.2977 e^{-7.4204t} + 1.6667$$

where $x(t)$ is measured in meters and time t in seconds. This is the analytical solution to the problem.

EXAMPLE 2-5 Obtain the inverse Laplace transform of

$$F(s) = \frac{s^5 + 8s^4 + 23s^3 + 35s^2 + 28s + 3}{s^3 + 6s^2 + 8s}$$

[Use MATLAB to find the partial-fraction expansion of $F(s)$.]
The following MATLAB program will produce the partial-fraction expansion of $F(s)$:

```
>> num = [1  8  23  35  28  3];
>> den = [1  6  8  0];
>> [r,p,k] = residue(num,den)

r =

    0.3750
    0.2500
    0.3750

p =

   -4
   -2
    0

k =

    1  2  3
```

Note that $k = \begin{bmatrix} 1 & 2 & 3 \end{bmatrix}$ means that $F(s)$ involves $s^2 + 2s + 3$ as follows:

$$F(s) = s^2 + 2s + 3 + \frac{0.375}{s+4} + \frac{0.25}{s+2} + \frac{0.375}{s}$$

Hence, the inverse Laplace transform of $F(s)$ is given by

$$f(t) = \frac{d^2}{dt^2}\delta(t) + 2\frac{d}{dt}\delta(t) + 3\delta(t) + 0.375e^{-4t} + 0.25e^{-2t} + 0.375, \quad \text{for } t > 0-$$

EXAMPLE 2-6 Given the zero(s), pole(s), and gain K of $B(s)/A(s)$, use MATLAB to obtain the function $B(s)/A(s)$. Consider the following three cases:

1. There is no zero. Poles are at $-1 + 2j$ and $-1 - 2j$. $K = 10$.
2. A zero is at 0. Poles are at $-1 + 2j$ and $-1 - 2j$. $K = 10$.
3. A zero is at -1. Poles are at -2, -4 and -8. $K = 12$.

MATLAB programs to obtain $B(s)/A(s)$ = num/den for the three cases are as follows:

```
>> z = [];
>> p = [-1+2*j;-1-2*j];
>> K = 10;
>> [num,den] = zp2tf(z,p,K);
>> printsys(num,den)

num/den =
        10
   -----------
   s^2 + 2s + 5
```

```
>> z = [0];
>> p = [-1+2*j; -1-2*j];
>> K = 10;
>> [num,den] = zp2tf(z,p,K);
>> printsys(num,den)

num/den =
        10s
   -----------
   s^2 + 2s + 5
```

```
>> z = [-1];
>> p = [-2; -4; -8];
>> K = 12;
>> [num,den] = zp2tf(z,p,K);
>> printsys(num,den)

num/den =
          12s + 12
   -----------------------
   s^3 + 14s^2 + 56s + 64
```

EXAMPLE 2-7 Solve the differential equation

$$\ddot{x} + 2\dot{x} + 10x = t^2, \quad x(0) = 0, \quad \dot{x}(0) = 0$$

Since the initial conditions are zeros, the Laplace transform of the equation becomes

$$s^2 X(s) + 2s X(s) + 10 X(s) = \frac{2}{s^3}$$

Hence,

$$X(s) = \frac{2}{s^3(s^2 + 2s + 10)}$$

We need to find the partial-fraction expansion of $X(s)$. Since the denominator involves a triple pole, it is simpler to use MATLAB to obtain the partial-fraction expansion. The following MATLAB program may be used:

```
>> num = [2];
>> den = [1  2  10  0  0  0];
>> [r,p,k] = residue(num,den)

r =

      0.0060 - 0.0087i
      0.0060 + 0.0087i
     -0.0120
     -0.0400
      0.2000

p =

     -1.0000 + 3.0000i
     -1.0000 - 3.0000i
            0
            0
            0

k =

     []
```

From the MATLAB output, we find that

$$X(s) = \frac{0.006 - 0.0087j}{s + 1 - 3j} + \frac{0.006 + 0.0087j}{s + 1 + 3j} + \frac{-0.012}{s} + \frac{-0.04}{s^2} + \frac{0.2}{s^3}$$

Combining the first two terms on the right-hand side of the equation, we get

$$X(s) = \frac{0.012(s + 1) + 0.0522}{(s + 1)^2 + 3^2} - \frac{0.012}{s} - \frac{0.04}{s^2} + \frac{0.2}{s^3}$$

The inverse Laplace transform of $X(s)$ gives

$$x(t) = 0.012e^{-t}\cos 3t + 0.0174e^{-t}\sin 3t - 0.012 - 0.04t + 0.1t^2, \quad \text{for } t \geq 0$$

Note that if only the roots of a polynomial such as

$$d(s) = s^3 + 2s^2 + 3s + 4$$

are desired, we may use the command

$$r = \text{roots}(d)$$

(See MATLAB Program 2–3.) The command poly(r) produces the original polynomial such that

$$d = \text{poly}(r)$$

(See also MATLAB Program 2–3.)

MATLAB Program 2–3

```
>> d = [1   2   3   4];
>> r = roots(d)

r =

   -1.6506
   -0.1747 + 1.5469i
   -0.1747 - 1.5469i

>> d = poly(r)

d =

   1.0000   2.0000   3.0000   4.0000
```

2-2 TRANSFORMATION OF MATHEMATICAL MODELS OF DYNAMIC SYSTEMS

MATLAB has useful commands to transform a mathematical model of a linear time-invariant system to another model. The following are linear system transformations that are useful for solving control engineering problems:

    Transfer-function to state-space conversion   (tf2ss)
    State-space to transfer-function conversion   (ss2tf)
    State-space to zero-pole conversion   (ss2zp)
    Zero-pole to state-space conversion   (zp2ss)
    Transfer-function to zero-pole conversion   (tf2zp)
    Zero-pole to transfer-function conversion   (zp2tf)
    Continuous-time to discrete-time conversion   (c2d)

We begin our discussion with the transformation from the transfer function to state space.

Transfer function to state space. The command

$$[A,B,C,D] = \text{tf2ss(num,den)}$$

converts a system from the transfer function form

$$\frac{Y(s)}{U(s)} = \frac{\text{num}}{\text{den}} = \mathbf{C}(s\mathbf{I} - \mathbf{A})^{-1}\mathbf{B} + D$$

to the state-space form

$$\dot{\mathbf{x}} = \mathbf{A}\mathbf{x} + \mathbf{B}u$$
$$y = \mathbf{C}\mathbf{x} + Du$$

It is important to note that the state-space representation of any system is not unique: There are many (indeed, infinitely many) state-space representations of the same system. The MATLAB command gives one possible such representation.

Consider the transfer-function system

$$\frac{Y(s)}{U(s)} = \frac{10s + 10}{s^3 + 6s^2 + 5s + 10} \tag{2-6}$$

There are many (again, infinitely many) possible state-space representations of this system. One possible such representation is

$$\begin{bmatrix} \dot{x}_1 \\ \dot{x}_2 \\ \dot{x}_3 \end{bmatrix} = \begin{bmatrix} 0 & 1 & 0 \\ 0 & 0 & 1 \\ -10 & -5 & -6 \end{bmatrix} \begin{bmatrix} x_1 \\ x_2 \\ x_3 \end{bmatrix} + \begin{bmatrix} 0 \\ 10 \\ -50 \end{bmatrix} u$$

$$y = \begin{bmatrix} 1 & 0 & 0 \end{bmatrix} \begin{bmatrix} x_1 \\ x_2 \\ x_3 \end{bmatrix} + [0]u$$

Another possible state-space representation (among infinitely many alternatives) is

$$\begin{bmatrix} \dot{x}_1 \\ \dot{x}_2 \\ \dot{x}_3 \end{bmatrix} = \begin{bmatrix} -6 & -5 & -10 \\ 1 & 0 & 0 \\ 0 & 1 & 0 \end{bmatrix} \begin{bmatrix} x_1 \\ x_2 \\ x_3 \end{bmatrix} + \begin{bmatrix} 1 \\ 0 \\ 0 \end{bmatrix} u \tag{2-7}$$

$$y = \begin{bmatrix} 0 & 10 & 10 \end{bmatrix} \begin{bmatrix} x_1 \\ x_2 \\ x_3 \end{bmatrix} + [0]u \tag{2-8}$$

MATLAB Program 2–4 transforms the transfer function given by Equation (2–6) into the state-space representation given by Equations (2–7) and (2–8).

```
MATLAB Program 2–4
>> num = [10   10];
>> den = [1  6  5  10];
>> [A,B,C,D] = tf2ss(num,den)

A =
       -6   -5   -10
        1    0     0
        0    1     0
B =
        1
        0
        0
C =
        0   10   10
D =
        0
```

EXAMPLE 2-8 Consider the transfer function system

$$\frac{Y(s)}{U(s)} = \frac{25.04s + 5.008}{s^3 + 5.03247s^2 + 25.1026s + 5.008}$$

Obtain a state-space representation of this system with MATLAB.
MATLAB command

$$[A,B,C,D] = \text{tf2ss(num,den)}$$

will produce a state-space representation of the system. (See MATLAB Program 2–5.)

MATLAB Program 2–5

```
>> num = [25.04   5.008];
>> den = [1   5.03247   25.1026   5.008];
>> [A,B,C,D] = tf2ss(num,den)

A =

    -5.0325   -25.1026    -5.0080
     1.0000         0           0
          0    1.0000           0

B =

     1
     0
     0

C =

     0   25.0400    5.0080

D =

     0
```

This is the MATLAB representation of the following state-space equations:

$$\begin{bmatrix} \dot{x}_1 \\ \dot{x}_2 \\ \dot{x}_3 \end{bmatrix} = \begin{bmatrix} -5.0325 & -25.1026 & -5.008 \\ 1 & 0 & 0 \\ 0 & 1 & 0 \end{bmatrix} \begin{bmatrix} x_1 \\ x_2 \\ x_3 \end{bmatrix} + \begin{bmatrix} 1 \\ 0 \\ 0 \end{bmatrix} u$$

$$y = \begin{bmatrix} 0 & 25.04 & 5.008 \end{bmatrix} \begin{bmatrix} x_1 \\ x_2 \\ x_3 \end{bmatrix} + [0]u$$

Transformation from state space to transfer function. To obtain the transfer function from state-space equations, use the command

$$[\text{num,den}] = \text{ss2tf}(A,B,C,D,iu)$$

iu must be specified for systems with more than one input. For example, if the system has three inputs ($u1, u2, u3$), then iu must be either 1, 2, or 3, where 1 implies $u1$, 2 implies $u2$, and 3 implies $u3$.

If the system has only one input, then either

$$[\text{num,den}] = \text{ss2tf}(A,B,C,D)$$

or

$$[\text{num,den}] = \text{ss2tf}(A,B,C,D,1)$$

may be used. (See Example 2–9 and MATLAB Program 2–6.)

EXAMPLE 2-9 Obtain the transfer function of the system defined by the following state-space equations:

$$\begin{bmatrix} \dot{x}_1 \\ \dot{x}_2 \\ \dot{x}_3 \end{bmatrix} = \begin{bmatrix} 0 & 1 & 0 \\ 0 & 0 & 1 \\ -5.008 & -25.1026 & -5.03247 \end{bmatrix} \begin{bmatrix} x_1 \\ x_2 \\ x_3 \end{bmatrix} + \begin{bmatrix} 0 \\ 25.04 \\ -121.005 \end{bmatrix} u$$

$$y = \begin{bmatrix} 1 & 0 & 0 \end{bmatrix} \begin{bmatrix} x_1 \\ x_2 \\ x_3 \end{bmatrix}$$

MATLAB Program 2–6 will produce the transfer function of the given system, namely,

$$\frac{Y(s)}{U(s)} = \frac{25.04s + 5.008}{s^3 + 5.0325s^2 + 25.1026s + 5.008}$$

MATLAB Program 2–6

```
>> A = [0  1  0;0  0  1;-5.008  -25.1026  -5.03247];
>> B = [0;25.04; -121.005];
>> C = [1  0  0];
>> D = [0];
>> [num,den] = ss2tf(A,B,C,D)

num =

        0  -0.0000   25.0400   5.0080

den =

   1.0000   5.0325   25.1026   5.0080

>> % *****The same result can be obtained by entering the following command *****
>> [num,den] = ss2tf(A,B,C,D,1)

num =

        0  -0.0000   25.0400   5.0080

den =

   1.0000   5.0325   25.1026   5.0080
```

If the system involves one output but more than one input, use the command

$$[\text{num},\text{den}] = \text{ss2tf}(A,B,C,D,iu)$$

This command converts the state-space system

$$\dot{x} = Ax + Bu$$
$$y = Cx + Du$$

to the transfer function system

$$\frac{Y(s)}{U_i(s)} = i\text{th element of } [C(sI - A)^{-1}B + D]$$

Note that the scalar iu is an index into the inputs of the system and specifies which input is to be used for the response.

EXAMPLE 2-10 Consider the system

$$\begin{bmatrix} \dot{x}_1 \\ \dot{x}_2 \end{bmatrix} = \begin{bmatrix} 0 & 1 \\ -2 & -3 \end{bmatrix} \begin{bmatrix} x_1 \\ x_2 \end{bmatrix} + \begin{bmatrix} 1 & 0 \\ 0 & 1 \end{bmatrix} \begin{bmatrix} u_1 \\ u_2 \end{bmatrix}$$

$$y = \begin{bmatrix} 1 & 0 \end{bmatrix} \begin{bmatrix} x_1 \\ x_2 \end{bmatrix} + \begin{bmatrix} 0 & 0 \end{bmatrix} \begin{bmatrix} u_1 \\ u_2 \end{bmatrix}$$

which has two inputs and one output.

Two transfer functions may be obtained for this system. One relates the output y and input u_1, and the other relates the output y and input u_2. (When considering input u_1, we assume that input u_2 is zero, and vice versa.) MATLAB Program 2–7 produces the transfer functions in question.

MATLAB Program 2–7

```
>> A = [0   1;-2  -3];
>> B = [1   0;0   1];
>> C = [1   0];
>> D = [0   0];
>> [num,den] = ss2tf(A, B, C, D, 1)

num =

         0    1.0000    3.0000

den =

         1    3    2

>> [num,den] = ss2tf(A, B, C, D, 2)

num =

         0    0.0000    1.0000

den =

         1    3    2
```

From the MATLAB output, we have

$$\frac{Y(s)}{U_1(s)} = \frac{s+3}{s^2+3s+2}$$

and

$$\frac{Y(s)}{U_2(s)} = \frac{1}{s^2+3s+2}$$

EXAMPLE 2-11 Consider a system with multiple inputs and multiple outputs. When the system has more than one output, the command

$$[\text{NUM,den}] = \text{ss2tf}(A,B,C,D,iu)$$

produces transfer functions for all outputs to each input. (The numerator coefficients are returned to matrix NUM with as many rows as there are outputs.)

Now consider the system defined by

$$\begin{bmatrix} \dot{x}_1 \\ \dot{x}_2 \end{bmatrix} = \begin{bmatrix} 0 & 1 \\ -25 & -4 \end{bmatrix} \begin{bmatrix} x_1 \\ x_2 \end{bmatrix} + \begin{bmatrix} 1 & 1 \\ 0 & 1 \end{bmatrix} \begin{bmatrix} u_1 \\ u_2 \end{bmatrix}$$

$$\begin{bmatrix} y_1 \\ y_2 \end{bmatrix} = \begin{bmatrix} 1 & 0 \\ 0 & 1 \end{bmatrix} \begin{bmatrix} x_1 \\ x_2 \end{bmatrix} + \begin{bmatrix} 0 & 0 \\ 0 & 0 \end{bmatrix} \begin{bmatrix} u_1 \\ u_2 \end{bmatrix}$$

This system involves two inputs and two outputs. Four transfer functions are involved: $Y_1(s)/U_1(s)$, $Y_2(s)/U_1(s)$, $Y_1(s)/U_2(s)$, and $Y_2(s)/U_2(s)$. (When considering input u_1, we assume that input u_2 is zero, and vice versa.) See the output of MATLAB Program 2-8.

MATLAB Program 2-8

```
>> A = [0    1;-25  -4];
>> B = [1    1;0    1];
>> C = [1    0;0    1];
>> D = [0    0;0    0];
>> [NUM,den] = ss2tf(A,B,C,D,1)

NUM =

     0    1     4
     0    0   -25

den =

     1    4    25

>> [NUM,den] = ss2tf(A,B,C,D,2)

NUM =

     0    1.0000     5.0000
     0    1.0000   -25.0000

den =

     1    4    25
```

This is the MATLAB representation of the following four transfer functions:

$$\frac{Y_1(s)}{U_1(s)} = \frac{s+4}{s^2+4s+25}, \qquad \frac{Y_1(s)}{U_2(s)} = \frac{s+5}{s^2+4s+25}$$

$$\frac{Y_2(s)}{U_1(s)} = \frac{-25}{s^2+4s+25}, \qquad \frac{Y_2(s)}{U_2(s)} = \frac{s-25}{s^2+4s+25}$$

Conversion from continuous time to discrete time. The command

$$[G,H] = c2d(A,B,Ts)$$

where Ts is the sampling period in seconds, converts the state-space model from continuous time to discrete time, assuming a zero-order hold on the inputs. That is, with that command,

$$\dot{\mathbf{x}} = \mathbf{A}\mathbf{x} + \mathbf{B}\mathbf{u}$$

is converted to

$$\mathbf{x}(k+1) = \mathbf{G}\mathbf{x}(k) + \mathbf{H}\mathbf{u}(k)$$

Consider, for example, the following continuous-time system:

$$\begin{bmatrix} \dot{x}_1 \\ \dot{x}_2 \end{bmatrix} = \begin{bmatrix} 0 & 1 \\ -25 & -4 \end{bmatrix} \begin{bmatrix} x_1 \\ x_2 \end{bmatrix} + \begin{bmatrix} 0 \\ 1 \end{bmatrix} u$$

An equivalent discrete-time system can be obtained with the command [G,H] = c2d(A,B,Ts). If the sampling period Ts is assumed to be 0.05 sec, we have the result shown in MATLAB Program 2–9.

MATLAB Program 2–9

```
>> A = [0    1;-25   -4];
>> B = [0;1];
>> [G, H] = c2d(A, B, 0.05)

G =

     0.9709    0.0448
    -1.1212    0.7915

H =

     0.0012
     0.0448
```

Notice that if we used format long, we would have a more accurate **H** matrix. (See MATLAB Program 2–10.)

```
MATLAB Program 2-10

>> A = [0   1;-25   -4];
>> B = [0;1];
>> format long
>> [G, H] = c2d(A, B, 0.05)

G =

        0.97088325381929      0.04484704238264
       -1.12117605956599      0.79149508428874

H =

        0.00116466984723
        0.04484704238264
```

The discrete-time state-space equation when Ts = 0.05 sec is then given by

$$\begin{bmatrix} x_1(k+1) \\ x_2(k+1) \end{bmatrix} = \begin{bmatrix} 0.9709 & 0.04485 \\ -1.1212 & 0.7915 \end{bmatrix} \begin{bmatrix} x_1(k) \\ x_2(k) \end{bmatrix} + \begin{bmatrix} 0.001165 \\ 0.04485 \end{bmatrix} u(k)$$

If we used format short or if we did not specify format, we would have gotten

$$\mathbf{H} = \begin{bmatrix} 0.0012 \\ 0.0448 \end{bmatrix}$$

which is not quite accurate. (See MATLAB Program 2-9.) In transforming a continuous-time system to a discrete-time system, it is recommended that we use format long.

2-3 MATLAB REPRESENTATION OF SYSTEMS IN BLOCK DIAGRAM FORM

In this section, we shall review MATLAB representations of systems given in the form of block diagrams.

Figure 2-2(a) shows a block with a transfer function. Such a block represents a system or an element of a system. To simplify our presentation, we shall call the

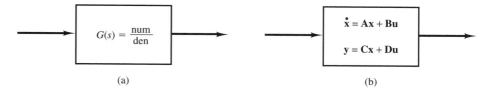

Figure 2-2
Block diagrams of systems.
(a) System is given by transfer function; (b) system is defined by state-space form.

block with a transfer function a system. MATLAB uses sys to represent such a system. The statement

$$\text{sys} = \text{tf(num,den)}$$

represents the system. Figure 2–2(b) shows a block with a state-space representation. The MATLAB representation of such a system is given by

$$\text{sys} = \text{ss(A,B,C,D)}$$

A physical system may involve many interconnected blocks. In what follows, we shall consider series-connected blocks, parallel-connected blocks, and feedback-connected blocks. Any linear time-invariant system may be represented by combinations of series-connected blocks, parallel-connected blocks, and feedback-connected blocks.

Series-connected blocks. In the system shown in Figure 2–3, G_1 and G_2 are series connected. System G_1 (which is represented by sys1) and system G_2 (which is represented by sys2) are respectively defined by

$$\text{sys1} = \text{tf(num1, den1)}$$
$$\text{sys2} = \text{tf(num2, den2)}$$

provided that the two systems are defined in terms of transfer functions. Then the series-connected system (sys) can be given by

$$\text{sys} = \text{series(sys1, sys2)}$$

The system's numerator and denominator can also be given by

$$[\text{num, den}] = \text{series(num1,den1,num2,den2)}$$

If systems G_1 and G_2 are given in state-space form, then their MATLAB representations are, respectively,

$$\text{sys1} = \text{ss(A1, B1, C1, D1)}$$
$$\text{sys2} = \text{ss(A2, B2, C2, D2)}$$

The series-connected system G_1G_2 is given by

$$\text{sys} = \text{series(sys1, sys2)}$$

Parallel-connected blocks. Figures 2–4(a) and (b) show parallel-connected systems. In Figure 2–4(a), the outputs from two systems G_1 and G_2 are added, while

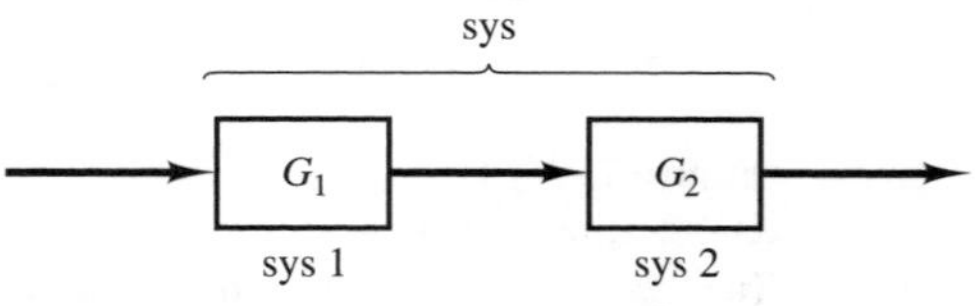

Figure 2–3
Series-connected blocks.

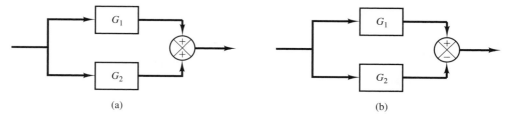

Figure 2-4
Parallel-connected systems.
(a) $G_1 + G_2$; (b) $G_1 - G_2$.

in Figure 2-4(b) the output from system G_2 is subtracted from the output of system G_1. If G_1 and G_2 are defined in terms of transfer functions, then

$$\text{sys1} = \text{tf(num1, den1)}$$
$$\text{sys2} = \text{tf(num2, den2)}$$

and the parallel-connected system $G_1 + G_2$ is given by

$$\text{sys} = \text{parallel(sys1, sys2)}$$

If G_1 and G_2 are in state-space form, then

$$\text{sys1} = \text{ss(A1, B1, C1, D1)}$$
$$\text{sys2} = \text{ss(A2, B2, C2, D2)}$$

and the parallel-connected system $G_1 + G_2$ is given by

$$\text{sys} = \text{parallel(sys1, sys2)}$$

If the parallel-connected system is $G_1 - G_2$, as shown in Figure 2-4(b), then we define sys1 and sys2 as before, but change sys2 to $-$sys2 in the expressions for sys; that is,

$$\text{sys} = \text{parallel(sys1, } -\text{sys2)}$$

Feedback-connected blocks. Figure 2-5(a) shows a negative-feedback system and Figure 2-5(b) shows a positive-feedback system.

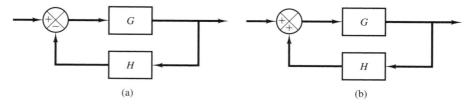

Figure 2-5
(a) Negative-feedback system; (b) positive-feedback system.

If G and H are defined in terms of transfer functions, then

$$\text{sysg} = [\text{numg, deng}]$$
$$\text{sysh} = [\text{numh, denh}]$$

and the entire feedback system is given by

$$\text{sys} = \text{feedback(sysg, sysh)}$$

or

$$[\text{num, den}] = \text{feedback(numg, deng, numh, denh)}$$

If the system has a unity feedback function, then $H = [1]$ and sys can be given by

$$\text{sys} = \text{feedback(sysg, [1])}$$

Note that, in treating the feedback system, MATLAB assumed that the feedback was negative. If the system involves a positive feedback, we need to add +1 in the argument of feedback as follows:

$$\text{sys} = \text{feedback(sysg, sysh, +1)}$$

Alternatively, we can use −sysh in the statement sys; that is,

$$\text{sys} = \text{feedback(sysg, −sysh)}$$

for the positive feedback system.

The following diagram summarizes the statement sys for feedback systems:

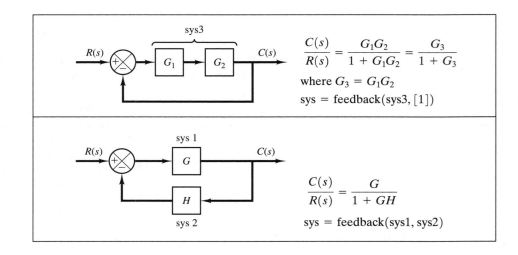

76   Chapter 2 / Preliminary Study of MATLAB Analysis of Dynamic Systems

EXAMPLE 2-12 Obtain the transfer functions of the cascaded system, parallel system, and feedback system shown in Figure 2–6. Assume that $G_1(s)$ and $G_2(s)$ are as follows:

$$G_1(s) = \frac{\text{num1}}{\text{den1}} = \frac{10}{s^2 + 2s + 10}, \quad G_2(s) = \frac{\text{num2}}{\text{den2}} = \frac{5}{s + 5}$$

To obtain the transfer functions of the cascaded system, parallel system, or feedback (closed-loop) system, we use the following commands:

[num, den] = series(num1,den1,num2,den2)
[num, den] = parallel(num1,den1,num2,den2)
[num, den] = feedback(num1,den1,num2,den2)

MATLAB Program 2–11 gives $C(s)/R(s) = \text{num/den}$ for each arrangement of $G_1(s)$ and $G_2(s)$. Note that the command

printsys(num,den)

displays the num/den [that is, the transfer function $C(s)/R(s)$] of the system considered.

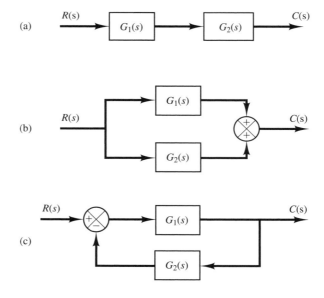

Figure 2–6
(a) Cascaded system;
(b) parallel system;
(c) feedback system.

MATLAB Program 2–11

```
>> num1 = [10];
>> den1 = [1   2   10];
>> num2 = [5];
>> den2 = [1   5];
>> [num, den] = series(num1,den1,num2,den2);
>> printsys(num,den)
```

num/den =

$$\frac{50}{s^3 + 7s^2 + 20s + 50}$$

```
>> [num, den] = parallel(num1,den1,num2,den2);
>> printsys(num,den)
```

num/den =

$$\frac{5s^2 + 20s + 100}{s^3 + 7s^2 + 20s + 50}$$

```
>> [num, den] = feedback(num1,den1,num2,den2);
>> printsys(num,den)
```

num/den =

$$\frac{10s + 50}{s^3 + 7s^2 + 20s + 100}$$

EXAMPLE 2–13 Consider the system shown in Figure 2–7. Obtain the closed-loop transfer function $Y(s)/U(s)$ with MATLAB.

MATLAB Program 2–12 generates the closed-loop transfer function $Y(s)/U(s)$.

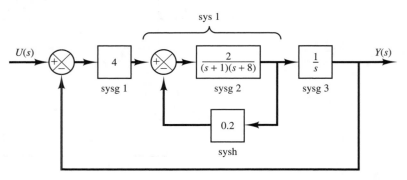

Figure 2–7
Closed-loop system.

MATLAB Program 2–12

```
>> sysg1 = [4];
>> numg2 = [2]; deng2 = [1   9   8]; sysg2 = tf(numg2,deng2);
>> numg3 = [1]; deng3 = [1   0]; sysg3 = tf(numg3,deng3);
>> sysh = [0.2];
>> sys1 = feedback (sysg2,sysh);
>> sys2 = series (sys1,sysg3);
>> sys3 = series (sysg1,sys2);
>> sys = feedback (sys3,[1])
```

Transfer function:
$$\frac{8}{s^3 + 9 s^2 + 8.4 s + 8}$$

EXAMPLE 2–14 Consider the system shown in Figure 2–8. Obtain the closed-loop transfer function $Y(s)/U(s)$ with MATLAB. Also, obtain a state-space representation of the system.

MATLAB Program 2–13 produces the closed-loop transfer function $Y(s)/U(s)$ and also the state-space representation of the closed-loop system.

Note that in MATLAB Program 2–13 the state-space representation of the transfer-function system was obtained with the command

$$sys\_ss = ss(sys)$$

where sys is given in terms of transfer functions.

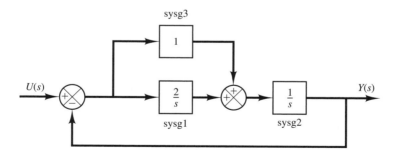

Figure 2–8
Closed-loop system.

> **MATLAB Program 2–13**
>
> ```
> >> numg1 = [2]; deng1 = [1  0]; sysg1 = tf(numg1,deng1);
> >> numg2 = [1]; deng2 = [1  0]; sysg2 = tf(numg2,deng2);
> >> sysg3 = [1];
> >> sys1 = parallel(sysg1,sysg3);
> >> sys2 = series(sys1,sysg2);
> >> sys = feedback(sys2, [1])
> ```
>
> Transfer function:
>
> $$\frac{s + 2}{s^2 + s + 2}$$
>
> ```
> >> sys_ss = ss(sys)
>
> a =
>        x1   x2
>   x1   -1   -1
>   x2    2    0
>
> b =
>        u1
>   x1    1
>   x2    0
>
> c =
>        x1   x2
>   y1    1    1
>
> d =
>        u1
>   y1    0
> ```
>
> Continuous-time model.

The state-space equations obtained are

$$\begin{bmatrix} \dot{x}_1 \\ \dot{x}_2 \end{bmatrix} = \begin{bmatrix} -1 & -1 \\ 2 & 0 \end{bmatrix} \begin{bmatrix} x_1 \\ x_2 \end{bmatrix} + \begin{bmatrix} 1 \\ 0 \end{bmatrix} u \qquad (2\text{–}9)$$

$$y = \begin{bmatrix} 1 & 1 \end{bmatrix} \begin{bmatrix} x_1 \\ x_2 \end{bmatrix} + [0]u \qquad (2\text{–}10)$$

Note that this state-space representation is not unique. If we use the statement

$$[A,B,C,D] = \text{tf2ss(num,den)}$$

we obtain the following state-space equations:

$$\begin{bmatrix} \dot{x}_1 \\ \dot{x}_2 \end{bmatrix} = \begin{bmatrix} -1 & -2 \\ 1 & 0 \end{bmatrix} \begin{bmatrix} x_1 \\ x_2 \end{bmatrix} + \begin{bmatrix} 1 \\ 0 \end{bmatrix} u \qquad (2\text{–}11)$$

$$y = \begin{bmatrix} 1 & 2 \end{bmatrix} \begin{bmatrix} x_1 \\ x_2 \end{bmatrix} + [0]u \qquad (2\text{--}12)$$

(See MATLAB Program 2–14.)

MATLAB Program 2–14

```
>> num = [1  2]; den = [1  1  2];
>> [A,B,C,D] = tf2ss(num,den)

A =
     -1    -2
      1     0
B =
      1
      0
C =
      1     2
D =
      0
```

The state-space equations [Equations (2–9) and (2–10)] can be transformed into those given by Equations (2–11) and (2–12) by means of the transformation matrix

$$\mathbf{P} = \begin{bmatrix} 1 & 0 \\ 0 & 2 \end{bmatrix}$$

If we use another transformation matrix

$$\mathbf{P} = \begin{bmatrix} 0 & 0.5 \\ 1 & -0.5 \end{bmatrix}$$

then the state-space equations [Equations (2–9) and (2–10)] become

$$\begin{bmatrix} \dot{x}_1 \\ \dot{x}_2 \end{bmatrix} = \begin{bmatrix} -1 & 1 \\ -2 & 0 \end{bmatrix} \begin{bmatrix} x_1 \\ x_2 \end{bmatrix} + \begin{bmatrix} 1 \\ 2 \end{bmatrix} u$$

$$y = \begin{bmatrix} 1 & 0 \end{bmatrix} \begin{bmatrix} x_1 \\ x_2 \end{bmatrix} + [0]u$$

In fact, there are infinitely many state-space representations of the system shown in Figure 2–8.

EXAMPLE 2–15 Consider the system shown in Figure 2–9(a). Obtain a state-space representation of the closed-loop system.

One approach to obtaining a state-space representation of the system is to obtain the transfer-function expression for the closed-loop system and then transform that expression to a state-space expression. Another approach is to obtain state-space representations of individual blocks and then obtain the state-space representation of the entire closed-loop system.

In this example, we shall use the second approach. Note that, to obtain a state-space expression of the transfer function in a block, it is necessary that the order of the numerator be less than or equal to the order of the denominator. Therefore, we redraw the system block diagram as shown in Figure 2–9(b).

MATLAB Program 2–15 produces the state-space representation of the closed-loop system shown in Figure 2–9(b). To obtain the corresponding transfer function, we use the statement

$$\text{sys\_tf} = \text{tf(sys\_ss)}$$

shown in MATLAB Program 2–15.

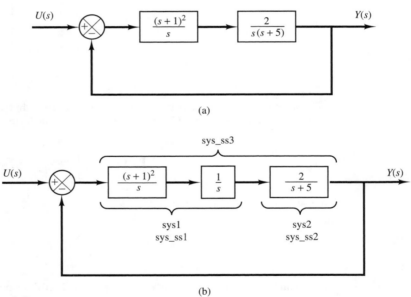

Figure 2–9
(a) Closed-loop system; (b) redrawn closed-loop system.

MATLAB Program 2–15

```
>> num1 = [1   2   1]; den1 = [1   0   0]; sys1 = tf(num1,den1);
>> num2 = [2]; den2 = [1   5]; sys2 = tf(num2,den2);
>> sys_ss1 = ss(sys1);
>> sys_ss2 = ss(sys2);
>> sys_ss3 = series(sys_ss1,sys_ss2);
>> sys_ss = feedback(sys_ss3, [1])
```

a =

	x1	x2	x3
x1	−7	1	1
x2	−4	0	0
x3	0	0.5	0

b =

	u1
x1	1
x2	2
x3	0

c =

	x1	x2	x3
y1	2	0	0

d =

	u1
y1	0

Continuous-time model.
```
>>
>> sys_tf = tf(sys_ss)
```

Transfer function:

$$\frac{2 s^2 + 4 s + 2}{s^3 + 7 s^2 + 4 s + 2}$$

Pole–zero cancellation. When we simplify a block diagram, the resulting closed-loop transfer function may involve pole–zero cancellation. If so, we would like to check the expression. If cancellation does occur, the cancelled expression can be obtained by use of the command

minreal

This command performs the pole–zero cancellation if there are common factors between the numerator and denominator and produces the minimal-order transfer function.

As an example, consider the following transfer function:

$$\frac{Y(s)}{U(s)} = \frac{(s+1)(s+2)}{(s+3)(s+4)(s+1)}$$

$$= \frac{s^2 + 3s + 2}{s^3 + 8s^2 + 19s + 12}$$

Clearly, $(s+1)$ in the numerator and $(s+1)$ in the denominator cancel each other. To obtain the minimal-order transfer function for $Y(s)/U(s)$, we may use the command minreal. (See MATLAB Program 2–16.)

MATLAB Program 2–16

```
>> num = [1   3   2]; den = [1   8   19   12]; sys = tf(num,den);
>> sys
```

Transfer function:

$$\frac{s^2 + 3s + 2}{s^3 + 8s^2 + 19s + 12}$$

```
>> sys_min = minreal(sys)
```

Transfer function:

$$\frac{s + 2}{s^2 + 7s + 12}$$

The minimal-order transfer function (the canceled transfer function) is

$$\frac{Y(s)}{U(s)} = \frac{s + 2}{s^2 + 7s + 12}$$

Transient-Response Analysis

3–1 INTRODUCTION

This chapter presents the MATLAB approach to obtaining system responses when the inputs are time-domain inputs such as the step, impulse, and ramp functions. The chapter also discusses system responses to initial conditions and arbitrary time functions. System responses to the frequency-domain inputs are presented in Chapter 5.

Section 3–2 discusses step responses and Section 3–3 deals with impulse responses. Section 3–4 treats ramp responses and Section 3–5 examines responses to arbitrary inputs. Section 3–6 obtains system responses to arbitrary initial conditions. Finally, Section 3–7 discusses three-dimensional plots of transient responses.

3–2 STEP RESPONSE

MATLAB description of standard second-order system. A second-order system of the form

$$G(s) = \frac{\omega_n^2}{s^2 + 2\zeta\omega_n s + \omega_n^2}$$

is called the *standard second-order system*. Given ω_n and ζ, the command

$$\text{printsys(num,den)} \quad \text{or} \quad \text{printsys(num,den,s)}$$

prints num/den as a ratio of polynomials in *s*.

Consider, for example, the case where ω_n = 5 rad/sec and ζ = 0.4. MATLAB Program 3–1 generates the standard second-order system when ω_n = 5 rad/sec and ζ = 0.4.

MATLAB Program 3–1

```
>> wn = 5;
>> damping_ratio = 0.4;
>> [num0,den] = ord2(wn,damping_ratio);
>> num = 5^2*num0;
>> printsys(num,den,'s')
num/den =
              25
         -----------
         s^2 + 4s + 25
```

Step response. If num and den (the numerator and denominator of a transfer function) are known, we may define the system by

$$\text{sys = tf(num,den)}$$

Then a command such as

$$\text{step(sys)} \quad \text{or} \quad \text{step(num,den)}$$

will generate a plot of a unit-step response and will display a response curve on the screen. The computation interval Δt and the time span of the response are determined by MATLAB.

If we wish MATLAB to compute the response every Δt seconds and plot the response curve for $0 \leq t \leq T$ (where *T* is an integer multiple of Δt), we enter the statement

$$t = 0{:}\Delta t{:}T;$$

in the program and use the command

$$\text{step(sys,t)} \quad \text{or} \quad \text{step(num,den,t)}$$

where t is the user-specified time.

If step commands have left-hand arguments, such as

$$\text{y = step(sys,t)} \quad \text{or} \quad \text{y = step(num,den,t)}$$

and

$$\text{[y,t] = step(sys,t)} \quad \text{or} \quad \text{[y,t] = step(num,den,t)}$$

MATLAB produces the unit-step response of the system, but displays no plot on the screen. It is necessary to use a plot command to see response curves.

MATLAB representation of linear systems. The transfer function of a system is represented by two arrays of numbers. For example, the system

$$\frac{C(s)}{R(s)} = \frac{2s + 25}{s^2 + 4s + 25}$$

can be represented as two arrays, each containing the coefficients of the polynomials in decreasing powers of s as follows:

$$\text{num} = [2 \quad 25]$$
$$\text{den} = [1 \quad 4 \quad 25]$$

An alternative representation is

$$\text{num} = [0 \quad 2 \quad 25]$$
$$\text{den} = [1 \quad 4 \quad 25]$$

In this expression, a zero is padded. Note that if zeros are padded, the dimensions of the num vector and den vector become the same. An advantage of padding zeros is that num and den vector can be directly added. For instance,

$$\text{num} + \text{den} = [0 \quad 2 \quad 25] + [1 \quad 4 \quad 25]$$
$$= [1 \quad 6 \quad 50]$$

In writing the numerator, either expression,

$$\text{num} = [2 \quad 25] \quad \text{or} \quad \text{num} = [0 \quad 2 \quad 25]$$

may be used. As you have already seen in Chapter 2, we use both expressions in solving MATLAB problems. MATLAB Program 3–2 produces the transfer function of the system.

MATLAB Program 3–2

```
>> num = [2   25];
>> den = [1   4   25];
>> sys = tf(num,den)
```

Transfer function:
$$\frac{2s + 25}{s^2 + 4s + 25}$$

For a control system defined in state-space form, where state matrix **A**, control matrix **B**, output matrix **C**, and direct transmission matrix **D** of state-space equations are known, the command

$$\text{step}(A,B,C,D)$$

will generate plots of unit-step responses. The time vector is automatically determined when t is not explicitly included in the step commands.

Note that the command step(sys) may be used to obtain the unit-step response of a system. First, we define the system by

$$sys = tf(num,den)$$

or

$$sys = ss(A,B,C,D)$$

Then, to obtain, for example, the unit-step response, we enter

$$step(sys)$$

into the computer.

When step commands have left-hand arguments, such as

$$[y,x,t] = step(num,den,t)$$
$$[y,x,t] = step(A,B,C,D,iu)$$
$$[y,x,t] = step(A,B,C,D,iu,t) \quad (3\text{--}1)$$

no plot is shown on the screen. Hence, it is necessary to use a plot command to see the response curves. The matrices y and x contain the output and state response of the system, respectively, evaluated at the computation time points t. (y has as many columns as outputs and one row for each element in t; x has as many columns as states and one row for each element in t.)

Note in Equation (3–1) that the scalar iu is an index into the inputs of the system and specifies which input is to be used for the response; t is the user-specified time. If the system involves multiple inputs and multiple outputs, the step command, as given by Equation (3–1), produces a series of step response plots, one for each input and output combination of

$$\dot{\mathbf{x}} = \mathbf{A}\mathbf{x} + \mathbf{B}\mathbf{u}$$
$$\mathbf{y} = \mathbf{C}\mathbf{x} + \mathbf{D}\mathbf{u}$$

(For details, see Example 3–5.)

EXAMPLE 3–1 Consider the spring–mass–dashpot system mounted on a cart, as shown in Figure 3–1. Here, u is the input displacement and y is the output displacement. To derive the transfer

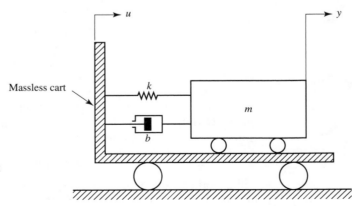

Figure 3–1
Spring–mass–dashpot system mounted on a cart.

function of the system, note that, for translational systems, Newton's second law states that

$$ma = \sum F$$

where m is a mass, a is the acceleration of the mass, and $\sum F$ is the sum of the forces acting on the mass in the direction of the acceleration. Applying Newton's second law to the present system and noting that the cart is massless, we obtain

$$m\frac{d^2y}{dt^2} = -b\left(\frac{dy}{dt} - \frac{du}{dt}\right) - k(y - u)$$

or

$$m\frac{d^2y}{dt^2} + b\frac{dy}{dt} + ky = b\frac{du}{dt} + ku$$

The latter equation represents a mathematical model of the system under consideration. Taking the Laplace transform of the equation, assuming zero initial conditions, gives

$$(ms^2 + bs + k)Y(s) = (bs + k)U(s)$$

Taking the ratio of $Y(s)$ to $U(s)$, we find the transfer function of the system to be

$$\text{Transfer function} = \frac{Y(s)}{U(s)} = \frac{bs + k}{ms^2 + bs + k}$$

Assuming that $m = 10$ kg, $b = 20$ N-s/m, $k = 100$ N/m, and the input $u(t)$ is a unit-step input (a step input of 1 m), obtain the response curve $y(t)$.

Substituting the given numerical values into the transfer function, we have

$$\frac{Y(s)}{U(s)} = \frac{20s + 100}{10s^2 + 20s + 100} = \frac{2s + 10}{s^2 + 2s + 10}$$

MATLAB Program 3–3 will produce the unit-step response $y(t)$. The resulting unit-step response curve is shown in Figure 3–2.

MATLAB Program 3–3

```
>> num = [2   10];
>> den = [1   2   10];
>> sys = tf(num,den);
>> step(sys)
>> grid
```

In this plot, the duration of the response is automatically determined by MATLAB. The title and axis labels are also automatically determined by MATLAB.

If we wish to compute and plot the curve every 0.01 sec over the interval $0 \leq t \leq 8$, we need to enter the following statement in the MATLAB program:

$$t = 0:0.01:8;$$

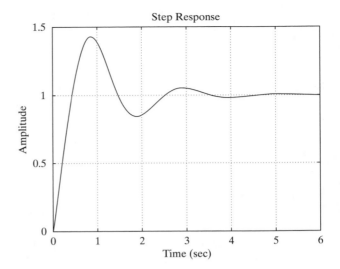

Figure 3–2
Unit-step response curve.

Also, if we wish to change the title and axis labels, we enter the desired title and desired labels as shown in MATLAB Program 3–4.

```
MATLAB Program 3–4
>> t = 0:0.01:8;
>> num = [2   10];
>> den = [1   2   10];
>> sys = tf(num,den);
>> step(sys,t)
>> grid
>> title('Unit-Step Response', 'Fontsize', 20')
>> xlabel('t','Fontsize', 20')
>> ylabel('Output y','Fontsize', 20')
```

Note that if we did not enter the desired title and desired axis labels in the program, the title, x-axis label, and y-axis label on the plot would have been "Step Response", "Time (sec)", and "Amplitude", respectively. (This statement does not apply to older versions of MATLAB.) When we enter the desired title and axis labels, as we have done in MATLAB Program 3–4, MATLAB erases the predetermined title and axis labels, except "(sec)" in the x-axis label, and replaces them with the ones we have specified. If the font sizes are too small, they can be made larger. For example, entering

'Fontsize', 20

in the title, xlabel, and ylabel variables as shown in MATLAB Program 3–4 results in that size of text appearing in those places. Figure 3–3 is a plot of the response curve obtained with MATLAB Program 3–4.

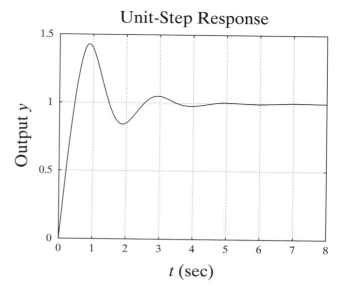

Figure 3–3
Unit-step response curve. Font sizes for title, xlabel, and ylabel are enlarged.

EXAMPLE 3–2 Consider the mechanical system shown in Figure 3–4. The transfer function $X(s)/P(s)$ is

$$\frac{X(s)}{P(s)} = \frac{b_2 s + k_2}{m b_2 s^3 + m k_2 s^2 + (k_1 + k_2) b_2 s + k_1 k_2} \quad (3\text{–}2)$$

(For the derivation of this transfer function, see Example 2–4.)

The transfer function $Y(s)/X(s)$ is obtained as

$$\frac{Y(s)}{X(s)} = \frac{k_2}{b_2 s + k_2}$$

Hence,

$$\frac{Y(s)}{P(s)} = \frac{Y(s)}{X(s)} \frac{X(s)}{P(s)} = \frac{k_2}{m b_2 s^3 + m k_2 s^2 + (k_1 + k_2) b_2 s + k_1 k_2} \quad (3\text{–}3)$$

Assuming that $m = 0.1$ kg, $b_2 = 0.4$ N-s/m, $k_1 = 6$ N/m, $k_2 = 4$ N/m, and $p(t)$ is a step force of magnitude 10 N, obtain the responses $x(t)$ and $y(t)$.

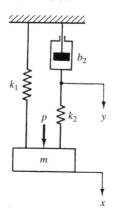

Figure 3–4
Mechanical system.

Substituting the numerical values for m, b_2, k_1, and k_2 into the transfer functions given by Equations (3–2) and (3–3), we obtain

$$\frac{X(s)}{P(s)} = \frac{0.4s + 4}{0.04s^3 + 0.4s^2 + 4s + 24}$$

$$= \frac{10s + 100}{s^3 + 10s^2 + 100s + 600} \quad (3\text{–}4)$$

and

$$\frac{Y(s)}{P(s)} = \frac{4}{0.04s^3 + 0.4s^2 + 4s + 24}$$

$$= \frac{100}{s^3 + 10s^2 + 100s + 600} \quad (3\text{–}5)$$

Since $p(t)$ is a step force of magnitude 10 N, we may define $p(t) = 10u(t)$, where $u(t)$ is a unit-step input of magnitude 1 N. Then Equations (3–4) and (3–5) can be written as

$$\frac{X(s)}{U(s)} = \frac{100s + 1000}{s^3 + 10s^2 + 100s + 600} \quad (3\text{–}6)$$

and

$$\frac{Y(s)}{U(s)} = \frac{1000}{s^3 + 10s^2 + 100s + 600} \quad (3\text{–}7)$$

Since $u(t)$ is a unit-step input, $x(t)$ and $y(t)$ can be obtained from Equations (3–6) and (3–7) with the use of a step command. (Step commands assume that the input is the unit-step input.)

In this example, we shall demonstrate the use of the commands

$$y = \text{step(sys,t)}$$

and

$$\text{plot(t,y)}$$

MATLAB Program 3–5 produces the responses $x(t)$ and $y(t)$ of the system on one diagram.

MATLAB Program 3–5

```
>> t = 0:0.01:5;
>> num1 = [100   1000];
>> num2 = [1000];
>> den = [1   10   100   600];
>> sys1 = tf(num1,den);
>> sys2 = tf(num2,den);
>> y1 = step(sys1,t);
>> y2 = step(sys2,t);
>> plot(t,y1,t,y2)
>> grid
>> title('Unit-Step Responses')
>> xlabel('t (sec)')
>> ylabel('x(t) and y(t)')
>> text(0.07,2.8,'x(t)')
>> text(0.7,2.35,'y(t)')
```

The response curves $x(t)$ and $y(t)$ are shown in Figure 3–5.

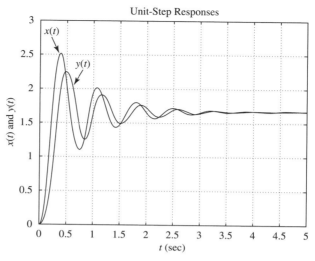

Figure 3–5
Step response curves $x(t)$ and $y(t)$.

Writing text on the graphics screen. To write text on the graphics screen, we enter, for example, the following statements:

$$\text{text}(0.07, 2.8, 'x(t)')$$

and

$$\text{text}(0.7, 2.35, 'y(t)')$$

The first statement tells the computer to write $x(t)$ beginning at the coordinates $x = 0.07$, $y = 2.8$. Similarly, the second statement tells the computer to write $y(t)$ beginning at the coordinates $x = 0.7$, $y = 2.35$. [See MATLAB Program 3–5 and Figure 3–5.]

Another way to write text in the plot is to use the gtext command. The syntax is

$$\text{gtext}('text')$$

When gtext is executed, the computer waits until the cursor is positioned (with a mouse) at the desired location on the screen. When the left mouse button is pressed, the text enclosed in simple quotes is written on the plot at the cursor's position. Any number of gtext commands can be used in a plot. (See, for example, MATLAB Program 3–18.)

EXAMPLE 3-3 Obtain both analytically and computationally the unit-step response of the following higher-order system:

$$\frac{C(s)}{R(s)} = \frac{3s^3 + 25s^2 + 72s + 80}{s^4 + 8s^3 + 40s^2 + 96s + 80}$$

[Obtain the partial-fraction expansion of $C(s)$ with MATLAB when $R(s)$ is a unit-step function.]

MATLAB Program 3–6 yields the unit-step response curve shown in Figure 3–6. It also yields the partial-fraction expansion of $C(s)$ as follows:

$$C(s) = \frac{3s^3 + 25s^2 + 72s + 80}{s^4 + 8s^3 + 40s^2 + 96s + 80} \frac{1}{s}$$

$$= \frac{-0.2813 - j0.1719}{s + 2 - j4} + \frac{-0.2813 + j0.1719}{s + 2 + j4}$$

$$+ \frac{-0.4375}{s + 2} + \frac{-0.375}{(s + 2)^2} + \frac{1}{s}$$

$$= \frac{-0.5626(s + 2)}{(s + 2)^2 + 4^2} + \frac{(0.3438) \times 4}{(s + 2)^2 + 4^2}$$

$$- \frac{0.4375}{s + 2} - \frac{0.375}{(s + 2)^2} + \frac{1}{s}$$

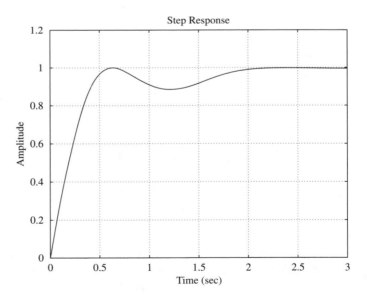

Figure 3–6
Unit-step response curve.

MATLAB Program 3–6

```
>> % —— Unit-Step Response of C(s)/R(s) and Partial-Fraction Expansion of C(s) ——
>>
>> num = [3   25   72   80];
>> den = [1   8   40   96   80];
>> step(num,den);
>> v = [0   3   0   1.2]; axis(v), grid
>>
>> % To obtain the partial-fraction expansion of C(s), enter commands
>> %     num1 = [3   25   72   80];
>> %     den1 = [1   8   40   96   80   0];
>> %     [r,p,k] = residue(num1,den1)
>>
>> num1 = [3   25   72   80];
>> den1 = [1   8   40   96   80   0];
>> [r,p,k] = residue(num1,den1)

r =

  −0.2813 − 0.1719i
  −0.2813 + 0.1719i
  −0.4375
  −0.3750
   1.0000

p =

  −2.0000 + 4.0000i
  −2.0000 − 4.0000i
  −2.0000
  −2.0000
   0

k =

  []
```

Hence, the time response can be given by

$$c(t) = -0.5626e^{-2t}\cos 4t + 0.3438e^{-2t}\sin 4t$$
$$-0.4375e^{-2t} - 0.375te^{-2t} + 1$$

The fact that the response curve consists of damped sinusoidal curves superimposed on an exponential curve can be seen from Figure 3–6.

EXAMPLE 3–4 Obtain both analytical and computational solutions of the unit-step response of a unity-feedback system whose open-loop transfer function is

$$G(s) = \frac{5(s + 20)}{s(s + 4.5883)(s^2 + 3.4118s + 16.346)}$$

The closed-loop transfer function is

$$\frac{C(s)}{R(s)} = \frac{5(s + 20)}{s(s + 4.5883)(s^2 + 3.4118s + 16.3460) + 5(s + 20)}$$

$$= \frac{5s + 100}{s^4 + 8s^3 + 32s^2 + 80s + 100}$$

$$= \frac{5(s + 20)}{(s^2 + 2s + 10)(s^2 + 6s + 10)}$$

The unit-step response of this system is then

$$C(s) = \frac{5(s + 20)}{s(s^2 + 2s + 10)(s^2 + 6s + 10)}$$

$$= \frac{1}{s} + \frac{\frac{3}{8}(s + 1) - \frac{17}{8}}{(s + 1)^2 + 3^2} + \frac{-\frac{11}{8}(s + 3) - \frac{13}{8}}{(s + 3)^2 + 1^2}$$

The time response $c(t)$ can be found by taking the inverse Laplace transform of $C(s)$ as follows:

$$c(t) = 1 + \frac{3}{8}e^{-t}\cos 3t - \frac{17}{24}e^{-t}\sin 3t - \frac{11}{8}e^{-3t}\cos t - \frac{13}{8}e^{-3t}\sin t, \quad \text{for } t \geq 0$$

A MATLAB program to obtain the unit-step response of this system is shown in MATLAB Program 3–7. The resulting unit-step response curve is shown in Figure 3–7.

MATLAB Program 3–7

```
>> % ——— Unit-step response ———
>>
>> num = [5   100];
>> den = [1   8   32   80   100];
>> step(num,den)
>> grid
>> title('Unit-Step Response of C(s)/R(s) = (5s + 100)/(s^4 + 8s^3 + 32s^2 + 80s + 100)')
```

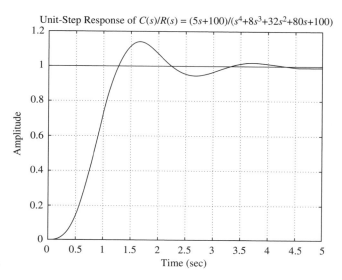

Figure 3–7
Unit-step response curve.

EXAMPLE 3-5 Obtain the unit-step response curves of the system

$$\begin{bmatrix} \dot{x}_1 \\ \dot{x}_2 \end{bmatrix} = \begin{bmatrix} -1 & -1 \\ 6.5 & 0 \end{bmatrix} \begin{bmatrix} x_1 \\ x_2 \end{bmatrix} + \begin{bmatrix} 1 & 1 \\ 1 & 0 \end{bmatrix} \begin{bmatrix} u_1 \\ u_2 \end{bmatrix}$$

$$\begin{bmatrix} y_1 \\ y_2 \end{bmatrix} = \begin{bmatrix} 1 & 0 \\ 0 & 1 \end{bmatrix} \begin{bmatrix} x_1 \\ x_2 \end{bmatrix} + \begin{bmatrix} 0 & 0 \\ 0 & 0 \end{bmatrix} \begin{bmatrix} u_1 \\ u_2 \end{bmatrix}$$

Although it is not necessary to obtain the transfer matrix expression for the system to obtain the unit-step response curves with MATLAB, we shall derive such an expression for reference. For the system defined by

$$\dot{\mathbf{x}} = \mathbf{A}\mathbf{x} + \mathbf{B}\mathbf{u}$$
$$\mathbf{y} = \mathbf{C}\mathbf{x} + \mathbf{D}\mathbf{u}$$

the transfer matrix $\mathbf{G}(s)$ is a matrix that relates $\mathbf{Y}(s)$ and $\mathbf{U}(s)$ as follows:

$$\mathbf{Y}(s) = \mathbf{G}(s)\mathbf{U}(s)$$

Taking Laplace transforms of the state-space equations, we obtain

$$s\mathbf{X}(s) - \mathbf{x}(0) = \mathbf{A}\mathbf{X}(s) + \mathbf{B}\mathbf{U}(s) \quad (3\text{–}8)$$

$$\mathbf{Y}(s) = \mathbf{C}\mathbf{X}(s) + \mathbf{D}\mathbf{U}(s) \quad (3\text{–}9)$$

In deriving the transfer matrix, we assume that $\mathbf{x}(0) = \mathbf{0}$. Then, from Equation (3–8), we get

$$\mathbf{X}(s) = (s\mathbf{I} - \mathbf{A})^{-1}\mathbf{B}\mathbf{U}(s) \quad (3\text{–}10)$$

Substituting Equation (3–10) into Equation (3–9), we obtain

$$Y(s) = [C(sI - A)^{-1}B + D]U(s)$$

Thus, the transfer matrix $G(s)$ is given by

$$G(s) = C(sI - A)^{-1}B + D$$

The transfer matrix $G(s)$ for the given system then becomes

$$G(s) = C(sI - A)^{-1}B$$

$$= \begin{bmatrix} 1 & 0 \\ 0 & 1 \end{bmatrix} \begin{bmatrix} s+1 & 1 \\ -6.5 & s \end{bmatrix}^{-1} \begin{bmatrix} 1 & 1 \\ 1 & 0 \end{bmatrix}$$

$$= \frac{1}{s^2 + s + 6.5} \begin{bmatrix} s & -1 \\ 6.5 & s+1 \end{bmatrix} \begin{bmatrix} 1 & 1 \\ 1 & 0 \end{bmatrix}$$

$$= \frac{1}{s^2 + s + 6.5} \begin{bmatrix} s-1 & s \\ s+7.5 & 6.5 \end{bmatrix}$$

Hence,

$$\begin{bmatrix} Y_1(s) \\ Y_2(s) \end{bmatrix} = \begin{bmatrix} \dfrac{s-1}{s^2+s+6.5} & \dfrac{s}{s^2+s+6.5} \\ \dfrac{s+7.5}{s^2+s+6.5} & \dfrac{6.5}{s^2+s+6.5} \end{bmatrix} \begin{bmatrix} U_1(s) \\ U_2(s) \end{bmatrix}$$

Since this system involves two inputs and two outputs, four transfer functions may be defined, depending on which signals are considered as input and output. Note that, when considering the signal u_1 as the input, we assume that signal u_2 is zero, and vice versa. The four transfer functions are

$$\frac{Y_1(s)}{U_1(s)} = \frac{s-1}{s^2+s+6.5}, \quad \frac{Y_1(s)}{U_2(s)} = \frac{s}{s^2+s+6.5}$$

$$\frac{Y_2(s)}{U_1(s)} = \frac{s+7.5}{s^2+s+6.5}, \quad \frac{Y_2(s)}{U_2(s)} = \frac{6.5}{s^2+s+6.5}$$

The four individual unit-step response curves can be plotted with the command

$$\text{step(A,B,C,D)}$$

MATLAB Program 3–8 produces four such step response curves, shown in Figure 3–8.

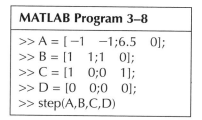

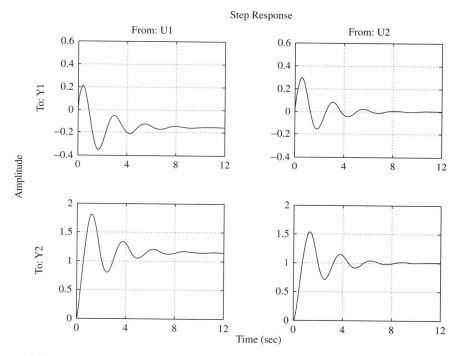

Figure 3–8
Unit-step response curves.

To plot two step response curves for the input u_1 in one diagram and two step response curves for the input u_2 in another diagram, we may use the commands

$$\text{step}(A,B,C,D,1)$$

and

$$\text{step}(A,B,C,D,2)$$

Section 3-2 / Step Response

respectively. MATLAB Program 3–9 plots the desired step-response curves. Figure 3–9 shows the resulting two diagrams, each consisting of two step response curves.

MATLAB Program 3–9

```
>> % ——— Step response curves for system defined in state
>> % space ———
>>
>> % ***** In this program, we plot step response curves of a system
>> % having two inputs (u1 and u2) and two outputs (y1 and y2) *****
>>
>> % ***** First we shall plot step response curves when the input is
>> % u1. Then we shall plot step response curves when the input is
>> % u2 *****
>> % ***** Enter matrices A, B, C, and D *****
>>
>> A = [−1   −1;6.5   0];
>> B = [1    1;1   0];
>> C = [1    0;0   1];
>> D = [0    0;0   0];
>>
>> % ***** To plot step response curves when the input is u1, enter
>> % the command 'step(A,B,C,D,1)' *****
>>
>> step(A,B,C,D,1)
>> grid
>> title('Step Response Plots: Input = u_1 (u_2 = 0)')
>> text(3.4, −0.05, 'Y1')
>> text(3.4, 1.4,'Y2')
>>
>> % ***** Next, we shall plot step response curves when the input
>> % is u2. Enter the command 'step(A,B,C,D,2)' *****
>>
>> step(A,B,C,D,2)
>> grid
>> title('Step Response Plots: Input = u_2 (u_1 = 0)')
>> text(3,0.14,'Y1')
>> text(2.8,1.1,'Y2')
```

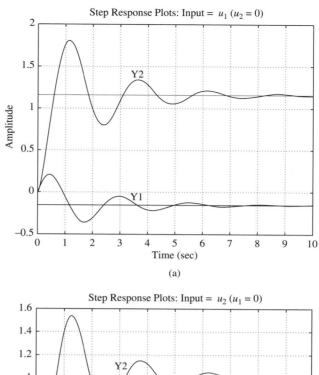

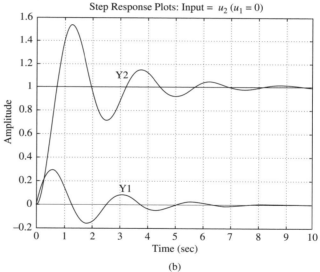

Figure 3–9
Unit-step response curves. (a) u_1 is the input ($u_2 = 0$); (b) u_2 is the input ($u_1 = 0$).

EXAMPLE 3-6 Consider the system shown in Figure 3–10. Obtain the output $c_r(t)$ when the input $r(t)$ is a unit-step displacement. Then obtain the output $c_d(t)$ when the disturbance input $d(t)$ is a unit-step torque. Assume that when $r(t) = 1(t)$, $d(t) = 0$, and vice versa.

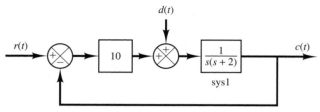

Figure 3–10
Control system.

MATLAB Program 3–10 produces the unit-step response $c_r(t)$ shown in Figure 3–11.

MATLAB Program 3–10

```
>> t = 0:0.01:5;
>> num1 = [1]; den1 = [1   2   0]; sys1 = tf(num1,den1);
>> sys = feedback(10*sys1, [1]);
>> [c_r,t] = step(sys,t);
>> plot(t,c_r)
>> grid
>> title('Unit-Step Response C_r(s)/R(s)')
>> xlabel('t sec')
>> ylabel('Output to Unit-Step Reference Input')
```

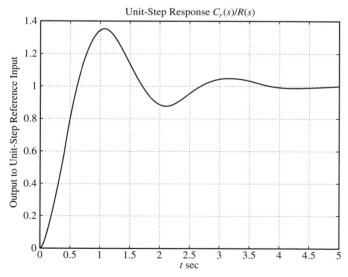

Figure 3–11
Unit-step response $c_r(t)$.

To find the response $c_d(t)$ when the disturbance input $d(t)$ is a unit-step torque, we redraw the block diagram as shown in Figure 3–12.

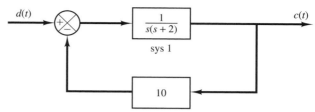

Figure 3–12
Control system when $r(t) = 0$.

MATLAB Program 3–11 will produce the output $c_d(t)$. The output curve is shown in Figure 3–13.

```
>> t = 0:0.01:5;
>> num1 = [1]; den1 = [1   2   0]; sys1 = tf(num1,den1);
>> sys = feedback(sys1, [10]);
>> [c_d,t] = step(sys,t);
>> plot(t,c_d)
>> grid
>> title('Unit-Step Response C_d(s)/D(s)')
>> xlabel('t sec')
>> ylabel('Output to Unit-Step Disturbance Input')
```

MATLAB Program 3–11

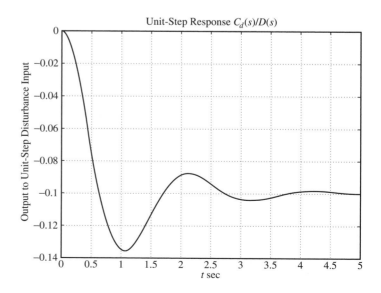

Figure 3–13
Unit-step response $c_d(t)$.

EXAMPLE 3–7 For the system shown in Figure 3–14, obtain the response $c(t)$ to a unit-step input $r(t)$.

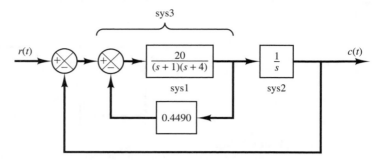

Figure 3–14
Control system.

MATLAB Program 3–12 produces the unit-step response curve $c(t)$, which is shown in Figure 3–15.

```
MATLAB Program 3–12

>> t = 0:0.01:10;
>> num1 = [20]; den1 = [1   5   4]; sys1 = tf(num1,den1);
>> num2 = [1]; den2 = [1   0]; sys2 = tf(num2,den2);
>> sys3 = feedback(sys1,0.4490);
>> sys4 = series(sys3,sys2);
>> sys = feedback(sys4,1);
>> step(sys,t)
>> grid
>> title('Unit-Step Response','Fontsize', 20)
>> xlabel('t sec','Fontsize', 20)
>> ylabel('Output c(t)','Fontsize', 20)
```

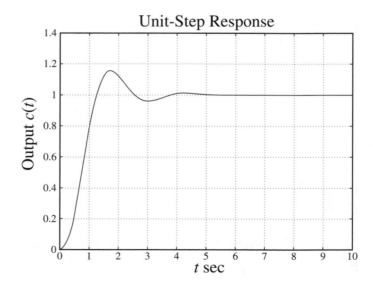

Figure 3–15
Unit-step response curve $c(t)$.

EXAMPLE 3–8 Write a MATLAB program to obtain the unit-step response of the control system shown in Figure 3–16.

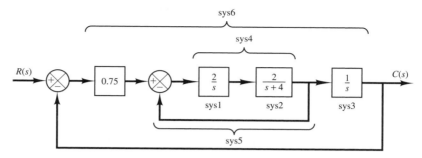

Figure 3–16
Control system.

MATLAB Program 3–13 produces the unit-step response $c(t)$ shown in Figure 3–17.

MATLAB Program 3–13

```
>> t = 0:0.02:12;
>> num1 = [2]; den1 = [1   0]; sys1 = tf(num1,den1);
>> num2 = [2]; den2 = [1   4]; sys2 = tf(num2,den2);
>> num3 = [1]; den3 = [1   0]; sys3 = tf(num3,den3);
>> sys4 = series(sys1,sys2);
>> sys5 = feedback(sys4,1);
>> sys6 = series(0.75*sys5,sys3);
>> sys = feedback(sys6,1);
>> [c,t] = step(sys,t);
>> plot(t,c)
>> grid
>> title('Unit-Step Response C(s)/R(s)')
>> xlabel('t sec')
>> ylabel('Output c(t)')
```

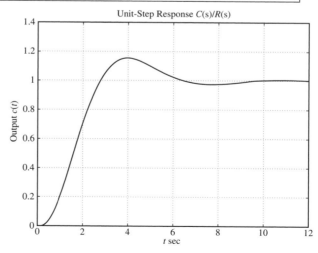

Figure 3–17
Unit-step response $c(t)$.

Simplification of MATLAB programs when repetitive computations are involved. When a MATLAB program involves repetitive computation, we use a *for loop* to execute a series of statements iteratively. The MATLAB syntax for a for loop is

>   for   expression (loop variable)
>   statements
>   end

Note that expression assigns an initial value of the loop variable. Each time the loop variable passes through the loop, it is incremented by a fixed value specified in expression. The statements are executed once during each pass, using the current value of the loop variable. The keyword end denotes the end of the statements. The looping process continues until the loop variable exceeds the ending value. Then MATLAB executes any program listed after end. Note that the loops can be nested and the statements can continue across lines. Any command must terminate with its own end statement. If k commands are involved, there must be k end statements. The following example is illustrative:

```
>> t = 0:0.01:20;
>> K = [5   10   20];
>>    for n = 1:3;                                     % expression
          numg = [K(n)]; deng = [1   11   10   0];     % statements
          sysg = tf(numg,deng);                        % statements
          sys = feedback(sysg, [1]);                   % statements
          y(:,n) = step(sys,t);                        % statements
      end                                              % end
>> plot (t, y)                                         % program
```

In this program, the expression n = 1:3 assigns the initial value 1 to the loop variable n. As n passes through the loop, it is incremented by 1. The statements are executed once during each pass. The looping process continues for n = 2 and n = 3. The looping process ends when n exceeds 3. MATLAB then executes the statement plot(t,y) following end. The statement plot(t,y) plots the outputs y(:,1), y(:,2), and y(:,3) versus t. In the plot, K = 20, K = 10, and K = 5 are entered by means of text statements.

Note that the execution of a set of commands can be carried out conditionally with the use of if, else. The syntax is

>   if statement 1
>   else if statement 2
>   else statement 3
>   end

If statement 1 is true, then statement 1 is executed. If statement 1 is not true, then statement 2 is executed if it is true. Otherwise statement 3 is executed.

EXAMPLE 3-9 Consider the control system shown in Figure 3–18. Draw unit-step response curves for $K = 5, 10,$ and 20.

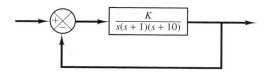

Figure 3–18
Control system.

MATLAB program 3–14 produces unit-step response curves for the three cases $K = 5, K = 10,$ and $K = 20$. The response curves are shown in Figure 3–19.

MATLAB Program 3–14

```
>> t = 0:0.01:20;
>> K = [5   10   20];
>>    for n = 1:3;
           numg = [K(n)]; deng = [1   11   10   0]; sysg = tf(numg,deng);
           sys = feedback(sysg, [1]);
           y(:,n) = step(sys,t);
      end
>> plot(t,y)
>> grid
>> title('Unit-Step Response Curves for K = 5, 10, 20')
>> xlabel('t sec')
>> ylabel('Outputs')
>> text(4.6,1.33,'K = 20')
>> text(6.5,1.12,'K = 10')
>> text(4.4,0.72,'K = 5')
>>
>> % Note that if we want to plot individual curves separately, we use
>> % the commands plot(t,y(:,1)), plot(t,y(:,2)), and plot(t,y(:,3)).
>>
>> subplot(2,2,1), plot(t,y(:,1)), grid, xlabel('t sec'), ylabel('Output')
>> text(10.5,1.27, 'K = 5')
>> subplot(2,2,2), plot(t,y(:,2)), grid, xlabel('t sec'), ylabel('Output')
>> text(10.2,1.27, 'K = 10')
>> subplot(2,2,3), plot(t,y(:,3)), grid, xlabel('t sec'), ylabel('Output')
>> text(10.1,1.27,'K = 20')
```

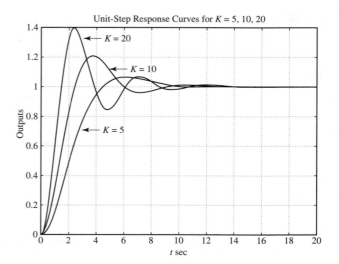

Figure 3–19
Unit-step response curves for $K = 5, 10,$ and 20.

If we want to plot individual response curves separately, we use the commands plot(t,y(:,1)), plot(t,y(:,2)), and plot(t,y(:,3)). (See the last eight lines of MATLAB Program 3–14.) The three individual response curves are shown in Figure 3–20.

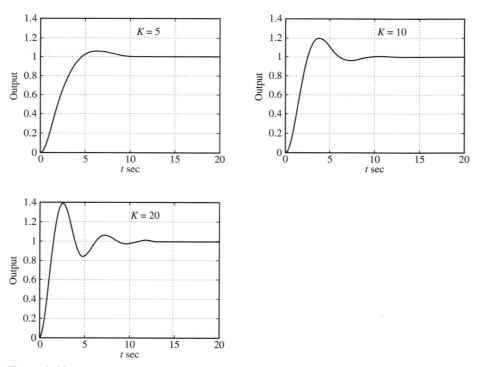

Figure 3–20
Unit-step response curves for $K = 5$, $K = 10$, and $K = 20$ plotted separately.

Chapter 3 / Transient-Response Analysis

EXAMPLE 3-10 Consider the tachometer feedback control system shown in Figure 3–21. Obtain the unit-step response curves when the tachometer feedback variable k assumes five values: $k = 0.1, 0.2, 0.3, 0.4,$ and 0.5.

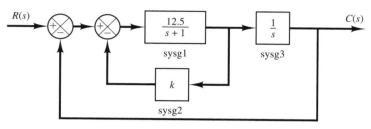

Figure 3–21
Tachometer feedback control system.

MATLAB Program 3–15 produces the five unit-step response curves as shown in Figure 3–22.

```
MATLAB Program 3–15

>> t = 0:0.01:5;
>> k = [0.1   0.2   0.3   0.4   0.5];
>>      for n = 1:5
        numg1 = [12.5]; deng1 = [1   1]; sysg1 = tf(numg1,deng1);
        sysg2 = [k(n)];
        numg3 = [1]; deng3 = [1   0]; sysg3 = tf(numg3,deng3);
        sys1 = feedback(sysg1,sysg2);
        sys2 = series(sys1,sysg3);
        sys = feedback(sys2,1);
        y(:,n) = step(sys,t);
        end
>> plot(t,y)
>> grid
>> title('Unit-Step Response Curves for k = 0.1, 0.2, 0.3, 0.4, and 0.5')
>> xlabel('t sec')
>> ylabel('Outputs')
>> text(1.51,1.28, '\zeta = 0.1')
>> text(1.51,1.16, '\zeta = 0.2')
>> text(1.51,0.75, '\zeta = 0.3')
>> text(1.51,0.67, '\zeta = 0.4')
>> text(1.01,0.53, '\zeta = 0.5')
```

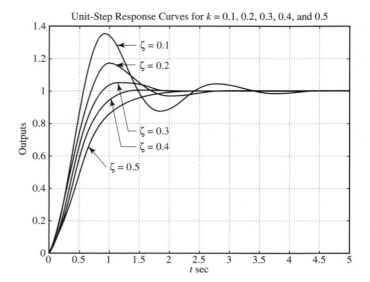

Figure 3–22
Unit-step response curves for $k = 0.1$, 0.2, 0.3, 0.4, and 0.5.

EXAMPLE 3–11 Consider a third-order system defined by

$$\frac{Y(s)}{R(s)} = \frac{6}{s^3 + 6s^2 + 11s + 6} \quad (3\text{–}11)$$

Find a second-order system of the form

$$\frac{Y(s)}{U(s)} = \frac{\omega_n^2}{s^2 + 2\zeta\omega_n s + \omega_n^2} \quad (3\text{–}12)$$

that closely approximates the unit-step response of the third-order system given by Equation (3–11).

We choose the search region to be

$$0.65 \leq \zeta \leq 1.00$$
$$0.5 \leq \omega_n \leq 4$$

We also arbitrarily choose the computation steps for ζ and ω_n to be 0.05 and 0.5, respectively. Then, for $p = 1, 2, 3, \ldots, 8$, we have

$$\zeta = 0.6 + 0.05p$$

and for $q = 1, 2, 3, \ldots, 8$, we have

$$\omega_n = 0.5q$$

We desire the error squared at any time t, namely,

$$[y_1(t) - y_2(t)]^2$$

to be smaller than 0.05, where $y_1(t)$ is the unit-step response of the third-order system defined by Equation (3–11) and $y_2(t)$ is the unit-step response of the second-order system given by Equation (3–12).

In MATLAB Program 3–16, the (ζ, ω_n) pairs that satisfy the condition are shown in the

MATLAB Program 3–16

```
>> t = 0:0.01:6;
>> k = 0;
>> for p = 1:8;
        zeta (p) = 0.6+0.05*p;
        for q = 1:8;
           w(q) = 0.5*q;
           num1 = [6]; den1 = [1   6   11   6];
           sys1 = tf(num1,den1);
           num2 = [w(q).^2]; den2 = [1   2*zeta (p)*w(q)   w(q).^2];
           sys2 = tf(num2,den2);
        [y1,t] = step(sys1,t);
        [y2,t] = step(sys2,t);
           error_square = (y1 − y2).^2;
           emax = max(error_squared);
            if emax < 0.05;
           k = k+1;
           solution (k,:) = [k   zeta (p)   w(q)   emax];
        end
      end
    end
>> solution

solution =

    1.0000    0.6500    1.0000    0.0204
    2.0000    0.7000    1.0000    0.0117
    3.0000    0.7500    1.0000    0.0059
    4.0000    0.8000    1.0000    0.0024
    5.0000    0.8500    1.0000    0.0014
    6.0000    0.9000    1.0000    0.0012
    7.0000    0.9500    1.0000    0.0015
    8.0000    0.9500    1.5000    0.0456
    9.0000    1.0000    1.0000    0.0037
   10.0000    1.0000    1.5000    0.0384
```

form of a table. From this table, we find that the combination of $\zeta = 0.90$ and $\omega_n = 1.0$ produces the minimum error squared and the combination of $\zeta = 0.85$ and $\omega_n = 1.0$ produces the second minimum error squared. Note that this conclusion is correct only if ζ and ω_n are chosen from

$$\zeta = 0.65, 0.70, 0.75, 0.80, 0.85, 0.90, 0.95, 1.0$$
$$\omega_n = 0.5, 1.0, 1.5, 2.0, 2.5, 3.0, 3.5, 4.0$$

If the ranges of ζ and ω_n are different from those just given or the step sizes are different, the best combination of ζ and ω_n may also be different.

Let us define the system given by Equation (3–11) as System 1. We then define the system having $\zeta = 0.90$ and $\omega_n = 1$ to be System 2_1, or

$$\text{System } 2_1: \frac{1}{s^2 + 1.8s + 1}$$

Also, define the system having $\zeta = 0.85$ and $\omega_n = 1$ to be System 2_2, or

$$\text{System } 2_2: \frac{1}{s^2 + 1.7s + 1}$$

MATLAB Program 3–17 produces the unit-step responses of System 1 and System 2_1 in one diagram and the unit-step responses of System 1 and System 2_2 in another diagram. These curves are shown in Figures 3–23 and 3–24. Notice that both System 2_1 and System 2_2 closely approximate the unit-step response of System 1.

MATLAB Program 3–17

```
>> sys1 = tf([6], [1  6  11  6]);
>> sys2_1 = tf([1], [1  2*0.9  1]);
>> sys2_2 = tf([1], [1  2*0.85  1]);
>> [y1,t] = step(sys1,t);
>> [y2_1,t] = step(sys2_1,t);
>> [y2_2,t] = step(sys2_2,t);
>>
>> plot(t,y1,t,y2_1)   % See Figure 3–23.
>> grid
>> title('Unit-Step Responses of sys1 and sys2_1')
>> xlabel('t sec')
>> ylabel('Outputs y1 and y2_1')
>> text(2.05,0.87, 'sys1')
>> text(3.3,0.72, sys2_1')
>> text(1.45,0.14, sys1')
>> text(0.16,0.25, sys2_1')
>>
>> plot(t,y1,t,y2_2)   % See Figure 3–24.
>> grid
>> title('Unit-Step Responses of sys1 and sys2_2')
>> xlabel('t sec')
>> ylabel('Outputs y1 and y2_2')
>> text(4.1,0.85,'sys1')
>> text(3.05,1.1,'sys2_2')
>> text(1.4,0.08,'sys1')
>> text(0.1,0.28,'sys2_2')
```

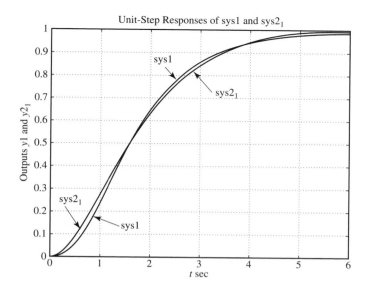

Figure 3–23
Unit-step response curves of System 1 and System 2_1.

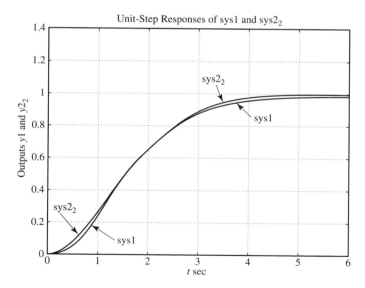

Figure 3–24
Unit-step response curves of System 1 and System 2_2.

Section 3–2 / Step Response

EXAMPLE 3-12 Consider the control system defined by

$$\frac{C(s)}{R(s)} = \frac{\omega_n^2}{s^2 + 2\zeta\omega_n s + \omega_n^2}$$

Using a for loop, write a MATLAB program to obtain the unit-step response of this system for the following four cases:

Case 1: $\zeta = 0.3$, $\omega_n = 1$
Case 2: $\zeta = 0.5$, $\omega_n = 2$
Case 3: $\zeta = 0.7$, $\omega_n = 4$
Case 4: $\zeta = 0.8$, $\omega_n = 6$

Define $\omega_n^2 = a$ and $2\zeta\omega_n = b$. Then a and b each have four elements as follows:

$$a = [1 \quad 4 \quad 16 \quad 36]$$
$$b = [0.6 \quad 2 \quad 5.6 \quad 9.6]$$

Using vectors a and b, MATLAB Program 3–18 will produce the unit-step response curves as shown in Figure 3–25.

MATLAB Program 3–18

```
>> a = [1   4   16   36];
>> b = [0.6   2   5.6   9.6];
>> t = 0:0.1:8;
>> y = zeros(81,4);
>>      for i = 1:4;
        num = [a(i)];
        den = [1   b(i)   a(i)];
        y(:,i) = step(num,den,t);
        end
>> plot(t,y(:,1),'o',t,y(:,2),'x',t,y(:,3),'-',t,y(:,4),'-.')
>> grid
>> title('Unit-Step Response Curves for Four Cases')
>> xlabel('t Sec')
>> ylabel('Outputs')
>> gtext('1')
>> gtext('2')
>> gtext('3')
>> gtext('4')
```

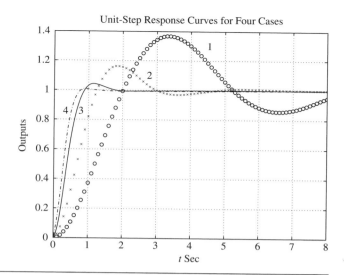

Figure 3–25
Unit-step response curves for four cases.

Obtaining rise time, peak time, maximum overshoot, and settling time with MATLAB. MATLAB can conveniently be used to obtain the rise time, peak time, maximum overshoot, and settling time of the unit-step response curve. Consider the system defined by

$$\frac{C(s)}{R(s)} = \frac{25}{s^2 + 6s + 25}$$

MATLAB Program 3–19 yields the rise time, peak time, maximum overshoot, and settling time of the unit-step response curve. A unit-step response curve for the system is given in Figure 3–26 to verify the results obtained with MATLAB Program 3–19. The program is applied to a higher-order system in the next example.

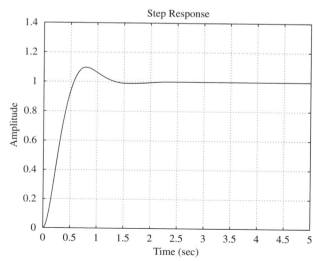

Figure 3–26
Unit-step response curve.

Section 3–2 / Step Response

MATLAB Program 3-19

```
>> % ——— This is a MATLAB program to find the rise time, peak time,
>> % maximum overshoot, and settling time of the unit-step response
>> % curve of the second-order system and higher-order system ———
>>
>> % ——— In this example, we assume that zeta = 0.6 and wn = 5 ———
>> num = [25];
>> den = [1  6  25];
>> t = 0:0.005:5;
>> [y,x,t] = step(num,den,t);
>> r =1; while y(r) < 1.0001; r = r + 1; end;
>> rise_time = (r − 1)*0.005
rise_time =

    0.5550

>> [ymax,tp] = max(y);
>> peak_time = (tp − 1)*0.005
peak_time =

    0.7850

>> max_overshoot = ymax − 1
max_overshoot =

    0.0948

>> s = 1001; while y(s) > 0.98 & y(s) < 1.02; s = s − 1; end;
>> settling_time = (s − 1)*0.005
settling _time =

    1.1850
```

EXAMPLE 3-13 Consider a higher-order system defined by

$$\frac{C(s)}{R(s)} = \frac{6.3223s^2 + 18s + 12.811}{s^4 + 6s^3 + 11.3223s^2 + 18s + 12.811}$$

Using MATLAB, plot the unit-step response curve of this system and obtain the rise time, peak time, maximum overshoot, and settling time.

MATLAB Program 3–20 plots the unit-step response curve, as well as gives the rise time, peak time, maximum overshoot, and settling time. The unit-step response curve is shown in Figure 3–27.

```
MATLAB Program 3–20

>> % ——— This program plots the unit-step response curve and finds the
>> % rise time, peak time, maximum overshoot, and settling time.
>> % In this program the rise time is calculated as the time required for the
>> % response to rise from 10% to 90% of its final value. ———
>>
>> num = [6.3223   18   12.811];
>> den = [1   6   11.3223   18   12.811];
>> t = 0:0.02:20;
>> [y,x,t] = step(num,den,t);
>> plot(t,y)
>> grid
>> title('Unit-Step Response')
>> xlabel('t (sec)')
>> ylabel('Output y(t)')
>> r1 = 1; while y(r1) < 0.1, r1 = r1+1; end;
>> r2 = 1; while y(r2) < 0.9, r2 = r2+1; end;
>> rise_time = (r2 – r1)*0.02
rise_time =

    0.5800

>> [ymax,tp] = max(y);
>> peak_time = (tp – 1)*0.02
peak_time =

    1.6600

>> max_overshoot = ymax – 1
max_overshoot =

    0.6182

>> s = 1001; while y(s) > 0.98 & y(s) < 1.02; s = s – 1; end;
>> settling_time = (s – 1)*0.02
settling_time =

    10.0200
```

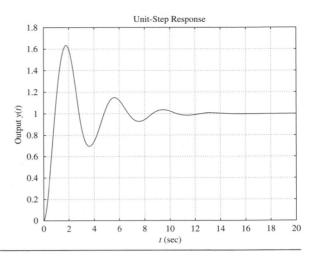

Figure 3–27
Unit-step response curve.

3–3 IMPULSE RESPONSE

The unit-impulse response of a control system may be obtained with one of the following MATLAB commands:

    impulse(num,den)   or   impulse(sys)
    impulse(A,B,C,D)
    impulse(num,den,t)   or   impulse(sys,t)
    y = impulse(num,den)   or   y = impulse(sys)
    [y,t] = impulse(num,den,t)   or   [y,t] = impulse(sys,t)   (3–13)
    [y,x,t] = impulse(num,den)
    [y,x,t] = impulse(num,den,t)   (3–14)
    [y,x,t] = impulse(A,B,C,D)
    [y,x,t] = impulse(A,B,C,D,iu)   (3–15)
    [y,x,t] = impulse(A,B,C,D,iu,t)   (3–16)

The command impulse(num,den) plots the unit-impulse response on the screen. The command impulse(A,B,C,D) produces a series of unit-impulse-response plots, one for each input and output combination of the system

$$\dot{x} = Ax + Bu$$
$$y = Cx + Du$$

with the time vector automatically determined. In equations such as Equations (3–13), (3–14) and (3–16), t is a user-supplied time vector that specifies the times at which the impulse response is to be computed. In Equations (3–15) and (3–16), the scalar iu is an index into the inputs of the system and specifies which input is to be used for the impulse response.

If MATLAB is invoked with the left-hand argument [y,x,t], as in the case of [y,x,t] = impulse(A,B,C,D), the command returns the output and state responses of the system and the time vector t. No plot is drawn on the screen. The matrices y

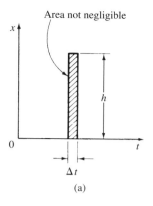
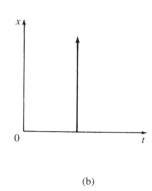

Figure 3-28
Impulse inputs.   (a)   (b)

and x contain the output and state responses of the system evaluated at the time points t. (y has as many columns as outputs and one row for each element in t. x has as many columns as state variables and one row for each element in t.)

Before discussing computational solutions of problems involving impulse inputs, we present some necessary background material.

Impulse input. The impulse response of a mechanical system can be observed when the system is subjected to a very large force for a very short time—for instance, when the mass of a spring–mass–dashpot system is hit by a hammer or a bullet. Mathematically, such an impulse input can be expressed by an impulse function.

The impulse function is a mathematical function without any actual physical counterpart. However, as shown in Figure 3-28(a), if the actual input lasts for a short time (Δt s), but has a large amplitude (h), so that the area ($h\Delta t$) in a time plot is not negligible, it can be approximated by an impulse function. The impulse input is usually denoted by a vertical arrow, as shown in Figure 3-28(b), to indicate that it has a very short duration and a very large height.

In handling impulse functions, only the magnitude (or area) of the function is important; its actual shape is immaterial. In other words, an impulse of amplitude $2h$ and duration $\Delta t/2$ can be considered the same as an impulse of amplitude h and duration Δt, as long as Δt approaches zero and $h\Delta t$ is finite.

EXAMPLE 3-14 Obtain the unit-impulse response of the following system:

$$\begin{bmatrix} \dot{x}_1 \\ \dot{x}_2 \end{bmatrix} = \begin{bmatrix} 0 & 1 \\ -1 & -1 \end{bmatrix} \begin{bmatrix} x_1 \\ x_2 \end{bmatrix} + \begin{bmatrix} 0 \\ 1 \end{bmatrix} u$$

$$y = \begin{bmatrix} 1 & 0 \end{bmatrix} \begin{bmatrix} x_1 \\ x_2 \end{bmatrix} + [0]u$$

A possible MATLAB program is shown in MATLAB Program 3-21. The resulting response curve is shown in Figure 3-29.

MATLAB Program 3–21

```
>> A = [0   1;-1   -1];
>> B = [0;1];
>> C = [1   0];
>> D = [0];
>> impulse(A,B,C,D);
>> grid;
>> title('Unit-Impulse Response')
```

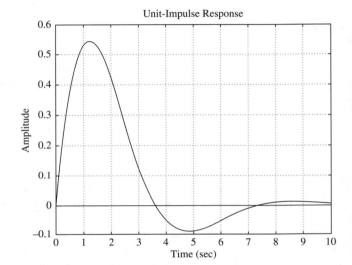

Figure 3–29
Unit-impulse response curve.

EXAMPLE 3–15 Obtain the unit-impulse response of the system

$$\frac{C(s)}{R(s)} = G(s) = \frac{1}{s^2 + 0.2s + 1}$$

MATLAB Program 3–22 will produce the unit-impulse response. The resulting plot is shown in Figure 3–30.

MATLAB Program 3–22

```
>> num = [1];
>> den = [1   0.2   1];
>> impulse(num,den);
>> grid
>> title('Unit-Impulse Response of G(s) = 1/(s^2 + 0.2s + 1)')
```

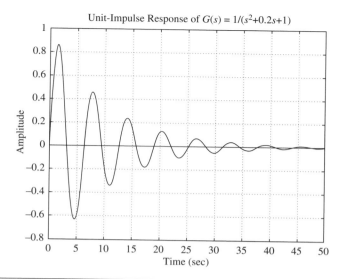

Figure 3–30
Unit-impulse response curve.

Alternative approach to obtaining the impulse response. Note that when the initial conditions are zero the unit-impulse response of $G(s)$ is the same as the unit-step response of $sG(s)$.

Consider the unit-impulse response of the system in Example 3–15. Since $R(s) = 1$ for the unit-impulse input, we have

$$\frac{C(s)}{R(s)} = C(s) = G(s) = \frac{1}{s^2 + 0.2s + 1}$$

$$= \frac{s}{s^2 + 0.2s + 1} \frac{1}{s}$$

We can thus convert the unit-impulse response of $G(s)$ to the unit-step response of $sG(s)$.

If we enter

$$\text{num} = [0 \quad 1 \quad 0]$$
$$\text{den} = [1 \quad 0.2 \quad 1]$$

into MATLAB and use the step-response command as given in MATLAB Program 3–23, we obtain a plot of the unit-impulse response of the system as shown in Figure 3–31. Notice that the curves shown in Figures 3–30 and 3–31 are identical.

MATLAB Program 3–23

```
>> num = [1  0];
>> den = [1  0.2  1];
>> step(num,den);
>> grid
>> title('Unit-Step Response of sG(s) = s/(s^2 + 0.2s + 1)')
```

Section 3–3 / Impulse Response

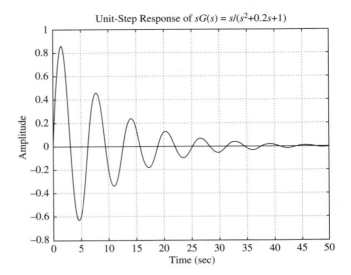

Figure 3–31
Unit-impulse response curve obtained as the unit-step response of $sG(s) = s/(s^2 + 0.2s + 1)$.

3–4 RAMP RESPONSE

There is no ramp command in MATLAB. Therefore, we need to use the step command or lsim command (presented in Section 3–5) to obtain the ramp response. Specifically, to obtain the ramp response of the transfer-function system $G(s)$ with zero initial condition, divide $G(s)$ by s and use the step-response command. For example, consider the closed-loop system

$$\frac{C(s)}{R(s)} = \frac{1}{s^2 + s + 1}$$

For a unit-ramp input, $R(s) = 1/s^2$. Hence,

$$C(s) = \frac{1}{s^2 + s + 1}\frac{1}{s^2} = \frac{1}{(s^2 + s + 1)s}\frac{1}{s}$$

To obtain the unit-ramp response of this system, enter the following numerator and denominator into the MATLAB program:

$$\text{num} = [0 \ \ 0 \ \ 0 \ \ 1];$$
$$\text{den} = [1 \ \ 1 \ \ 1 \ \ 0];$$

Then use the step response command. (See MATLAB Program 3–24.) The plot obtained with this program is shown in Figure 3–32.

MATLAB Program 3–24

```
>> % ——— Unit-ramp response ———
>>
>> % ***** The unit-ramp response is obtained as the unit-step
>> % response of G(s)/s *****
>> % ***** Enter the numerator and denominator of G(s)/s *****
>>
>> num = [1];
>> den = [1   1   1   0];
>>
>> % ***** Specify the computing time points (such as t = 0:0.1:7)
>> % and then enter step-response command: c = step(num,den,t) *****
>>
>> t = 0:0.1:7;
>> c = step(num,den,t);
>>
>> % ***** In plotting the ramp-response curve, add the reference
>> % input to the plot. The reference input is t. Add to the
>> % argument of the plot command with the following: t,t, '-'. Thus,
>> % the plot command becomes plot(t,c,'o',t,t,'-') *****
>> % ***** Add grid, title, xlabel, and ylabel *****
>>
>> plot(t,c,'o',t,t,'-')
>> grid
>> title('Unit-Ramp Response Curve for System G(s) = 1/(s^2 + s + 1)')
>> xlabel('t Sec')
>> ylabel('Input and Output')
```

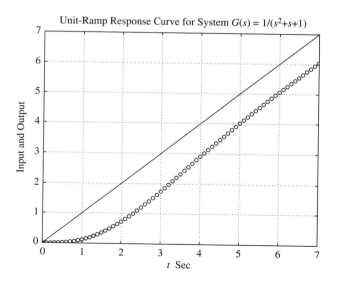

Figure 3–32
Unit-ramp response curve.

Unit-ramp response of a system defined in state space. Next, we shall treat the unit-ramp response of the system in state-space form. Consider the system described by

$$\dot{\mathbf{x}} = \mathbf{A}\mathbf{x} + \mathbf{B}u$$
$$y = \mathbf{C}\mathbf{x} + Du$$

where u is the unit-ramp function. We assume the initial condition to be zero. In what follows, we shall consider a simple example to explain the method. Suppose

$$\mathbf{A} = \begin{bmatrix} 0 & 1 \\ -1 & -1 \end{bmatrix}, \quad \mathbf{B} = \begin{bmatrix} 0 \\ 1 \end{bmatrix}, \quad \mathbf{x}(0) = \mathbf{0}$$
$$\mathbf{C} = \begin{bmatrix} 1 & 0 \end{bmatrix}, \quad D = \begin{bmatrix} 0 \end{bmatrix}$$

When the initial conditions are zeros, the unit-ramp response is the integral of the unit-step response. Hence the unit-ramp response can be given by

$$z = \int_0^t y\, dt \tag{3-17}$$

From Equation (3–17), we obtain

$$\dot{z} = y = x_1 \tag{3-18}$$

Let us define

$$z = x_3$$

Then Equation (3–18) becomes

$$\dot{x}_3 = x_1 \tag{3-19}$$

Combining Equation (3–19) with the original state-space equation, we obtain

$$\begin{bmatrix} \dot{x}_1 \\ \dot{x}_2 \\ \dot{x}_3 \end{bmatrix} = \begin{bmatrix} 0 & 1 & 0 \\ -1 & -1 & 0 \\ 1 & 0 & 0 \end{bmatrix} \begin{bmatrix} x_1 \\ x_2 \\ x_3 \end{bmatrix} + \begin{bmatrix} 0 \\ 1 \\ 0 \end{bmatrix} u \tag{3-20}$$

$$z = \begin{bmatrix} 0 & 0 & 1 \end{bmatrix} \begin{bmatrix} x_1 \\ x_2 \\ x_3 \end{bmatrix} \tag{3-21}$$

where u appearing in Equation (3–20) is the unit-step function. Equations (3–20) and (3–21) can be written as

$$\dot{\mathbf{x}} = \mathbf{AA}\mathbf{x} + \mathbf{BB}u$$
$$z = \mathbf{CC}\mathbf{x} + DDu$$

where

$$\mathbf{AA} = \begin{bmatrix} 0 & 1 & 0 \\ -1 & -1 & 0 \\ 1 & 0 & 0 \end{bmatrix} = \begin{bmatrix} \mathbf{A} & \mathbf{0} \\ \hline \mathbf{C} & 0 \end{bmatrix}$$

$$\mathbf{BB} = \begin{bmatrix} 0 \\ 1 \\ 0 \end{bmatrix} = \begin{bmatrix} \mathbf{B} \\ 0 \end{bmatrix}, \quad \mathbf{CC} = \begin{bmatrix} 0 & 0 & 1 \end{bmatrix}, \quad DD = \begin{bmatrix} 0 \end{bmatrix}$$

Note that x_3 is the third element of **x**. A plot of the unit-ramp response curve $z(t)$ can be obtained by entering MATLAB Program 3–25 into the computer. A plot of the unit-ramp response curve obtained from this MATLAB program is shown in Figure 3–33.

MATLAB Program 3–25

```
>> % ————— Unit-ramp response —————
>>
>> % ***** The unit-ramp response is obtained by adding a new
>> % state variable x3. The dimension of the state equation
>> % is enlarged by 1 *****
>> % ***** Enter matrices A, B, C, and D of the original state
>> % equation and output equation *****
>>
>> A = [0    1;-1   -1];
>> B = [0;   1];
>> C = [1    0];
>> D = [0];
>>
>> % ***** Enter matrices AA, BB, CC, and DD of the new,
>> % enlarged state equation and output equation *****
>>
>> AA = [A   zeros(2,1);C   0];
>> BB = [B;0];
>> CC = [0   0    1];
>> DD = [0];
>>
>> % ***** Enter step-response command: [z,x,t] = step(AA,BB,CC,DD) *****
>>
>> [z,x,t] = step(AA,BB,CC,DD);
>>
>> % ***** In plotting x3, add the unit-ramp input t in the plot
>> % by entering the following command: plot(t,x3,'o',t,t,'-') *****
>>
>> x3 = [0   0    1]*x'; plot(t,x3,'o',t,t,'-')
>> grid
>> title('Unit-Ramp Response')
>> xlabel('t Sec')
>> ylabel('Input and Output')
```

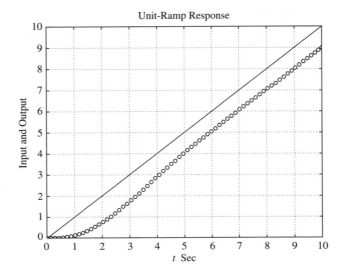

Figure 3-33
Unit-ramp response curve.

EXAMPLE 3-16 When the closed-loop system involves numerator dynamics, the unit-step response curve may exhibit a large overshoot. Obtain the unit-step response of the following system with MATLAB:

$$\frac{C(s)}{R(s)} = \frac{10s + 4}{s^2 + 4s + 4}$$

Obtain also the unit-ramp response with MATLAB.

MATLAB Program 3-26 produces the unit-step response as well as the unit-ramp response of the system. The unit-step response curve and unit-ramp response curve, together with the unit-ramp input, are shown in Figures 3-34(a) and (b), respectively.

Notice that the unit-step response curve exhibits over 115% of overshoot. The unit-ramp response curve leads the input curve. These phenomena occurred because of the presence of a large derivative term in the numerator.

MATLAB Program 3–26

```
>> num = [10   4];
>> den = [1   4   4];
>> t = 0:0.02:10;
>> y = step(num,den,t);
>> plot(t,y)
>> grid
>> title('Unit-Step Response')
>> xlabel('t (sec)')
>> ylabel('Output')
>>
>> num1 = [10   4];
>> den1 = [1   4   4   0];
>> y1 = step(num1,den1,t);
>> plot(t,t,'--',t,y1)
>> v = [0   10   0   10]; axis(v);
>> grid
>> title('Unit-Ramp Response')
>> xlabel('t (sec)')
>> ylabel('Unit-Ramp Input and Output')
>> text(6.1,5.0,'Unit-Ramp Input')
>> text(3.5,7.1,'Output')
```

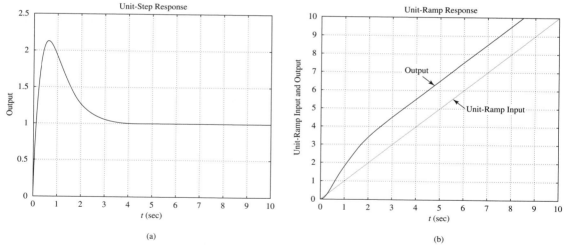

Figure 3–34
(a) Unit-step response curve; (b) unit-ramp response curve plotted with unit-ramp input.

3-5 RESPONSE TO ARBITRARY INPUT

Obtaining the response to arbitrary input. To obtain the response of a time-invariant system to an arbitrary input, the command lsim may be used. Commands such as

$$\begin{array}{ll}
\text{lsim(num,den,u,t),} & \text{lsim(A,B,C,D,u,t)} \\
\text{lsim(sys,u,t),} & \text{y = lsim(num,den,u,t)} \\
\text{y = lsim(A,B,C,D,u,t),} & \text{y = lsim(sys,u,t)} \\
\text{[y,t] = lsim(sys,u,t)} &
\end{array}$$

will generate the response to the input time function u if the initial conditions are zero. If t is given as

$$t = 0: \Delta t : T$$

then the response is computed every Δt sec, beginning at $t = 0$ and ending at T, where T is a positive integer multiple of Δt. Note that the command with a left-hand argument such as

$$y = \text{lsim(sys,u,t)}$$

returns the output response y. The columns of the matrix y are outputs and the number of its rows equals the length of t. No plot is drawn. To plot the response curve, it is necessary to use the command plot(t,y).

If the initial conditions are nonzero in a state-space model, the command

$$\text{lsim(sys,u,t,}x_0\text{)}$$

where x_0 is the initial state, produces the response of the system subjected to the input u and the initial condition x_0. (This case will be discussed in Section 3-6.)
Note also that the command

$$\text{lsim(sys1, sys2,\ldots, u,t)}$$

plots the responses of systems (sys1, sys2, ...) on a single diagram.

EXAMPLE 3-17 Using the lsim command, obtain the unit-ramp response of the following system:

$$\frac{C(s)}{R(s)} = \frac{1}{s^2 + s + 1}$$

We may enter MATLAB Program 3-27 into the computer to obtain the unit-ramp response. The resulting plot is shown in Figure 3-35.

MATLAB Program 3–27

```
>> % ———— Ramp Response ————
>>
>> num = [1];
>> den = [1   1   1];
>> t = 0:0.1:8;
>> r = t;
>> y = lsim(num,den,r,t);
>> plot(t,r,'-', t,y,'o')
>> grid
>> title('Unit-Ramp Response Obtained by Use of Command "lsim"')
>> xlabel('t Sec')
>> ylabel('Unit-Ramp Input and System Output')
>> text(2.1,4.65,'Unit-Ramp Input')
>> text(4.5,2.0,'Output')
```

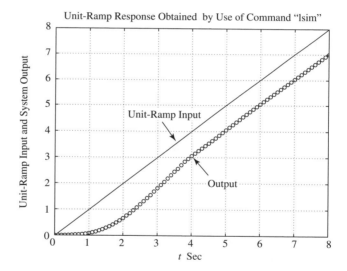

Figure 3–35
Unit-ramp response.

EXAMPLE 3–18 Consider the system

$$\begin{bmatrix} \dot{x}_1 \\ \dot{x}_2 \end{bmatrix} = \begin{bmatrix} -1 & 0.5 \\ -1 & 0 \end{bmatrix} \begin{bmatrix} x_1 \\ x_2 \end{bmatrix} + \begin{bmatrix} 0 \\ 1 \end{bmatrix} u$$

$$y = \begin{bmatrix} 1 & 0 \end{bmatrix} \begin{bmatrix} x_1 \\ x_2 \end{bmatrix}$$

Section 3–5 / Response to Arbitrary Input

Using MATLAB, obtain the response curves $y(t)$ when the input u is given by

1. $u = $ unit-step input
2. $u = e^{-t}$

Assume that the initial state is $\mathbf{x}(0) = \mathbf{0}$.

A possible MATLAB program to produce the responses of this system to the unit-step input $[u = 1(t)]$ and the exponential input $[u = e^{-t}]$ is shown in MATLAB Program 3–28. The resulting response curves are shown in Figures 3–36(a) and (b), respectively.

MATLAB Program 3–28

```
>> t = 0:0.1:12;
>> A = [-1   0.5;-1   0];
>> B = [0;1];
>> C = [1   0];
>> D = [0];
>>
>> % For the unit-step input u = 1(t), use the command "y = step(A,B,C,D,1,t)".
>>
>> y = step(A,B,C,D,1,t);
>> plot(t,y)
>> grid
>> title('Unit-Step Response')
>> xlabel('t Sec')
>> ylabel('Output')
>>
>> % For the response to exponential input u = exp(-t), use the command
>> % z = lsim(A,B,C,D,u,t).
>>
>> u = exp(-t);
>> z = lsim(A,B,C,D,u,t);
>> plot(t,u,'-',t,z,'o')
>> grid
>> title('Response to Exponential Input u = exp^-^t')
>> xlabel('t Sec')
>> ylabel('Exponential Input and System Output')
>> text(2.3,0.49,'Exponential input')
>> text(6.4,0.28,'Output')
```

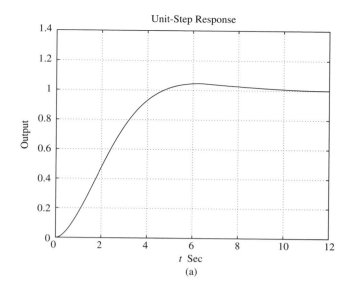

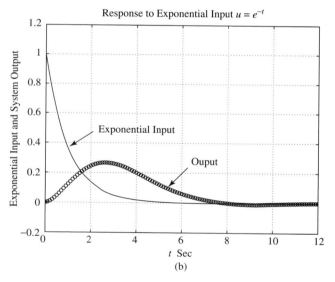

Figure 3–36
(a) Unit-step response; (b) response to input $u = e^{-t}$.

EXAMPLE 3–19 Using the lsim command, obtain the unit-ramp response of the closed-loop control system whose closed-loop transfer function is

$$\frac{C(s)}{R(s)} = \frac{s + 10}{s^3 + 6s^2 + 9s + 10}$$

Also, obtain the response of this system when the input is given by

$$r_1 = e^{-0.5t}$$

MATLAB Program 3–29 produces the unit-ramp response and the response to the exponential input $r_1 = e^{-0.5t}$. The resulting response curves are shown in Figures 3–37(a) and (b), respectively.

MATLAB Program 3–29

```
>> % —— Unit-Ramp Response ——
>>
>> num = [1   10];
>> den = [1   6   9   10];
>> t = 0:0.1:10;
>> r = t;
>> y = lsim(num,den,r,t);
>> plot(t,r,'-',t,y,'o')
>> grid
>> title('Unit-Ramp Response by Use of Command "lsim"')
>> xlabel('t Sec')
>> ylabel('Output')
>> text(3.2,6.5,'Unit-Ramp Input')
>> text(6.0,3.1,'Output')
>>
>> % —— Response to Input r1 = exp(-0.5t). ——
>>
>> num = [1   10];
>> den = [1   6   9   10];
>> t = 0:0.1:12;
>> r1 = exp(-0.5*t);
>> y1 = lsim(num,den,r1,t);
>> plot(t,r1,'-',t,y1,'o')
>> grid
>> title('Response to Input r_1 = exp^-^0^.^5^t')
>> xlabel('t Sec')
>> ylabel('Input and Output')
>> text(1.4,0.75,'Input r_1 = exp^-^0^.^5^t')
>> text(6.2,0.34,'Output')
```

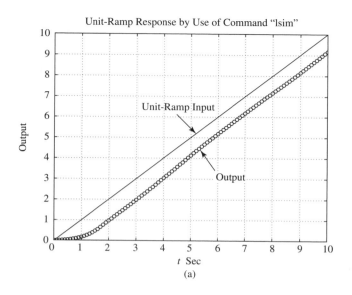

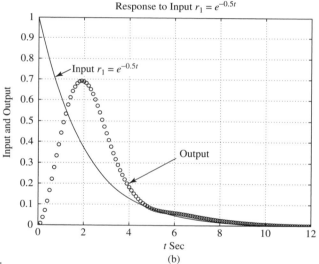

Figure 3-37
(a) Unit-ramp response curve; (b) response to exponential input $r_1 = e^{-0.5t}$.

EXAMPLE 3-20 Obtain the response of the closed-loop system defined by

$$\frac{C(s)}{R(s)} = \frac{5}{s^2 + s + 5}$$

when the input $r(t)$ is given by

$$r(t) = 2 + t$$

[The input $r(t)$ is a step input of magnitude 2 plus a unit-ramp input.]

Section 3-5 / Response to Arbitrary Input

A possible MATLAB program is shown in MATLAB Program 3–30. The resulting response curve, together with a plot of the input function, is shown in Figure 3–38.

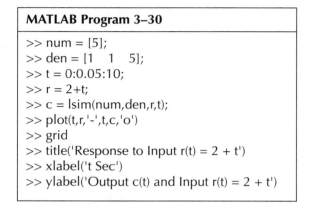

MATLAB Program 3–30

```
>> num = [5];
>> den = [1   1   5];
>> t = 0:0.05:10;
>> r = 2+t;
>> c = lsim(num,den,r,t);
>> plot(t,r,'-',t,c,'o')
>> grid
>> title('Response to Input r(t) = 2 + t')
>> xlabel('t Sec')
>> ylabel('Output c(t) and Input r(t) = 2 + t')
```

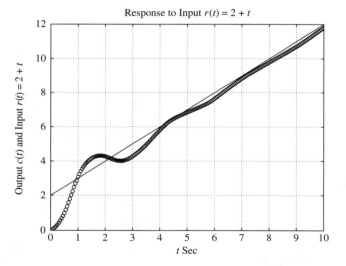

Figure 3–38
Response to input $r(t) = 2 + t$.

EXAMPLE 3–21 Obtain the response of the system shown in Figure 3–39 when the input $r(t)$ is given by

$$r(t) = \frac{1}{2}t^2$$

[The input $r(t)$ is the unit-acceleration input.]

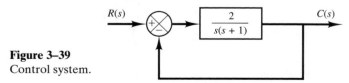

Figure 3–39
Control system.

The closed-loop transfer function is

$$\frac{C(s)}{R(s)} = \frac{2}{s^2 + s + 2}$$

MATLAB Program 3–31 produces the unit-accerelation response. The resulting response, together with the unit-acceleration input, is shown in Figure 3–40.

MATLAB Program 3–31

```
>> num = [2];
>> den = [1   1    2];
>> t = 0:0.2:10;
>> r = 0.5*t.^2;
>> y = lsim(num,den,r,t);
>> plot(t,r,'-',t,y,'o',t,y,'-')
>> grid
>> title('Unit-Acceleration Response')
>> xlabel('t Sec')
>> ylabel('Input and Output')
>> text(2.1,27.5,'Unit-Acceleration Input')
>> text(7.2,7.5,'Output')
```

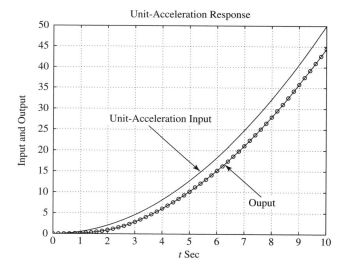

Figure 3–40
Response to unit-acceleration input.

EXAMPLE 3-22 Plot the response curve $y(t)$ versus t for the equation

$$\ddot{y} + 3\dot{y} + 5y = u(t), \qquad t \geq 0$$

when the input $u(t)$ is an acceleration input $t^2/2$. Assume that $y(0) = 0$ and $\dot{y}(0) = 0$.

To plot the response curve $y(t)$ for $0 \leq t \leq 10$ with a time increment of 0.01 sec, enter MATLAB Program 3–32 into the computer. The resulting response curve is shown in Figure 3–41.

MATLAB Program 3–32

```
>> t = 0:0.01:10;
>> u = 0.5*t.^2;
>> sys = tf([1], [1  3  5]);
>> [y,t] = lsim(sys,u,t);
>> plot(t,y)
>> grid
>> title('Plot of Solution Curve for Differential Equation')
>> xlabel('t sec')
>> ylabel('Output y(t)')
```

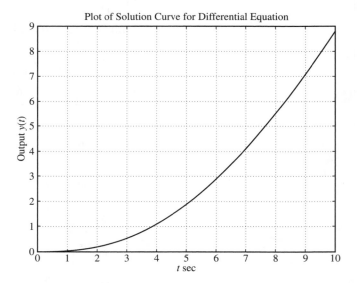

Figure 3–41
Plot of the solution curve.

3–6 RESPONSE TO ARBITRARY INITIAL CONDITION

In what follows, we shall present a few methods for obtaining the response to an arbitrary initial condition. Among the commands we may use are step and initial. First, we shall present a method for obtaining the response to an initial condition in a simple case. Then we shall discuss the response to initial condition when the system is given in state-space form. Finally, we shall present the command initial used to obtain the response of a system given in state-space form.

Laplace transform approach. Consider the mechanical system shown in Figure 3–42, where $m = 1$ kg, $b = 3$ N-sec/m, and $k = 2$ N/m. Assume that at $t = 0$ the mass m is pulled downward such that $x(0) = 0.1$ m and

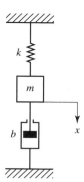

Figure 3-42
Mechanical system.

$\dot{x}(0) = 0.05$ m/sec. The displacement $x(t)$ is measured from the equilibrium position before the mass is pulled down. Obtain the motion of the mass subjected to the initial condition. (Assume that there is no external forcing function.)

The system equation is

$$m\ddot{x} + b\dot{x} + kx = 0$$

with the initial conditions $x(0) = 0.1$ m and $\dot{x}(0) = 0.05$ m/sec. (x is measured from the equilibrium position.) The Laplace transform of the system equation gives

$$m[s^2 X(s) - sx(0) - \dot{x}(0)] + b[sX(s) - x(0)] + kX(s) = 0$$

or

$$(ms^2 + bs + k)X(s) = mx(0)s + m\dot{x}(0) + bx(0)$$

Solving this last equation for $X(s)$ and substituting the given numerical values, we obtain

$$X(s) = \frac{mx(0)s + m\dot{x}(0) + bx(0)}{ms^2 + bs + k}$$

$$= \frac{0.1s + 0.35}{s^2 + 3s + 2}$$

This equation can be written as

$$X(s) = \frac{0.1s^2 + 0.35s}{s^2 + 3s + 2} \frac{1}{s}$$

Hence, the motion of the mass m may be obtained as the unit-step response of the following system:

$$G(s) = \frac{0.1s^2 + 0.35s}{s^2 + 3s + 2}$$

MATLAB Program 3–33 will give a plot of the motion of the mass. The plot is shown in Figure 3–43.

```
MATLAB Program 3–33
>> % ——— Response to initial condition ———
>>
>> % ***** System response to initial condition is converted to
>> % a unit-step response by modifying the numerator polynomial *****
>> % ***** Enter the numerator and denominator of the transfer
>> % function G(s) *****
>>
>> num = [0.1   0.35   0];
>> den = [1   3   2];
>>
>> % ***** Enter the following step-response command *****
>>
>> step(num,den)
>>
>> % ***** Enter grid and title of the plot *****
>>
>> grid
>> title('Response of Spring-Mass-Damper System to Initial Condition')
```

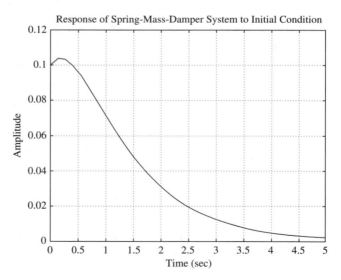

Figure 3–43
Response of the mechanical system to the initial condition.

138   Chapter 3 / Transient-Response Analysis

Response to initial condition (State-space approach, case 1). Consider the system defined by

$$\dot{\mathbf{x}} = \mathbf{A}\mathbf{x}, \qquad \mathbf{x}(0) = \mathbf{x}_0 \qquad (3\text{--}22)$$

Let us obtain the response $\mathbf{x}(t)$ when the initial condition $\mathbf{x}(0)$ is specified. (There is no external input function acting on this system.) Assume that $\mathbf{x}$ is an n-vector.

First, take Laplace transforms of both sides of Equation (3–22):

$$s\mathbf{X}(s) - \mathbf{x}(0) = \mathbf{A}\mathbf{X}(s)$$

This equation can be rewritten as

$$s\mathbf{X}(s) = \mathbf{A}\mathbf{X}(s) + \mathbf{x}(0) \qquad (3\text{--}23)$$

Taking the inverse Laplace transform of Equation (3–23), we obtain

$$\dot{\mathbf{x}} = \mathbf{A}\mathbf{x} + \mathbf{x}(0)\delta(t) \qquad (3\text{--}24)$$

(Notice that taking the Laplace transform of a differential equation and then taking the inverse Laplace transform of the Laplace-transformed equation generates a differential equation that involves the initial condition.)

Now define

$$\dot{\mathbf{z}} = \mathbf{x} \qquad (3\text{--}25)$$

Then Equation (3–24) can be written as

$$\ddot{\mathbf{z}} = \mathbf{A}\dot{\mathbf{z}} + \mathbf{x}(0)\delta(t) \qquad (3\text{--}26)$$

Integrating Equation (3–26) with respect to t, we obtain

$$\dot{\mathbf{z}} = \mathbf{A}\mathbf{z} + \mathbf{x}(0)1(t) = \mathbf{A}\mathbf{z} + \mathbf{B}u \qquad (3\text{--}27)$$

where

$$\mathbf{B} = \mathbf{x}(0), \qquad u = 1(t)$$

From Equation (3–25), the state $\mathbf{x}(t)$ is given by $\dot{\mathbf{z}}(t)$. Thus,

$$\mathbf{x} = \dot{\mathbf{z}} = \mathbf{A}\mathbf{z} + \mathbf{B}u \qquad (3\text{--}28)$$

The solution of Equations (3–27) and (3–28) gives the response to the initial condition.

Summarizing, the response of Equation (3–22) to the initial condition $\mathbf{x}(0)$ is obtained by solving the state-space equations

$$\dot{\mathbf{z}} = \mathbf{A}\mathbf{z} + \mathbf{B}u$$
$$\mathbf{x} = \mathbf{A}\mathbf{z} + \mathbf{B}u$$

where

$$\mathbf{B} = \mathbf{x}(0), \qquad u = 1(t)$$

MATLAB commands for obtaining the response curves in one diagram are as follows:

```
[x,z,t] = step(A,B,A,B,1,t);
x1 = [1  0  0...0]*x';
x2 = [0  1  0...0]*x';
  ⋮
xn = [0  0  0...1]*x';
plot(t,x1,t,x2,...,t,xn)
```

Response to initial condition (State-space approach, case 2). Consider the system defined by

$$\dot{\mathbf{x}} = \mathbf{A}\mathbf{x}, \quad \mathbf{x}(0) = \mathbf{x}_0 \quad (3\text{--}29)$$

$$\mathbf{y} = \mathbf{C}\mathbf{x} \quad (3\text{--}30)$$

(Assume that $\mathbf{x}$ is an n-vector and $\mathbf{y}$ is an m-vector.)

Similarly to case 1, by defining

$$\dot{\mathbf{z}} = \mathbf{x}$$

we obtain the equation

$$\dot{\mathbf{z}} = \mathbf{A}\mathbf{z} + \mathbf{x}(0)1(t) = \mathbf{A}\mathbf{z} + \mathbf{B}u \quad (3\text{--}31)$$

where

$$\mathbf{B} = \mathbf{x}(0), \quad u = 1(t)$$

Noting that $\mathbf{x} = \dot{\mathbf{z}}$, we can write Equation (3–30) as

$$\mathbf{y} = \mathbf{C}\dot{\mathbf{z}} \quad (3\text{--}32)$$

Substituting Equation (3–31) into Equation (3–32), we obtain

$$\mathbf{y} = \mathbf{C}(\mathbf{A}\mathbf{z} + \mathbf{B}u) = \mathbf{C}\mathbf{A}\mathbf{z} + \mathbf{C}\mathbf{B}u \quad (3\text{--}33)$$

The solution of Equations (3–31) and (3–33) gives the response of the system to a given initial condition. MATLAB commands to obtain the response curves (output curves y1 versus t, y2 versus t, ..., ym versus t) are as follows:

```
[y,z,t] = step(A,B,C*A,C*B,1,t);
y1 = [1  0  0...0]*y';
y2 = [0  1  0...0]*y';
  ⋮
ym = [0  0  0...1]*y';
plot(t,y1,t,y2,..., t,ym)
```

EXAMPLE 3–23 Obtain the response of the following system subjected to the given initial condition:

$$\begin{bmatrix} \dot{x}_1 \\ \dot{x}_2 \end{bmatrix} = \begin{bmatrix} 0 & 1 \\ -10 & -5 \end{bmatrix} \begin{bmatrix} x_1 \\ x_2 \end{bmatrix}, \quad \begin{bmatrix} x_1(0) \\ x_2(0) \end{bmatrix} = \begin{bmatrix} 2 \\ 1 \end{bmatrix}$$

or

$$\dot{\mathbf{x}} = \mathbf{A}\mathbf{x}, \quad \mathbf{x}(0) = \mathbf{x}_0$$

Obtaining the response of the system to the given initial condition becomes that of solving the unit-step response of the system

$$\dot{z} = Az + Bu$$
$$x = Az + Bu$$

where

$$B = x(0), \quad u = 1(t)$$

A possible MATLAB program for obtaining the response is shown in MATLAB Program 3–34. The resulting response curves are plotted in Figure 3–44.

MATLAB Program 3–34

```
>> t = 0:0.01:3;
>> A = [0  1;−10  −5];
>> B = [2;1];
>> [x,z,t] = step(A,B,A,B,1,t);
>> x1 = [1  0]*x';
>> x2 = [0  1]*x';
>> plot(t,x1,'x',t,x2,'-')
>> grid
>> title('Response to Initial Condition')
>> xlabel('t Sec')
>> ylabel('State Variables x_1 and x_2')
>> gtext('x_1')
>> gtext('x_2')
```

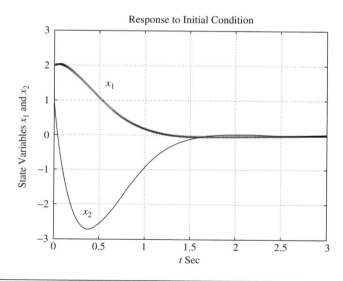

Figure 3–44
Response of system in Example 3–23 to initial condition.

Obtaining the response to an initial condition by use of the command initial. If the system is given in state-space form, then the command

$$\text{initial}(A,B,C,D,[\text{initial condition}],t)$$

will produce the response to the initial condition.

Suppose that we have the system defined by

$$\dot{\mathbf{x}} = \mathbf{A}\mathbf{x} + \mathbf{B}u, \qquad \mathbf{x}(0) = \mathbf{x}_0$$

$$y = \mathbf{C}\mathbf{x} + Du$$

where

$$\mathbf{A} = \begin{bmatrix} 0 & 1 \\ -10 & -5 \end{bmatrix}, \quad \mathbf{B} = \begin{bmatrix} 0 \\ 0 \end{bmatrix}, \quad \mathbf{C} = [0 \ 0], \quad D = 0$$

$$\mathbf{x}_0 = \begin{bmatrix} 2 \\ 1 \end{bmatrix}$$

Then the command initial can be used as shown in MATLAB Program 3–35 to obtain the response to the initial condition. The response curves $x_1(t)$ and $x_2(t)$ are shown in Figure 3–45 and are the same as those shown in Figure 3–44.

MATLAB Program 3–35

```
>> t = 0:0.05:3;
>> A = [0    1;-10   -5];
>> B = [0;0];
>> C = [0   0];
>> D = [0];
>> [y,x] = initial(A,B,C,D,[2;1],t);
>> x1 = [1   0]*x';
>> x2 = [0   1]*x';
>> plot(t,x1,'o',t,x1,t,x2,'x',t,x2)
>> grid
>> title('Response to Initial Condition')
>> xlabel('t Sec')
>> ylabel('State Variables x_1 and x_2')
>> gtext('x_1')
>> gtext('x_2')
```

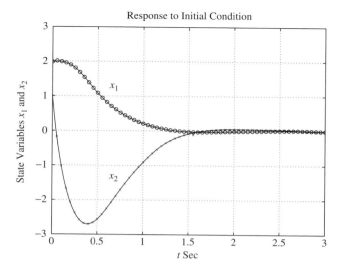

Figure 3–45
Response curves to initial condition.

EXAMPLE 3-24 Consider the following system that is subjected to the given initial condition:

$$\dddot{y} + 8\ddot{y} + 17\dot{y} + 10y = 0$$
$$y(0) = 2, \quad \dot{y}(0) = 1, \quad \ddot{y}(0) = 0.5$$

Assume that no external forcing function is present, and obtain the response $y(t)$ to the initial condition.

By defining the state variables as

$$x_1 = y$$
$$x_2 = \dot{y}$$
$$x_3 = \ddot{y}$$

we obtain the following state-space representation of the system:

$$\begin{bmatrix} \dot{x}_1 \\ \dot{x}_2 \\ \dot{x}_3 \end{bmatrix} = \begin{bmatrix} 0 & 1 & 0 \\ 0 & 0 & 1 \\ -10 & -17 & -8 \end{bmatrix} \begin{bmatrix} x_1 \\ x_2 \\ x_3 \end{bmatrix}, \quad \begin{bmatrix} x_1(0) \\ x_2(0) \\ x_3(0) \end{bmatrix} = \begin{bmatrix} 2 \\ 1 \\ 0.5 \end{bmatrix}$$

$$y = \begin{bmatrix} 1 & 0 & 0 \end{bmatrix} \begin{bmatrix} x_1 \\ x_2 \\ x_3 \end{bmatrix}$$

A possible MATLAB program to obtain the response $y(t)$ is given in MATLAB Program 3–36. The resulting response curve is shown in Figure 3–46.

MATLAB Program 3–36

```
>> t = 0:0.05:10;
>> A = [0  1  0;0  0  1;-10 -17 -8];
>> B = [0;0;0];
>> C = [1  0  0];
>> D = [0];
>> y = initial(A,B,C,D,[2;1;0.5],t);
>> plot(t,y)
>> grid
>> title('Response to Initial Condition')
>> xlabel('t (sec)')
>> ylabel('Output y')
```

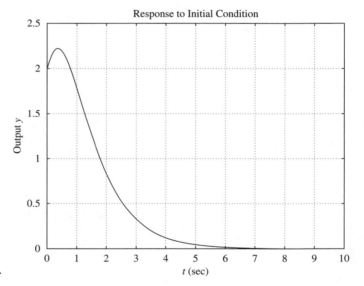

Figure 3–46
Response $y(t)$ to the given initial condition.

3–7 THREE-DIMENSIONAL PLOTS

Obtaining a three-dimensional plot of unit-step response curves with MATLAB. MATLAB enables us to plot three-dimensional plots easily. The command used here to obtain three-dimensional plots is mesh.

EXAMPLE 3–25 Consider the closed-loop system defined by

$$\frac{C(s)}{R(s)} = \frac{1}{s^2 + 2\zeta s + 1}$$

where $\zeta = 0$, 0.2, 0.4, 0.6, 0.8, and 1.0. Write a MATLAB program using a for loop to obtain the two-dimensional and three-dimensional plots of the system output. The input is the unit-step function.

MATLAB Program 3–37 is a possible program to obtain two-dimensional and three-dimensional plots. Figure 3–47(a) is the two-dimensional plot of the unit-step response curves for various values of ζ. Figure 3–47(b) is the three-dimensional plot obtained with the command mesh(y), and Figure 3–47(c) is obtained with the command mesh(y'). (These two three-dimensional plots are basically the same; the only difference is that x-axis and y-axis are interchanged.)

When a MATLAB program involves repetitive computations, many different MATLAB programs can be written. In this book, many different MATLAB programs using for loops are presented for illustration purposes. The reader is advised to study all those programs and try to improve them if possible.

MATLAB Program 3–37

```
>> t = 0:0.2:12;
>>      for n = 1:6;
        num = [1];
        den = [1   2*(n-1)*0.2   1];
        [y(1:61,n),x,t] = step(num,den,t);
        end
>> plot(t,y)
>> grid
>> title('Unit-Step Response Curves')
>> xlabel('t Sec')
>> ylabel('Outputs')
>> gtext('\zeta = 0'),
>> gtext('0.2')
>> gtext('0.4')
>> gtext('0.6')
>> gtext('0.8')
>> gtext('1.0')
>>
>> % To draw a three-dimensional plot, enter the following command: mesh(y) or mesh(y').
>> % We shall show two three-dimensional plots, one using "mesh(y)" and the other using
>> % "mesh(y')". These two plots are the same, except that the x-axis and y-axis are
>> % interchanged.
>>
>> mesh(y)
>> title('Three-Dimensional Plot of Unit-Step Response Curves Using Command "mesh(y)"')
>> xlabel('n, where n = 1,2,3,4,5,6')
>> ylabel('Computation Time Points')
>> zlabel('Outputs')
>>
>> mesh(y')
>> title('Three-Dimensional Plot of Unit-Step Response Curves Using Command
                                                           "mesh(y transpose)"')
>> xlabel('Computation Time Points')
>> ylabel('n, where n = 1,2,3,4,5,6')
>> zlabel('Outputs')
```

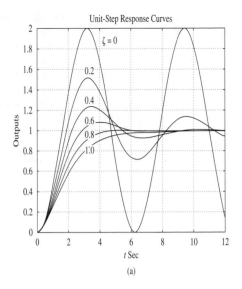

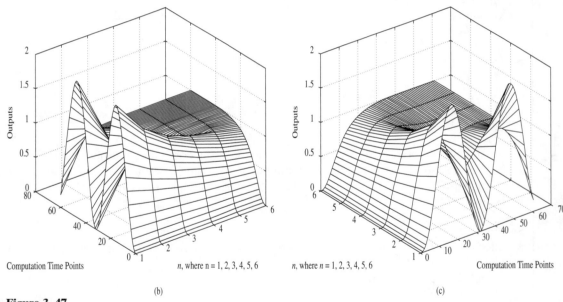

Figure 3–47
(a) Two-dimensional plot of unit-step response curves; (b) three-dimensional plot of unit-step response curves using command mesh(y); (c) three-dimensional plot of unit-step response curves using command mesh(y').

An alternative MATLAB program to plot the two-dimensional and three-dimensional diagrams is given in the next example. Either MATLAB program may be used in practice.

EXAMPLE 3–26 Consider again the problem presented in Example 3–25. The input to the system is the unit-step function. We would like to plot unit-step response curves for six different values of zeta.

MATLAB Program 3–38 is another possible program to obtain two-dimensional and three-dimensional plots of unit-step response curves.

MATLAB Program 3–38

```
>> t = 0:0.2:10;
>> zeta = [0   0.2   0.4   0.6   0.8   1];
>>      for n = 1:6;
            num = [1]; den = [1   2*zeta(n)   1]; sys = tf(num,den);
            y(:,n) = step(sys,t);
        end
>>
>> % To plot a two-dimensional diagram, enter the command plot(t,y).
>>
>> plot(t,y)
>> grid
>> title('Two-Dimensional Plot of Unit-Step Response Curves')
>> xlabel('t sec')
>> ylabel('Outputs')
>> text(4.1,1.86,'\zeta = 0')
>> text(3.5,1.5,'0.2')
>> text(3.5,1.24,'0.4')
>> text(3.5,1.08,'0.6')
>> text(3.5,0.95,'0.8')
>> text(3.5,0.86,'1.0')
>>
>> % To plot a three-dimensional diagram, enter the command mesh(zeta,t,y)
>> % or mesh(t,zeta,y').
>>
>> mesh(zeta,t,y)
>> title('Three-Dimensional Plot of Unit-Step Response Curves')
>> xlabel('\zeta')
>> ylabel('t sec')
>> zlabel('Outputs')
>>
>> mesh(t,zeta,y')
>> title('Three-Dimensional Plot of Unit-Step Response Curves')
>> xlabel('t sec')
>> ylabel('\zeta')
>> zlabel('Outputs')
```

Figure 3–48(a) is the resulting two-dimensional plot of the unit-step response curves for six different values of ζ. Figure 3–48(b) is the three-dimensional plot obtained with the command mesh(zeta,t,y). Figure 3–48(c) is obtained with the command mesh(t,zeta,y'). (These two three-dimensional plots are basically the same; the only difference is that x-axis and y-axis are interchanged.)

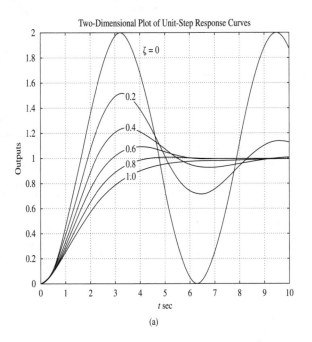

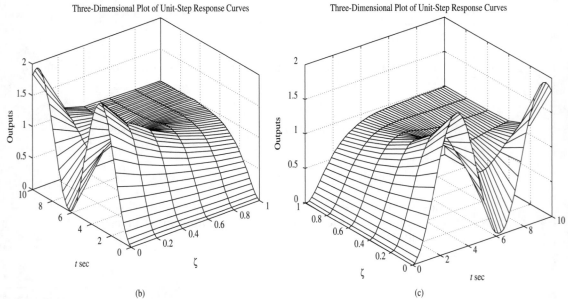

Figure 3–48
(a) Two-dimensional plot of the unit-step response curves; (b) three-dimensional plot obtained with the command mesh(zeta,t,y); (c) three-dimensional plot obtained with the command mesh(t,zeta,y').

EXAMPLE 3–27 Consider the normalized second-order system

$$\frac{\text{Output}}{\text{Input}} = \frac{1}{s^2 + 2\zeta s + 1}$$

Plot unit-impulse response curves for this system for ζ = 0, 0.2, 0.4, 0.6, 0.8, and 1.0. Plot a two-dimensional diagram and three-dimensional diagrams. Use the commands mesh(zeta,t,y) and mesh(t,zeta,y') to draw the latter diagrams.

MATLAB Program 3–39 produces the desired result.

MATLAB Program 3–39

```
>> t = 0:0.1:10;
>> zeta = [0   0.2   0.4   0.6   0.8   1];
>>      for n = 1:6;
             num = [1]; den = [1   2*zeta(n)   1]; sys = tf(num,den);
             y(:,n) = impulse(sys,t);
         end
>> % To plot a two-dimensional diagram, enter the command plot(t,y).
>>
>> plot(t,y)
>> grid
>> title('Two-Dimensional Plot of Unit-Impulse Response Curves')
>> xlabel('t sec')
>> ylabel('Outputs')
>> text(3.2,0.9, '\zeta = 0')
>> text(3.2,0.71, '0.2')
>> text(3.2,0.53, '0.4')
>> text(5.2,0.48, '0.6')
>> text(5.2,0.32, '0.8')
>> text(5.2,0.16, '1.0')
>>
>> % To plot a three-dimensional diagram, enter the command mesh(zeta,t,y)
>> % or mesh(t,zeta,y').
>>
>> mesh(zeta,t,y)
>> title('Three-Dimensional Plot of Unit-Impulse Response Curves')
>> xlabel('\zeta')
>> ylabel('t sec')
>> zlabel('Outputs')
>>
>> mesh(t,zeta,y')
>> title('Three-Dimensional Plot of Unit-Impulse Response Curves')
>> xlabel('t sec')
>> ylabel('\zeta')
>> zlabel('Outputs')
```

Figure 3–49(a) shows the two-dimensional plot. Figures 3–49(b) and (c) show three-dimensional plots using the commands mesh(zeta,t,y) and mesh(t,zeta,y'), respectively.

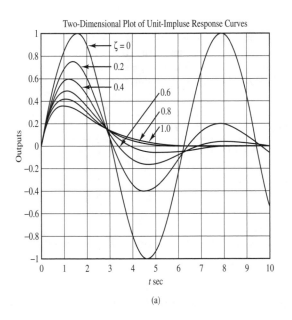

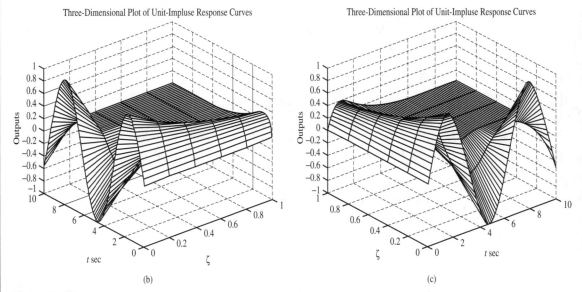

Figure 3–49
(a) Two-dimensional plot of unit-impulse response curves; (b) three-dimensional plot of unit-impulse response curves using the command mesh(zeta,t,y); (c) three-dimensional plot of unit-impulse response curves using the command mesh(t,zeta,y').

Root-Locus Analysis

4–1 INTRODUCTION

In this chapter, we present the MATLAB approach to the generation of root-locus plots and to finding relevant information from such plots.

Consider the system shown in Figure 4–1. The closed-loop transfer function is

$$\frac{C(s)}{R(s)} = \frac{G(s)}{1 + G(s)H(s)}$$

The characteristic equation for this closed-loop system is obtained by setting the denominator of the right-hand side equal to zero. That is,

$$1 + G(s)H(s) = 0 \qquad (4\text{–}1)$$

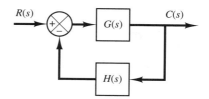

Figure 4–1
Closed-loop system.

In many cases, $G(s)H(s)$ involves a gain parameter K, and the characteristic equation may be written as

$$1 + \frac{K(s + z_1)(s + z_2)\cdots(s + z_m)}{(s + p_1)(s + p_2)\cdots(s + p_n)} = 0 \qquad (4\text{-}2)$$

or

$$1 + K\frac{\text{num}}{\text{den}} = 0$$

where num is the numerator polynomial and den is the denominator polynomial. That is,

$$\begin{aligned}
\text{num} &= (s + z_1)(s + z_2)\cdots(s + z_m) \\
&= s^m + (z_1 + z_2 + \cdots + z_m)s^{m-1} + \cdots + z_1 z_2 \cdots z_m \\
\text{den} &= (s + p_1)(s + p_2)\cdots(s + p_n) \\
&= s^n + (p_1 + p_2 + \cdots + p_n)s^{n-1} + \cdots + p_1 p_2 \cdots p_n
\end{aligned}$$

Note that both vectors num and den must be written in descending powers of s.

Equation (4–1) may be written as

$$G(s)H(s) = -1 \qquad (4\text{-}3)$$

Here, we assume that $G(s)H(s)$ is a ratio of polynomials in s. [Later, in Section 4–6, we extend the analysis to the case when $G(s)H(s)$ involves the transport lag e^{-Ts}.] Since $G(s)H(s)$ is a complex quantity, Equation (4–3) can be split into two equations by equating the angles and magnitudes of both sides, respectively, to obtain the following:

Angle condition:

$$\underline{/G(s)H(s)} = \pm 180°(2k + 1) \qquad (k = 0, 1, 2, \ldots) \qquad (4\text{-}4)$$

Magnitude condition:

$$|G(s)H(s)| = 1 \qquad (4\text{-}5)$$

The values of s that fulfill both the angle and magnitude conditions are the roots of the characteristic equation, or the closed-loop poles. A locus of the points in the complex plane satisfying the angle condition alone is the root locus. The roots of the characteristic equation (the closed-loop poles) corresponding to a given value of the gain can be determined from the angle and magnitude conditions. (For details of applying the angle condition and magnitude condition to obtain the closed-loop poles, see Chapter 6 of Reference 1.) Then, the root loci for the system are the loci of the closed-loop poles as the gain K is varied from zero to infinity.

A MATLAB command commonly used for plotting root loci is

rlocus(num,den)

This command draws the root-locus plot on the screen. The gain vector K is automatically determined. (The vector K contains all the gain values for which the closed-loop poles are to be computed.)

For the systems defined in state space, rlocus(A,B,C,D) plots the root locus of the system, with the gain vector automatically determined.

Note that commands

$$\text{rlocus(num,den,K)} \quad \text{and} \quad \text{rlocus(A,B,C,D,K)}$$

use the user-supplied gain vector K.

If invoked with left-hand arguments

$$[r,K] = \text{rlocus(num,den)}$$
$$[r,K] = \text{rlocus(num,den,K)}$$
$$[r,K] = \text{rlocus(A,B,C,D)}$$
$$[r,K] = \text{rlocus(A,B,C,D,K)}$$
$$[r,K] = \text{rlocus(sys)}$$

the screen will show the matrix r and gain vector K. [r has length K rows and length n columns (where n is the order of the denominator) containing the locations of the roots. Each row of the matrix corresponds to a gain from vector K.] The plot command

$$\text{plot(r,'-')}$$

plots the root loci.

If it is desired to plot the root loci with marks o or x, it is necessary to use the following command:

$$r = \text{rlocus(num,den)}$$
$$\text{plot(r,'o')} \quad \text{or} \quad \text{plot(r,'x')}$$

Plotting root loci with marks o or x is instructive, since each calculated closed-loop pole is shown graphically; in some portion of the root loci, those marks are densely placed; in another portion, they are sparsely placed. MATLAB supplies its own set of gain values used to calculate a root-locus plot. It does so by an internal adaptive step-size routine. Also, MATLAB uses the automatic axis-scaling feature of the plot command.

Finally, note that, since the gain vector is automatically determined, root-locus plots of

$$G(s)H(s) = \frac{K(s+1)}{s(s+2)(s+3)}$$

$$G(s)H(s) = \frac{10K(s+1)}{s(s+2)(s+3)}$$

$$G(s)H(s) = \frac{200K(s+1)}{s(s+2)(s+3)}$$

are all the same. The num and den set of the system is the same for all three systems:

$$\text{num} = [0 \quad 0 \quad 1 \quad 1]$$
$$\text{den} = [1 \quad 5 \quad 6 \quad 0]$$

EXAMPLE 4–1 Consider the system shown in Figure 4–2. Plot root loci with a square aspect ratio so that a line with slope 1 is a true 45° line. Choose the region of the root-locus plot to be

$$-6 \leq x \leq 6, \quad -6 \leq y \leq 6$$

where x and y are the real-axis coordinate and imaginary-axis coordinate, respectively.

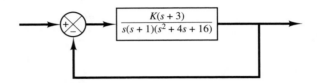

Figure 4–2
Control system.

To set the given plot region on the screen to be square, enter the command

$$v = [-6 \quad 6 \quad -6 \quad 6]; \text{ axis (v); axis('square')}$$

With this command, the region of the plot is as specified and a line with slope 1 is at a true 45°, not skewed by the irregular shape of the screen.

For this problem, the denominator is given as a product of first- and second-order terms. So we must multiply these terms to get a polynomial in s. The terms can be multiplied easily with the conv (convolution) command, shown next.

Define

$$a = s(s + 1): \quad\quad a = [1 \quad 1 \quad 0]$$
$$b = s^2 + 4s + 16: \quad\quad b = [1 \quad 4 \quad 16]$$

Then we use the command

$$c = \text{conv}(a, b)$$

Note that conv(a, b) gives the product of two polynomials a and b, shown in the following computer output:

```
>> a = [1   1   0];
>> b = [1   4   16];
>> c = conv (a,b)
c =
       1   5   20   16   0
```

The denominator polynomial is thus found to be

$$\text{den} = [1 \quad 5 \quad 20 \quad 16 \quad 0]$$

To find the complex-conjugate open-loop poles (the roots of $s^2 + 4s + 16 = 0$), we may enter the roots command as follows:

```
>> r = roots(b)
r =
   -2.0000 + 3.4641i
   -2.0000 - 3.4641i
```

Thus, the system has the following open-loop zero and open-loop poles:

Open-loop zero: $s = -3$
Open-loop poles: $s = 0$, $s = -1$, $s = -2 \pm j3.4641$

MATLAB Program 4–1 will plot the root-locus diagram for this system. The plot is shown in Figure 4–3.

MATLAB Program 4–1

```
>> clf % Clear figure
>> num = [1  3]; den = [1  5  20  16  0];
>> r = rlocus(num,den);
>> plot(r,'-'); v = [-6  6  -6  6]; axis(v); axis('square')
>> grid
>> title('Root-Locus Plot of G(s) = K(s+3)/[s(s+1)(s^2+4s+16)]')
>> xlabel('Real Axis'); ylabel('Imag Axis')
>> gtext('o') % Place 'o' mark on the open-loop zero.
>> gtext('x') % Place 'x' mark on each of 4 open-loop poles.
>> gtext('x')
>> gtext('x')
>> gtext('x')
```

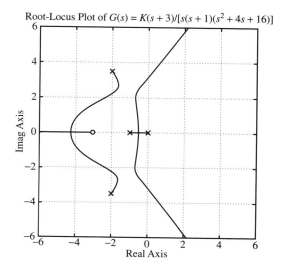

Figure 4–3
Root-locus plot.

Note that in MATLAB Program 4–1, instead of

$$\text{den} = [1 \quad 5 \quad 20 \quad 16 \quad 0]$$

we may enter

$$\text{den} = \text{conv}([1 \quad 1 \quad 0], [1 \quad 4 \quad 16])$$

The results are the same.

EXAMPLE 4–2 Consider the system whose open-loop transfer function $G(s)H(s)$ is

$$G(s)H(s) = \frac{K}{s(s + 0.5)(s^2 + 0.6s + 10)}$$

$$= \frac{K}{s^4 + 1.1s^3 + 10.3s^2 + 5s}$$

There are no open-loop zeros. Open-loop poles are located at $s = -0.3 + j3.1480$, $s = -0.3 - j3.1480$, $s = -0.5$, and $s = 0$.

Entering MATLAB Program 4–2 into the computer, we obtain the root-locus plot shown in Figure 4–4.

MATLAB Program 4–2

```
>> % _____ Root-locus plot _____
>>
>> num = [1];
>> den = [1   1.1   10.3   5   0];
>> r = rlocus(num,den);
>> plot(r,'o')
>> v = [-6   6   -6   6]; axis(v)
>> grid
>> title('Root-Locus Plot of G(s) = K/[s(s + 0.5)(s^2 + 0.6s +10)]')
>> xlabel('Real Axis')
>> ylabel('Imag Axis')
```

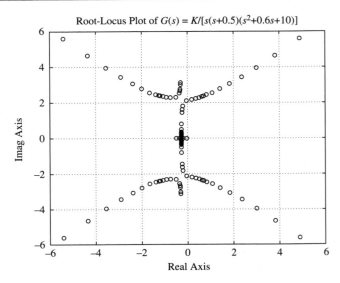

Figure 4–4
Root-locus plot.

Notice that in the regions near $x = -0.3$, $y = 2.3$ and $x = -0.3$, $y = -2.3$, two loci approach each other. We may wonder whether these two branches should touch or not. To explore this situation, we may plot the root loci, using smaller increments of K in the critical region.

By a conventional trial-and-error approach or using the command rlocfind to be presented later in this section, we find the particular region of interest to be $20 \leq K \leq 30$. By entering MATLAB Program 4–3, we obtain the root-locus plot shown in Figure 4–5. From this plot, it is clear that the two branches that approach in the upper half plane (or in the lower half plane) do not touch.

MATLAB Program 4–3

```
>> % _____ Root-locus plot _____
>>
>> num = [1];
>> den = [1   1.1   10.3   5   0];
>> K1 = 0:0.2:20;
>> K2 = 20:0.1:30;
>> K3 = 30:5:1000;
>> K = [K1   K2   K3];
>> r = rlocus(num,den,K);
>> plot(r,'o')
>> v = [-4   4   -4   4]; axis(v)
>> grid
>> title('Root-Locus Plot of G(s) = K/[s(s + 0.5)(s^2 + 0.6s + 10)]')
>> xlabel('Real Axis')
>> ylabel('Imag Axis')
```

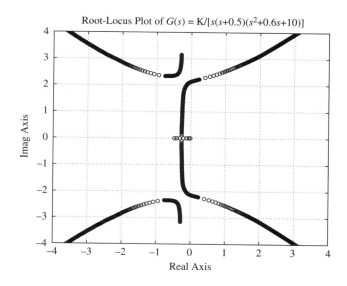

Figure 4–5
Root-locus plot.

EXAMPLE 4–3 Consider the system shown in Figure 4–6. The system equations are

$$\dot{x} = Ax + Bu$$
$$y = Cx + Du$$
$$u = r - y$$

In this problem, we shall obtain the root-locus diagram of the system defined in state space. As an example, let us consider the case where matrices **A**, **B**, **C**, and D are

$$\mathbf{A} = \begin{bmatrix} 0 & 1 & 0 \\ 0 & 0 & 1 \\ -160 & -56 & -14 \end{bmatrix}, \quad \mathbf{B} = \begin{bmatrix} 0 \\ 1 \\ -14 \end{bmatrix} \quad (4\text{--}6)$$

$$\mathbf{C} = \begin{bmatrix} 1 & 0 & 0 \end{bmatrix}, \quad D = [0]$$

The root-locus plot for this system can be obtained with MATLAB by use of the following command:

$$\text{rlocus}(A,B,C,D)$$

This command will produce the same root-locus plot that can be obtained with the rlocus (num,den) command, where num and den are obtained from

$$[\text{num,den}] = \text{ss2tf}(A,B,C,D)$$

as follows:

$$\text{num} = [0 \quad 0 \quad 1 \quad 0]$$
$$\text{den} = [1 \quad 14 \quad 56 \quad 160]$$

(The input and output of the transfer function are u and y, respectively.)

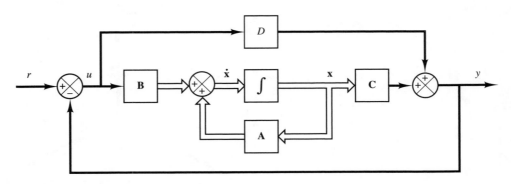

Figure 4–6
Closed-loop control system.

MATLAB Program 4–4 will generate the root-locus plot shown in Figure 4–7.

MATLAB Program 4–4

```
>> A = [0   1   0;0   0   1;−160   −56   −14];
>> B = [0;1;−14];
>> C = [1   0   0];
>> D = [0];
>> K = 0:0.1:400;
>> r = rlocus(A,B,C,D,K);
>> plot(r,'-'); v = [−20   20   −20   20]; axis(v)
>> grid
>> title('Root-Locus Plot of System Defined in State Space')
>> xlabel('Real Axis'); ylabel('Imag Axis')
>> gtext('o')   % Place 'o' mark on the open-loop zero.
>> gtext('x')   % Place 'x' mark on 3 open-loop poles.
>> gtext('x')
>> gtext('x')
```

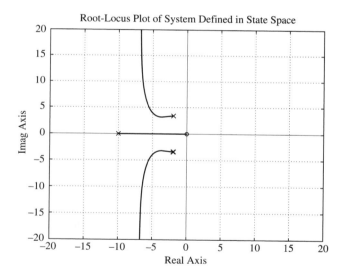

Figure 4–7
Root-locus plot of system defined in state space, where **A**, **B**, **C**, and *D* are as given by Equation (4–6).

EXAMPLE 4–4 Consider the system whose open-loop transfer function $G(s)H(s)$ is given by

$$G(s)H(s) = \frac{K}{s(s+1)(s+2)}$$

Using MATLAB, plot the root loci and their asymptotes.

We shall plot the root loci and asymptotes on one diagram. Since the open-loop transfer function is given by

$$G(s)H(s) = \frac{K}{s(s+1)(s+2)}$$
$$= \frac{K}{s^3 + 3s^2 + 2s}$$

the equation for the asymptotes may be obtained as follows: Noting that

$$\lim_{s \to \infty} \frac{K}{s^3 + 3s^2 + 2s} \doteq \lim_{s \to \infty} \frac{K}{s^3 + 3s^2 + 3s + 1} = \frac{K}{(s+1)^3}$$

we may find the equation for the asymptotes from

$$G_a(s)H_a(s) = \frac{K}{(s+1)^3}$$

Hence, for the given system, we have

$$\text{num} = [0 \quad 0 \quad 0 \quad 1]$$
$$\text{den} = [1 \quad 3 \quad 2 \quad 0]$$

and for the asymptotes,

$$\text{numa} = [0 \quad 0 \quad 0 \quad 1]$$
$$\text{dena} = [1 \quad 3 \quad 3 \quad 1]$$

In using the root-locus and plot commands

```
r = rlocus(num,den)
a = rlocus(numa,dena)
plot([r a])
```

the number of rows of r and of a must be the same. To ensure this, we include the gain constant K in the commands. For example,

```
K1 = 0:0.1:0.3;
K2 = 0.3:0.005:0.5:
K3 = 0.5:0.5:10;
K4 = 10:5:100;
K = [K1   K2   K3   K4]
r = rlocus(num,den,K)
a = rlocus(numa,dena,K)
y = [r   a]
plot(y,'-')
```

Including the gain K in the rlocus command ensures that the r matrix and the a matrix have the same number of rows. MATLAB Program 4–5 will generate a plot of root loci and their asymptotes. (See Figure 4–8.)

MATLAB Program 4-5

```
>> num = [1];
>> den = [1   3   2   0];
>> numa = [1];
>> dena = [1   3   3   1];
>> K1 = 0:0.1:0.3;
>> K2 = 0.3:0.005:0.5;
>> K3 = 0.5:0.5:10;
>> K4 = 10:5:100;
>> K = [K1   K2   K3   K4];
>> r = rlocus(num,den,K);
>> a = rlocus(numa,dena,K);
>> y = [r   a];
>> plot(y,'-')
>> v = [-4   4   -4   4]; axis(v)
>> grid
>> title('Root-Locus Plot of G(s) = K/[s(s + 1)(s + 2)] and Asymptotes')
>> xlabel('Real Axis'); ylabel('Imag Axis')
>> gtext('x')    %   Place 'x' mark on each of 3 open-loop poles.
>> gtext('x')
>> gtext('x')
```

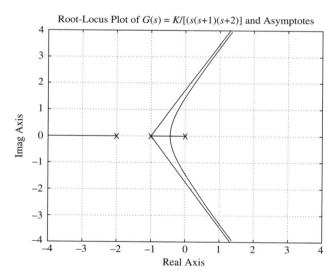

Figure 4-8
Root-locus plot.

Drawing two or more plots in one diagram can also be accomplished with the hold command, as is done in MATLAB Program 4-6. The resulting root-locus plot is shown in Figure 4-9.

MATLAB Program 4–6

```
>> % ——— Root-Locus Plots ———
>>
>> num = [1];
>> den = [1   3   2   0];
>> numa = [1];
>> dena = [1   3   3   1];
>> K1 = 0:0.1:0.3;
>> K2 = 0.3:0.005:0.5;
>> K3 = 0.5:0.5:10;
>> K4 = 10:5:100;
>> K = [K1   K2   K3   K4];
>> r = rlocus(num,den,K);
>> a = rlocus(numa,dena,K);
>> plot(r,'o')
>> v = [-4   4   -4   4]; axis(v)
>> hold
>> Current plot held
>> plot(a,'-')
>> grid
>> title('Root-Locus Plot of G(s) = K/[s(s+1)(s+2)] and Asymptotes')
>> xlabel('Real Axis')
>> ylabel('Imag Axis')
```

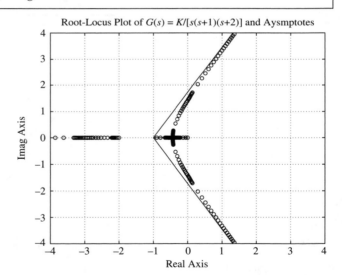

Figure 4–9
Root-locus plot.

EXAMPLE 4–5 Consider a unity-feedback system with the following feedforward transfer function $G(s)$:

$$G(s) = \frac{K(s + 2)^2}{(s^2 + 4)(s + 5)^2}$$

Plot the root loci of the system with MATLAB.

MATLAB Program 4–7 plots the root loci. The resulting root-locus plot is shown in Figure 4–10.

Notice that this is a special case where no root locus exists on the real axis. This means that, for any value of $K > 0$, the closed-loop poles of the system are two sets of complex-conjugate poles. (No real closed-loop poles exist.) For example, with $K = 25$, the characteristic equation for the system becomes

$$s^4 + 10s^3 + 54s^2 + 140s + 200$$
$$= (s^2 + 4s + 10)(s^2 + 6s + 20)$$
$$= (s + 2 + j2.4495)(s + 2 - j2.4495)(s + 3 + j3.3166)(s + 3 - j3.3166)$$

MATLAB Program 4–7

```
>> % ——— Root-Locus Plot ———
>>
>> num = [1   4   4];
>> den = [1   10   29   40   100];
>> r = rlocus(num,den);
>> plot(r,'o')
>> v = [-8   4   -6   6]; axis(v); axis('square')
>> hold
>> current plot held
>> plot(r,'-')
>> grid
>> title('Root-Locus Plot of G(s) = (s + 2)^2/[(s^2 + 4)(s + 5)^2]')
>> xlabel('Real Axis')
>> ylabel('Imag Axis')
```

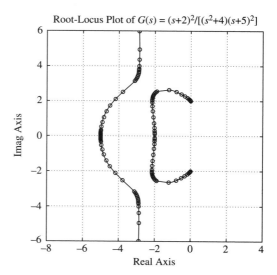

Figure 4–10
Root-locus plot.

Since no closed-loop poles exist in the right-half s plane, the system is stable for all values of $K > 0$.

EXAMPLE 4-6 Consider a unity-feedback control system with the following feedforward transfer function:

$$G(s) = \frac{K(s^2 + 25)s}{s^4 + 404s^2 + 1600}$$

Plot root loci for the system with MATLAB. Show that the system is stable for all values of $K > 0$.

MATLAB Program 4–8 gives a plot of root loci as shown in Figure 4–11. Since the root loci are entirely in the left-half s plane, the system is stable for all $K > 0$.

MATLAB Program 4–8

```
>> num = [1   0   25   0];
>> den = [1   0   404   0   1600];
>> K = 0:0.4:1000;
>> r = rlocus (num,den,K);
>> plot (r,'o');
>> v = [-30   20   -25   25]; axis(v); axis('square')
>> grid
>> title ('Root-Locus Plot of G(s) = K(s^2+25)s/(s^4+404s^2+1600)')
>> xlabel ('Real Axis'); ylabel ('Imag Axis')
```

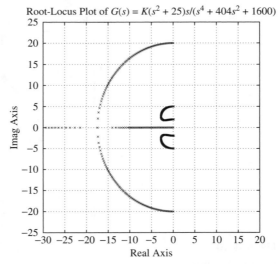

Figure 4–11
Root-locus plot.

4–2 ROOT LOCUS PLOTS WITH POLAR GRIDS

Constant ζ loci and constant ω_n loci. In the complex plane, the damping ratio ζ of a pair of complex-conjugate poles can be expressed in terms of the angle ϕ, which is measured from the negative real axis, as shown in Figure 4–12(a) with

$$\zeta = \cos \phi$$

In other words, lines of constant damping ratio ζ are radial lines passing through the origin as shown in Figure 4–12(b). For example, a damping ratio of 0.5 requires that the complex poles lie on the lines drawn through the origin making angles of $\pm 60°$ with the negative real axis. (If the real part of a pair of complex poles is positive, which means that the system is unstable, the corresponding ζ is negative.) the damping ratio determines the angular location of the poles, while the distance of the pole from the origin is determined by the undamped natural frequency ω_n. The constant ω_n loci are circles.

To draw constant ζ lines and constant ω_n circles on the root-locus diagram with MATLAB, use the command sgrid.

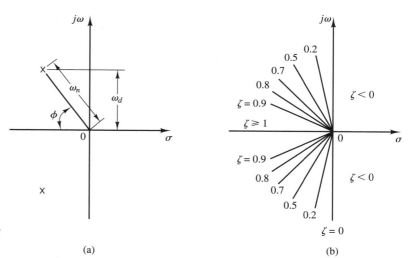

Figure 4–12
(a) Complex poles; (b) lines of constant damping ratio ζ.

Plotting polar grids in the root-locus diagram. The command

$$\text{sgrid}$$

overlays lines of constant damping ratio ($\zeta = 0 \sim 1$ with 0.1 increment) and circles of constant ω_n on the root-locus plot. (See MATLAB Program 4–9 and the resulting diagram shown in Figure 4–13.)

If only particular constant ζ lines (such as the $\zeta = 0.5$ line and $\zeta = 0.707$ line) and particular constant ω_n circles (such as the $\omega_n = 0.5$ circle, $\omega_n = 1$ circle, and $\omega_n = 2$ circle) are desired, use the following command:

$$\text{sgrid}([0.5, \quad 0.707], [0.5, \quad 1, \quad 2])$$

MATLAB Program 4–9

```
>> sgrid
>> v = [-3  3  -3  3]; axis(v); axis('square')
>> title('Constant \zeta Lines and Constant \omega_n Circles')
>> xlabel('Real Axis')
>> ylabel('Imag Axis')
```

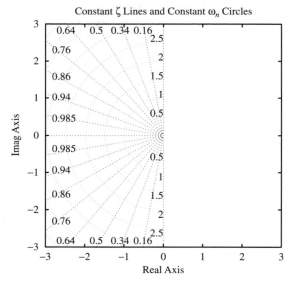

Figure 4–13
Constant ζ lines and constant ω_n circles.

If we wish to overlay lines of constant ζ and circles of constant ω_n on a root-locus plot of a system with

$$\text{num} = [0 \quad 0 \quad 0 \quad 1]$$
$$\text{den} = [1 \quad 4 \quad 5 \quad 0]$$

then we can enter MATLAB Program 4–10 into the computer. The resulting root-locus plot is shown in Figure 4–14.

MATLAB Program 4–10

```
>> num = [1];
>> den = [1   4   5   0];
>> K = 0:0.01:1000;
>> r = rlocus(num,den,K);
>> plot(r,'-'); v = [-3   1   -2   2]; axis(v); axis('square')
>> sgrid([0.5,0.707],[0.5,1,2])
>> grid
>> title('Root-Locus Plot with \zeta = 0.5 and 0.707 Lines and \omega_n = 0.5,1, and 2 Circles')
>> xlabel('Real Axis'); ylabel('Imag Axis')
>> gtext('\omega_n = 2')
>> gtext('\omega_n = 1')
>> gtext('\omega_n = 0.5')
>> gtext('x') % Place 'x' mark at each of 3 open-loop poles.
>> gtext('x')
>> gtext('x')
```

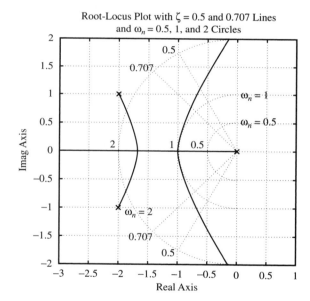

Figure 4–14
Constant ζ lines and constant ω_n circles superimposed on a root-locus plot.

If we want to omit either the entire constant ζ lines or entire constant ω_n circles, we may use empty brackets [ ] in the arguments of the sgrid command. For example, if we want to overlay only the constant damping ratio line corresponding to $\zeta = 0.5$ and no constant ω_n circles on the root-locus plot as shown in Figure 4–15, then we may use the command

$$\text{sgrid}(0.5,[\,])$$

See MATLAB Program 4–11 which produced Figure 4–15.

MATLAB Program 4–11
```
>> num = [1]; den = [1   4   5   0];
>> K1 = 0:0.0001:2; K2 = 2:0.01:1000; K = [K1   K2];
>> r = rlocus(num,den,K);
>> plot(r,'-'); v = [-3   1   -2   2]; axis(v); axis('square')
>> grid
>> sgrid(0.5,[])
>> title('Root-Locus Plot and \zeta = 0.5 Lines')
>> xlabel('Real Axis'); ylabel('Imag Axis')
>> gtext('x') % Place 'x' mark on each of 3 open-loop poles.
>> gtext('x')
>> gtext('x')
```

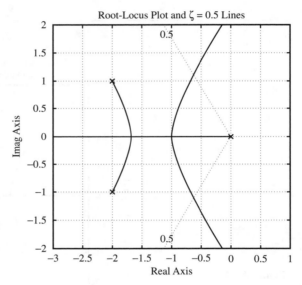

Figure 4–15
Root-locus plot with $\zeta = 0.5$ line.

EXAMPLE 4-7 Consider a unity-feedback control system with the following feedforward transfer function:

$$G(s) = \frac{s+2}{s^3 + 9s^2 + 8s}$$

Plot a root-locus diagram with MATLAB. Superimpose constant ζ lines and constant ω_n circles on the s plane.

MATLAB Program 4–12 produces the desired plot, as shown in Figure 4–16.

MATLAB Program 4–12

```
>> num = [1   2];
>> den = [1   9   8   0];
>> K = 0:0.02:200;
>> r = rlocus(num,den,K);
>> plot(r,'-'); v = [-10   2   -6   6]; axis(v);axis('square')
>> sgrid
>> title('Root-Locus Plot with Constant \zeta Lines and Constant \omega_n Circles')
>> xlabel('Real Axis')
>> ylabel('Imag Axis')
>> gtext('o')   %   Place 'o' mark on the open-loop zero.
>> gtext('x')   %   Place 'x' mark on 3 open-loop poles.
>> gtext('x')
>> gtext('x')
```

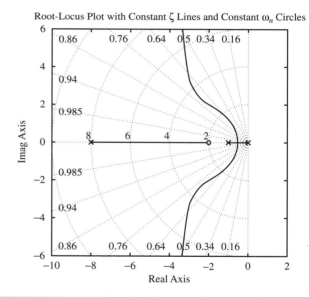

Figure 4–16
Root-locus plot with constant ζ lines and constant ω_n circles.

4–3 FINDING THE GAIN VALUE K AT AN ARBITRARY POINT ON THE ROOT LOCUS

Orthogonality of root loci and constant-gain loci. Consider the system whose open-loop transfer function is $G(s)H(s)$. In the $G(s)H(s)$ plane, the loci of $|G(s)H(s)|$ = constant are circles centered at the origin, and the loci corresponding to $\angle G(s)H(s) = \pm 180°(2k+1)(k = 0, 1, 2, \ldots)$ lie on the negative real axis of the $G(s)H(s)$ plane, as shown in Figure 4–17. [Note that the complex plane employed here is not the s plane, but the $G(s)H(s)$ plane.]

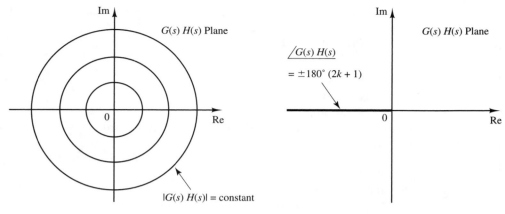

Figure 4–17
Plots of constant-gain and constant-phase loci in the $G(s)H(s)$ plane.

The root loci and constant-gain loci in the s plane are conformal mappings of the loci of $\angle G(s)H(s) = \pm 180°(2k+1)$ and of $|G(s)H(s)| = $ constant in the $G(s)H(s)$ plane.

Since the constant-phase and constant-gain loci in the $G(s)H(s)$ plane are orthogonal, the root loci and constant-gain loci in the s plane are orthogonal. Figure 4–18(a) shows the root loci and constant-gain loci for the following system:

$$G(s) = \frac{K(s+2)}{s^2 + 2s + 3}, \quad H(s) = 1$$

Notice that, since the pole–zero configuration is symmetrical about the real axis, the constant-gain loci are also symmetrical about the real axis.

Figure 4–18(b) shows the root loci and constant-gain loci for the system

$$G(s) = \frac{K}{s(s+1)(s+2)}, \quad H(s) = 1$$

Notice that, since the configuration of the poles in the s plane is symmetrical about the real axis and the line parallel to the imaginary axis passing through the point ($\sigma = -1, \omega = 0$), the constant-gain loci are symmetrical about the $\omega = 0$ line (real axis) and the line $\sigma = -1$.

Notice also, from Figures 4–18(a) and (b), that every point in the s plane has the corresponding K value. If we use a command rlocfind (presented next), MATLAB will give the K value of the specified point, as well as the nearest closed-loop poles corresponding to that K value.

Finding the gain value K at an arbitrary point on the root loci. In the MATLAB analysis of closed-loop systems, it is frequently desired to find the gain value K at an arbitrary point on the root locus. This can be accomplished by using either of the following rlocfind commands:

[K,r] = rlocfind(num, den)   or   [K, r] = rlocfind(A,B,C,D)

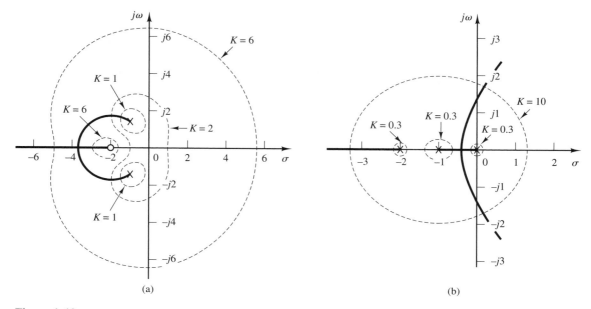

Figure 4–18
Plots of root loci and constant-gain loci. (a) System with $G(s) = K(s + 2)/(s^2 + 2s + 3)$, $H(s) = 1$; (b) system with $G(s) = K/[s(s + 1)(s + 2)]$, $H(s) = 1$.

Here, r gives the closed-loop poles. The rlocfind command, which must follow an rlocus command, overlays movable x–y coordinates on the screen. Using the mouse, we position the origin of the x–y coordinates over the desired point on the root locus and press the mouse button. Then MATLAB displays the coordinates of that point, the gain value at that point, and the closed-loop poles corresponding to the aforesaid gain value.

If the selected point is not on the root locus, the rlocfind command gives the coordinates of that selected point, the gain value of the point, and the locations of the closed-loop poles corresponding to that K value. [Note that every point on the s plane has a gain value; see, for example, Figures 4–18(a) and (b).]

EXAMPLE 4–8 Consider the unity-feedback control system with the following feedforward transfer function:

$$G(s) = \frac{K}{s(s^2 + 4s + 5)}$$

Plot the root loci with MATLAB. Determine the closed-loop poles that have the damping ratio 0.5. Find the gain value K at that point.

First we plot a root-locus diagram as shown in Figure 4–19. Then we enter the rlocfind command as shown in MATLAB Program 4–13. Next, we position the origin of the x–y coordinates over the intersection of the upper root-locus branch and the $\zeta = 0.5$ line. Then we press the button of the mouse. The screen shows the coordinates of that point, the gain value at the point, and the closed-loop poles corresponding to the aforesaid gain value.

The plot indicates the closed-loop poles by a plus sign (+). The three closed-loop poles obtained are

$$s = -2.7610, \quad s = -0.6195 + j1.0934, \quad s = -0.6195 - j1.0934$$

The gain K at this point is found to be 4.3602. Note that the three closed-loop poles are slightly off the exact locations obtained by the analytic method. The reason is that we cannot position the origin of the movable x–y coordinates exactly at the intersection of the upper root-locus branch and the $\zeta = 0.5$ line.

MATLAB Program 4–13

```
>> num = [1];
>> den = [1   4   5   0];
>> r = rlocus(num,den);
>> plot(r,'-'); v = [-3   1   -2   2]; axis(v); axis('square')
>> grid
>> sgrid(0.5,[])
>> [K,r] = rlocfind(num,den)
Select a point in the graphics window

selected_point =

    -0.6199 + 1.0936i

K =
    4.3602

r =

    -2.7610
    -0.6195 + 1.0934i
    -0.6195 - 1.0934i

>> xlabel('Real Axis')
>> ylabel('Imag Axis')
```

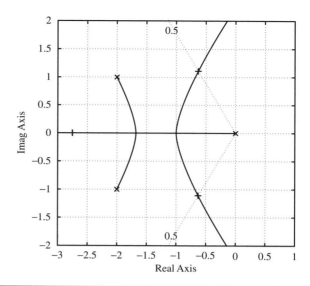

Figure 4–19
Root-locus plot
with $\zeta = 0.5$ lines.

4–4 ROOT-LOCUS PLOTS OF NON–MINIMUM-PHASE SYSTEMS

Non–minimum-phase systems. If all the poles and zeros of a system lie in the left-half s plane, then the system is called *minimum phase*. If a system has at least one pole or zero in the right-half s plane, then the system is called *non–minimum phase*. The term *non–minimum phase* comes from the phase-shift characteristics of such a system when subjected to sinusoidal inputs.

Consider the system shown in Figure 4–20(a), for which

$$G(s) = \frac{K(1 - T_a s)}{s(Ts + 1)} \quad (T_a > 0), \quad H(s) = 1$$

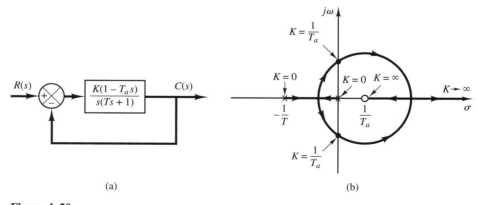

Figure 4–20
(a) Non–minimum-phase system; (b) root-locus plot.

This is a non–minimum-phase system, since there is one zero in the right-half s plane. For this system, the angle condition becomes

$$\angle G(s) = \angle \left[-\frac{K(T_a s - 1)}{s(Ts + 1)} \right]$$

$$= \angle \frac{K(T_a s - 1)}{s(Ts + 1)} + 180°$$

$$= \pm 180°(2k + 1) \qquad (k = 0, 1, 2, \ldots)$$

or

$$\angle \frac{K(T_a s - 1)}{s(Ts + 1)} = 0° \tag{4–7}$$

The root loci can be obtained from Equation (4–7). Figure 4–20(b) shows a root-locus plot for this system. From the diagram, we see that the system is stable if the gain K is less than $1/T_a$.

To obtain a root-locus plot with MATLAB, enter the numerator and denominator as usual. For example, if $T = 1$ sec and $T_a = 0.5$ sec, enter the following num and den in the program:

$$\text{num} = [0 \quad -0.5 \quad 1]$$
$$\text{den} = [1 \quad 1 \quad 0]$$

MATLAB Program 4–14 gives the plot of the root loci shown in Figure 4–21.

MATLAB Program 4–14

```
>> num = [-0.5  1];
>> den = [1  1  0];
>> K1 = 0:0.01:30;
>> K2 = 30:1:100;
>> K3 = 100:5:500;
>> K = [K1  K2  K3];
>> r = rlocus(num,den,K);
>> plot(r); v = [-2  6  -4  4]; axis(v); axis('square')
>> grid
>> title('Root-Locus Plot of G(s) = K(1-0.5s)/[s(s+1)]')
>> xlabel('Real Axis'); ylabel('Imag Axis')
```

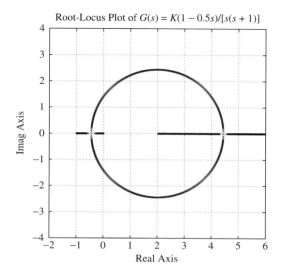

Figure 4–21
Root-locus plot of $G(s) = \dfrac{K(1 - 0.5s)}{s(s + 1)}$.

EXAMPLE 4–9 Consider the minimum-phase system and non–minimum-phase system shown in Figures 4–22(a) and (b), respectively. Obtain the unit-step responses of both systems.

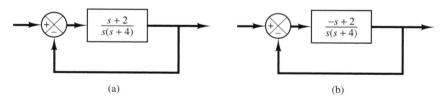

(a)   (b)

Figure 4–22
(a) Minimum-phase system; (b) non–minimum-phase system.

MATLAB Program 4–15 produces the step response curves shown in Figure 4–23. Notice that the non–minimum-phase system exhibits faulty behavior at the start of the response. (The response begins to move in the negative direction for a short time and then moves on to the positive direction.)

MATLAB Program 4–15

```
>> t = 0:0.1:10;
>> num = [1   2]; den = [1   4   0];
>> numn = [−1   2]; denn = [1   4   0];
>> sys = tf(num,den);
>> sysn = tf(numn,denn);
>> sys1 = feedback(sys, [1]);
>> sys2 = feedback(sysn, [1]);
>> [y1,t] = step(sys1,t);
>> [y2,t] = step(sys2,t);
>> plot(t,y1,t,y2,'o')
>> grid
>> title('Step Responses of Minimum- and Non–minimum-Phase Systems')
>> xlabel('t sec')
>> ylabel('Responses y_1 and y_2')
>> text(5.0,0.84, 'Minimum-Phase System')
>> text(2.3,0.3, 'Non–minimum-phase System')
```

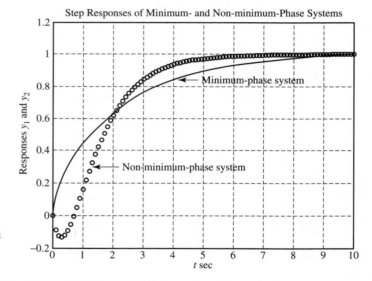

Figure 4–23
Step responses of minimum-phase system and non–minimum-phase system.

4–5 ROOT-LOCUS PLOTS OF CONDITIONALLY STABLE SYSTEMS

Consider the system shown in Figure 4–24. We can plot the root loci for this system by applying the general rules and procedure for constructing root loci, or we can use MATLAB to get root-locus plots. MATLAB Program 4–16 will plot the root-locus diagram for the system. The plot is shown in Figure 4–25.

It can be seen from the root-locus plot of Figure 4–25 that this system is stable only for limited ranges of the value of K—that is, $0 < K < 12$ and $73 < K < 154$.

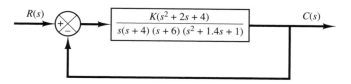

Figure 4–24
Control system.

MATLAB Program 4–16

```
>> num = [1   2   4];
>> den = conv(conv([1   4   0],[1   6]),[1   1.4   1]);
>> r = rlocus(num,den);
>> plot(r); v = [-7   3   -5   5];axis(v);axis('square')
>> grid
>> title('Root-Locus Plot of G(s) = K(s^2+2s+4)/[s(s+4)(s+6)(s^2+1.4s+1)]')
>> xlabel('Real Axis'); ylabel('Imag Axis')
>> text(1.0,0.5,'K = 12')
>> text(1.0,3.0,'K = 73')
>> text(1.0,4.15,'K = 154')
>> gtext('o')  % Place 'o' mark on each of 2 open-loop zeros.
>> gtext('o')
>> gtext('x')  % Place 'x' mark on each of 5 open-loop poles.
>> gtext('x')
>> gtext('x')
>> gtext('x')
>> gtext('x')
```

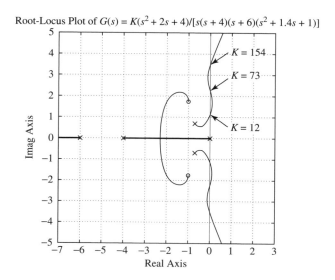

Figure 4–25
Root-locus plot of conditionally stable system.

The system becomes unstable for $12 < K < 73$ and $154 < K$. (If K assumes a value corresponding to unstable operation, the system may break down or may become nonlinear due to a saturation nonlinearity that may exist.) Such a system is called *conditionally stable*.

In practice, conditionally stable systems are not desirable. Conditional stability is dangerous, although it does occur in certain systems—in particular, a system that has an unstable feedforward path. Such a path may occur if the system has a minor loop. It is advisable to avoid such conditional stability, since, if the gain drops beyond the critical value for any reason, the system becomes unstable. Note that the addition of a proper compensating network will eliminate conditional stability. (The addition of a zero will cause the root loci to bend to the left; See Section 4–7.) Hence, conditional stability may be eliminated by adding proper compensation.

EXAMPLE 4–10 A simplified form of the open-loop transfer function of an airplane with an autopilot in the longitudinal mode is

$$G(s)H(s) = \frac{K(s+a)}{s(s-b)(s^2 + 2\zeta\omega_n s + \omega_n^2)}, \quad a > 0, \quad b > 0$$

Such a system involving an open-loop pole in the right-half s plane may be conditionally stable. Sketch the root loci when $a = b = 1$, $\zeta = 0.5$, and $\omega_n = 4$. Find the range of the gain K required for stability.

The open-loop transfer function for the system is

$$G(s)H(s) = \frac{K(s+1)}{s(s-1)(s^2 + 4s + 16)}$$

To obtain the root loci, we may enter MATLAB Program 4–17 into the computer. The resulting root-locus plot shown in Figure 4–26 indicates that the root loci cross the imaginary axis four times. The exact values of K where the root loci cross the imaginary axis can be determined from the Routh stability criterion as follows: Since the characteristic equation is

$$s^4 + 3s^3 + 12s^2 + (K - 16)s + K = 0$$

the Routh array becomes

s^4	1	12	K
s^3	3	$K - 16$	0
s^2	$\dfrac{52 - K}{3}$	K	0
s^1	$\dfrac{-K^2 + 59K - 832}{52 - K}$	0	
s^0	K		

MATLAB Program 4-17

```
>> num = [1  1]; den = conv([1  -1  0],[1  4  16]);
>> r = rlocus(num,den);
>> plot(r,'-'); v = [-4  4  -4  4]; axis(v); axis('square')
>> grid
>> title('Root-Locus Plot of G(s)H(s) = K(s+1)/[s(s-1)(s^2+4s+16)]')
>> xlabel('Real Axis'); ylabel('Imag Axis');
>> text(2,2.6,'K = 35.7')
>> text(2,1.6,'K = 23.3')
>> text(2,0.7,'K = 0')
>> text(-3.5,3.4,'K = 0')
>> text(-2,0.5,'K = \infty')
>> gtext('o') % Place 'o' mark on the open-loop zero.
>> gtext('x') % Place 'x' mark on each of 4 open-loop poles.
>> gtext('x')
>> gtext('x')
>> gtext('x')
```

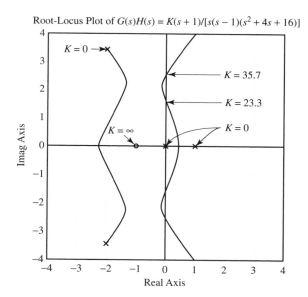

Figure 4-26
Root-locus plot.

The values of K that make the term at the interaction of the s^1 row and the first column equal to zero are $K = 35.7$ and $K = 23.3$.

The crossing points on the imaginary axis can be found by solving the auxiliary equation obtained from the s^2 row—that is, by solving the following equation for s:

$$\frac{52 - K}{3}s^2 + K = 0$$

The results are

$$s = \pm j2.56, \quad \text{for } K = 35.7$$
$$s = \pm j1.56, \quad \text{for } K = 23.3$$

The crossing points on the imaginary axis are thus $s = \pm j2.56$ and $s = \pm j1.56$. The range of the gain K required for stability is

$$23.3 < K < 35.7$$

4-6 ROOT LOCI FOR SYSTEMS WITH TRANSPORT LAG

Figure 4–27 shows a thermal system in which hot air is circulated to keep the temperature of a chamber constant. In this system, the measuring element is placed downstream a distance L ft from the furnace, the air velocity is v ft/sec, and $T = L/v$ sec would elapse before any change in the temperature of the furnace is sensed by the thermometer. Such a delay in measuring, in controller action, or in actuator operation, and the like, is called *transport lag* or *dead time*. Dead time is present in most process control systems.

The input $x(t)$ and the output $y(t)$ of a transport-lag or dead-time element are related by

$$y(t) = x(t - T)$$

where T is the dead time. The transfer function of transport lag or dead time is given by

$$\text{Transfer function of transport lag or dead time} = \frac{\mathscr{L}[x(t - T)1(t - T)]}{\mathscr{L}[x(t)1(t)]}$$
$$= \frac{X(s)e^{-Ts}}{X(s)} = e^{-Ts}$$

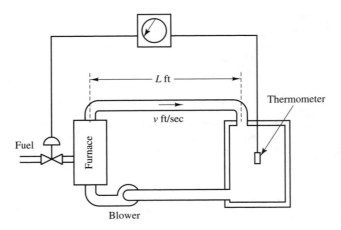

Figure 4–27
Thermal system.

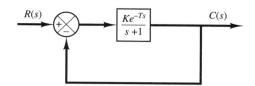

Figure 4-28
Block diagram of the system shown in Figure 4-27.

Suppose that the feedforward transfer function of this thermal system can be approximated by

$$G(s) = \frac{Ke^{-Ts}}{s+1}$$

as shown in Figure 4-28. Let us construct a root-locus plot for the system. The characteristic equation for this closed-loop system is

$$1 + \frac{Ke^{-Ts}}{s+1} = 0 \qquad (4-8)$$

For systems with transport lag, the rules of construction presented earlier need to be modified. For example, the number of root-locus branches is infinite, since the characteristic equation has an infinite number of roots. Similarly, the number of asymptotes is infinite. They are all parallel to the real axis of the s plane, as will be seen later.

From Equation (4-8), we obtain

$$\frac{Ke^{-Ts}}{s+1} = -1$$

Thus, the angle condition becomes

$$\underline{/\frac{Ke^{-Ts}}{s+1}} = \underline{/e^{-Ts}} - \underline{/s+1} = \pm 180°(2k+1) \qquad (k = 0, 1, 2, \ldots) \qquad (4-9)$$

To find the angle of e^{-Ts}, substitute $s = \sigma + j\omega$. Then we obtain

$$e^{-Ts} = e^{-T\sigma - j\omega T}$$

Since $e^{-T\sigma}$ is a real quantity, the angle of $e^{-T\sigma}$ is zero. Hence,

$$\underline{/e^{-Ts}} = \underline{/e^{-j\omega T}} = \underline{/\cos \omega T - j \sin \omega T}$$
$$= -\omega T \qquad \text{(radians)}$$
$$= -57.3°\omega T \qquad \text{(degrees)}$$

Since T is a given constant, the angle of e^{-Ts} is a function of ω only. The angle condition, Equation (4-9), then becomes

$$-57.3°\omega T - \underline{/s+1} = \pm 180°(2k+1)$$

We next determine the angle contribution due to e^{-Ts}, as given by Equation (4-9). For $k = 0$, the angle condition may be written

$$\underline{/s+1} = \pm 180° - 57.3°\omega T \qquad (4-10)$$

Since the angle contribution of e^{-Ts} is zero for $\omega = 0$, the real axis from -1 to $-\infty$ forms a part of the root loci. Now assume a value ω_1 for ω and compute $57.3°\omega_1 T$. At point -1 on the negative real axis, draw a line that makes an angle of $180° - 57.3°\omega_1 T$ with the real axis. Find the intersection of this line and the horizontal line $\omega = \omega_1$. This intersection, point P in Figure 4–29(a), is a point satisfying Equation (4–10) and hence is on a root locus. Continuing the same process, we obtain the root-locus plot shown in Figure 4–29(b).

Note that as s approaches minus infinity, the open-loop transfer function

$$\frac{Ke^{-Ts}}{s+1}$$

also approaches minus infinity, since

$$\lim_{s=-\infty} \frac{Ke^{-Ts}}{s+1} = \frac{\dfrac{d}{ds}(Ke^{-Ts})}{\dfrac{d}{ds}(s+1)}\bigg|_{s=-\infty}$$
$$= -KTe^{-Ts}\big|_{s=-\infty}$$
$$= -\infty$$

Therefore, $s = -\infty$ is a pole of the open-loop transfer function. Thus, root loci start from $s = -1$ or $s = -\infty$ and terminate at $s = \infty$, as K increases from zero to infinity. Since the right-hand side of the angle condition given by Equation (4–9) has an infinite number of values, there are an infinite number of root

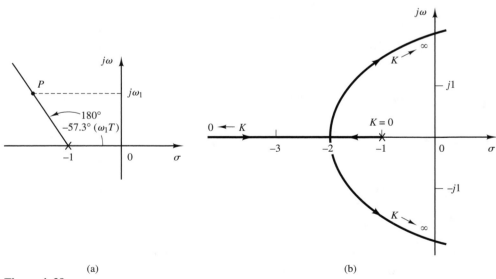

(a)   (b)

Figure 4–29
(a) Construction of the root locus; (b) root-locus plot.

loci as the value of $k(k = 0, 1, 2, \ldots)$ goes from zero to infinity. For example, if $k = 1$, the angle condition becomes

$$\underline{/s + 1} = \pm 540° - 57.3° \omega T \quad \text{(degrees)}$$
$$= \pm 3\pi - \omega T \quad \text{(radians)}$$

The construction of the root loci for $k = 1$ is the same as that for $k = 0$. A plot of root loci for $k = 0, 1$, and 2 when $T = 1$ sec is shown in Figure 4–30.

The magnitude condition states that

$$\left| \frac{Ke^{-Ts}}{s + 1} \right| = 1$$

Since the magnitude of e^{-Ts} is equal to that of $e^{-T\sigma}$, or

$$|e^{-Ts}| = |e^{-T\sigma}| \cdot |e^{-j\omega T}| = e^{-T\sigma}$$

the magnitude condition becomes

$$|s + 1| = Ke^{-T\sigma}$$

The root loci shown in Figure 4–30 are graduated in terms of K when $T = 1$ sec.

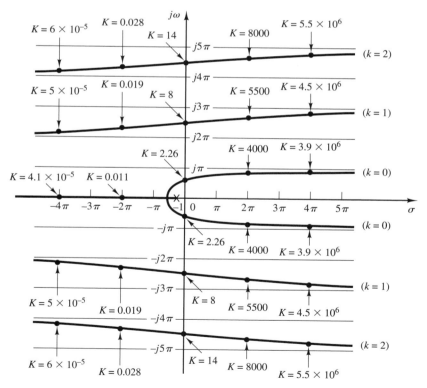

Figure 4–30
Root-locus plot for the system shown in Figure 4–28 ($T = 1$ sec).

Although there are an infinite number of root-locus branches, the primary branch that lies between $-j\pi$ and $j\pi$ is the most important. From Figure 4–30, the critical value of K at the primary branch is equal to 2.26, while the critical values of K at other branches are much higher $(8, 14, \ldots)$. Therefore, the critical value $K = 2.26$ on the primary branch is most significant from the viewpoint of stability. The transient response of the system is determined by the roots located closest to the $j\omega$-axis and lie on the primary branch. In sum, the root-locus branch corresponding to $k = 0$ is the dominant one; other branches, corresponding to $k = 1, 2, 3, \ldots$, are not so important and may be neglected.

This example illustrates the fact that dead time can cause instability even in a first-order system, because the root loci enter the right-half s plane for large values of K. Therefore, although the gain K of a first-order system can be set at a high value in the absence of dead time, it cannot be set too high if dead time is present. (For the system considered here, the value of the gain K must be considerably less than 2 for satisfactory operation.)

Approximation of transport lag or dead time. If the dead time T is very small, then e^{-Ts} is frequently approximated by

$$e^{-Ts} \doteq 1 - Ts$$

or

$$e^{-Ts} \doteq \frac{1}{Ts + 1}$$

Such approximations are good if the dead time is very small and, in addition, the input time function $f(t)$ to the dead-time element is smooth and continuous. [This means that the second- and higher-order derivatives of $f(t)$ are small.]

A more elaborate expression to approximate e^{-Ts} is

$$e^{-Ts} = \frac{1 - \dfrac{Ts}{2} + \dfrac{(Ts)^2}{8} - \dfrac{(Ts)^3}{48} + \cdots}{1 + \dfrac{Ts}{2} + \dfrac{(Ts)^2}{8} + \dfrac{(Ts)^3}{48} + \cdots}$$

If only the first two terms in the numerator and denominator are taken, then

$$e^{-Ts} \doteq \frac{1 - \dfrac{Ts}{2}}{1 + \dfrac{Ts}{2}} = \frac{2 - Ts}{2 + Ts}$$

This approximation is also used frequently.

MATLAB approximation of dead time. To handle the dead time e^{-sT}, MATLAB uses the Padé approximation, named after the French mathematician Henri Eugène Padé. For example, if $T = 0.1$ sec, then, using the third-order transfer function as an approximation to e^{-Ts}, we enter the following MATLAB program into the computer:

[num,den] = pade(0.1, 3);
printsys(num, den, 's')
num/den =

$$\frac{-1s^{\wedge}3 + 120s^{\wedge}2 - 6000s + 120000}{s^{\wedge}3 + 120s^{\wedge}2 + 6000s + 120000}$$

Similarly, the program for the fourth-order transfer function approximation with $T = 0.1$ sec is

[num,den] = pade(0.1, 4);
printsys(num, den, 's')
num/den =

$$\frac{s^{\wedge}4 - 200s^{\wedge}3 + 18000s^{\wedge}2 - 840000s + 16800000}{s^{\wedge}4 + 200s^{\wedge}3 + 18000s^{\wedge}2 + 840000s + 16800000}$$

Notice that the Padé approximation depends on the dead time T and the order desired for the approximating transfer function.

EXAMPLE 4–11 Consider the system shown in Figure 4–31, in which the dead time $T = 1$ sec. Suppose that we approximate the dead time by the second-order Padé approximation. The expression for this approximation can be obtained with MATLAB as follows:

```
>> [num,den] = pade(1, 2);
>> printsys(num, den, 's')
num/den =
   s^2 - 6s + 12
   s^2 + 6s + 12
```

Hence,

$$e^{-s} = \frac{s^2 - 6s + 12}{s^2 + 6s + 12} \qquad (4-11)$$

Using this approximation, determine the critical value of K (where $K > 0$) for stability. Since the characteristic equation for the system is

$$s + 1 + Ke^{-s} = 0$$

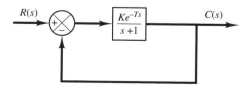

Figure 4–31
A control system with dead time.

by substituting Equation (4–11) into this characteristic equation, we obtain

$$s + 1 + K\frac{s^2 - 6s + 12}{s^2 + 6s + 12} = 0$$

or

$$s^3 + (7 + K)s^2 + (18 - 6K)s + 12(1 + K) = 0$$

Applying the Routh stability criterion, we get the following Routh table:

$$\begin{array}{ccc} s^3 & 1 & 18 - 6K \\ s^2 & 7 + K & 12(1 + K) \\ s^1 & \dfrac{-6K^2 - 36K + 114}{7 + K} & 0 \\ s^0 & 12(1 + K) & \end{array}$$

Hence, for stability, we require that

$$-6K^2 - 36K + 114 > 0$$

which can be written as

$$(K + 8.2915)(K - 2.2915) < 0$$

or

$$K < 2.2915$$

Since K must be positive, the range of K required for stability is

$$0 < K < 2.2915$$

Notice that, according to the present analysis, the upper limit of K required for stability is 2.2915. This value is greater than the exact upper limit of K. (Earlier, we obtained the exact upper limit of K to be 2, as shown in Figure 4–30.) This is because we approximated e^{-s} by the second-order Padé approximation. A higher-order Padé approximation will improve the accuracy of the analysis. However, the computations involved increase considerably.

4–7 ROOT-LOCUS APPROACH TO CONTROL SYSTEMS COMPENSATION

Basic design aspects of control systems are carried out mostly by analytical means, and only small parts may be carried out with MATLAB. However, once the system is designed, the response performance of the mathematical model must be checked throughly. MATLAB plays an important role in this regard.

Design and performance verification must work as a unit, and the latter can be done most effectively with MATLAB. In this section, we present series compensation techniques using the root-locus approach (such as lead compensation, lag compensation, and lag–lead compensation) and the parallel compensation technique using the root-locus approach. Sample problems illustrate each of these cases. Designs are carried out analytically, and the response characteristics of the designed systems are examined with MATLAB.

Lead compensation techniques based on the root-Locus approach.
The root-locus approach to design is very powerful when the specifications are given in terms of time-domain quantities, such as the damping ratio and undamped natural frequency of the desired dominant closed-loop poles, maximum overshoot, rise time, and settling time.

Consider a design problem in which the original system either is unstable for all values of the gain or is stable, but has undesirable transient-response characteristics. In such a case, reshaping of the root locus is necessary in the broad neighborhood of the $j\omega$-axis and the origin in order that the dominant closed-loop poles be at desired locations in the complex plane. This problem may be solved by inserting an appropriate lead compensator

$$G_c(s) = K_c \frac{s + \dfrac{1}{T}}{s + \dfrac{1}{\alpha T}} \quad (0 < \alpha < 1)$$

in cascade with the feedforward transfer function. The pole–zero configuration of the lead compensator is shown in Figure 4–32.

The procedures for designing a lead compensator for the system shown in Figure 4–33 by the root-locus method may be stated as follows:

1. From the performance specifications, determine the desired locations of the dominant closed-loop poles.
2. By drawing the root-locus plot of the uncompensated system (the original system), ascertain whether the gain adjustment alone can yield the desired closed-loop poles. If not, calculate the angle deficiency ϕ. This angle must be contributed by the lead compensator if the new root locus is to pass through the desired locations of the dominant closed-loop poles.
3. Assume that the lead compensator $G_c(s)$ is given by

$$G_c(s) = K_c \alpha \frac{Ts + 1}{\alpha Ts + 1} = K_c \frac{s + \dfrac{1}{T}}{s + \dfrac{1}{\alpha T}}, \quad (0 < \alpha < 1)$$

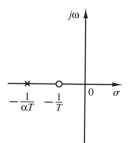

Figure 4–32
Pole–zero configuration of lead network.

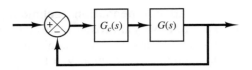

Figure 4-33
Control system.

where α and T are determined from the angle deficiency. K_c is determined from the requirement of the open-loop gain.

4. If no static error constants are specified, determine the locations of the pole and zero of the lead compensator so that the lead compensator will contribute the necessary angle ϕ. If no other requirements are imposed on the system, try to make the value of α as large as possible. A larger value of α generally results in a larger value of K_v, which is desirable. (If a particular static error constant is specified, it is generally simpler to use the frequency-response approach.)
5. Determine the open-loop gain of the compensated system from the magnitude condition.

Once a compensator has been designed, check to see whether all performance specifications have been met. If the compensated system does not meet the performance specifications, then repeat the design procedure by adjusting the compensator pole and zero until all such specifications are met. If a large static error constant is required, cascade a lag network or change the lead compensator to a lag–lead compensator.

Note that if the selected dominant closed-loop poles are not really dominant, it will be necessary to modify the locations of the pairs of such selected dominant closed-loop poles. (Closed-loop poles other than the dominant ones modify the response obtained from the dominant closed-loop poles alone. The amount of modification depends on the locations of these remaining closed-loop poles.) Also, closed-loop zeros affect the response if they are located near the origin.

EXAMPLE 4-12 Consider the system shown in Figure 4-34(a). The feedforward transfer function is

$$G(s) = \frac{4}{s(s+2)}$$

The root-locus plot for this system is shown in Figure 4-34(b). The closed-loop transfer function becomes

$$\frac{C(s)}{R(s)} = \frac{4}{s^2 + 2s + 4}$$

$$= \frac{4}{(s + 1 + j\sqrt{3})(s + 1 - j\sqrt{3})}$$

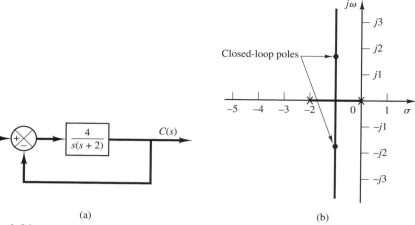

(a)   (b)

Figure 4-34
(a) Control system; (b) root-locus plot.

The closed-loop poles are located at

$$s = -1 \pm j\sqrt{3}$$

The damping ratio of the closed-loop poles is 0.5. The undamped natural frequency of the closed-loop poles is 2 rad/sec. The static velocity error constant K_v is 2 sec^{-1}.

It is desired to modify the closed-loop poles so that an undamped natural frequency $\omega_n = 4$ rad/sec is obtained, without changing the value of the damping ratio, $\zeta = 0.5$.

The damping ratio of 0.5 requires that the complex-conjugate poles lie on the lines drawn through the origin making angles of $\pm 60°$ with the negative real axis.

Since the damping ratio determines the angular location of the complex-conjugate closed-loop poles, while the distance of the pole from the origin is determined by the undamped natural frequency ω_n, the desired locations of the closed-loop poles of this sample problem are

$$s = -2 \pm j2\sqrt{3}$$

In some cases, after the root loci of the original system have been obtained, the dominant closed-loop poles may be moved to the desired location by simple gain adjustment. This is, however, not the case for the present system. Therefore, we shall insert a lead compensator in the feedforward path.

A general procedure for determining the lead compensator is as follows: First, find the sum of the angles at the desired location of one of the dominant closed-loop poles with the open-loop poles and zeros of the original system, and determine the necessary angle ϕ to be added so that the total sum of the angles is equal to $\pm 180°(2k + 1)$. The lead compensator must contribute this angle ϕ. (If the angle ϕ is quite large, then two or more lead networks may be needed rather than a single one.)

If the original system has the open-loop transfer function $G(s)$, then the compensated system will have the open-loop transfer function

$$G_c(s)G(s) = \left(K_c \frac{s + \dfrac{1}{T}}{s + \dfrac{1}{\alpha T}} \right) G(s)$$

where $G_c(s)$ is a lead compensator with the following form:

$$G_c(s) = K_c \alpha \frac{Ts + 1}{\alpha Ts + 1} = K_c \frac{s + \dfrac{1}{T}}{s + \dfrac{1}{\alpha T}}, \quad (0 < \alpha < 1)$$

Notice that there are many possible values for T and α that will yield the necessary angle contribution at the desired closed-loop poles.

The next step is to determine the locations of the zero and pole of the lead compensator. There are many possibilities for the choice of such locations. (See the comments at the end of this problem.) In what follows, we shall introduce a procedure to obtain the largest possible value for α. (Note that a larger value of α will produce a larger value of K_v. In most cases, the larger K_v is, the better is the system performance.) First, draw a horizontal line passing through point P, the desired location for one of the dominant closed-loop poles. This is shown as line PA in Figure 4–35. Also, draw a line connecting point P and the origin. Bisect the angle between the lines PA and PO, as shown in Figure 4–35. Draw two lines PC and PD that make angles $\pm \phi/2$ with the bisector PB. The intersections of PC and PD with the negative real axis give the necessary locations of the pole and zero of the lead network. The compensator thus designed will make point P a point on the root locus of the compensated system. The open-loop gain is determined from the magnitude condition.

In the present system, the angle of $G(s)$ at the desired closed-loop pole is

$$\left. \angle \frac{4}{s(s+2)} \right|_{s=-2+j2\sqrt{3}} = -210°$$

Thus, if we need to force the root locus to go through the desired closed-loop pole, the lead compensator must contribute $\phi = 30°$ at this point. By following the foregoing design procedure, we determine the zero and pole of the lead compensator, as shown in Figure 4–36, to be

$$\text{Zero at } s = -2.9, \quad \text{Pole at } s = -5.4$$

or

$$T = \frac{1}{2.9} = 0.345, \quad \alpha T = \frac{1}{5.4} = 0.185$$

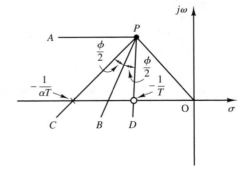

Figure 4–35
Determination of the pole and zero of a lead network.

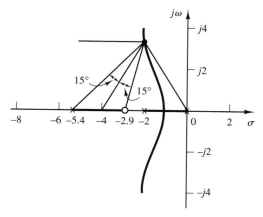

Figure 4-36
Root-locus plot of the compensated system.

Thus, $\alpha = 0.537$. The open-loop transfer function of the compensated system then becomes

$$G_c(s)G(s) = K_c \frac{s + 2.9}{s + 5.4} \frac{4}{s(s+2)} = \frac{K(s+2.9)}{s(s+2)(s+5.4)}$$

where $K = 4K_c$. The root-locus plot of the compensated system is shown in Figure 4-36. The gain K is evaluated from the magnitude condition as

$$\left| \frac{K(s+2.9)}{s(s+2)(s+5.4)} \right|_{s=-2+j2\sqrt{3}} = 1$$

or

$$K = 18.7$$

It follows that

$$G_c(s)G(s) = \frac{18.7(s+2.9)}{s(s+2)(s+5.4)}$$

The constant K_c of the lead compensator is

$$K_c = \frac{18.7}{4} = 4.68$$

Hence, $K_c \alpha = 2.51$. The lead compensator, therefore, has the transfer function

$$G_c(s) = 2.51 \frac{0.345s + 1}{0.185s + 1} = 4.68 \frac{s + 2.9}{s + 5.4}$$

The static velocity error constant K_v is obtained from the expression

$$K_v = \lim_{s \to 0} sG_c(s)G(s)$$
$$= \lim_{s \to 0} \frac{s18.7(s + 2.9)}{s(s + 2)(s + 5.4)}$$
$$= 5.02 \text{ sec}^{-1}$$

Note that the third closed-loop pole of the designed system is found by dividing the characteristic equation by the known factors as follows:

$$s(s + 2)(s + 5.4) + 18.7(s + 2.9) = \left(s + 2 + j2\sqrt{3}\right)\left(s + 2 - j2\sqrt{3}\right)(s + 3.4)$$

The foregoing compensation method enables us to place the dominant closed-loop poles at the desired points in the complex plane. The third pole, at $s = -3.4$, is close to the added zero at $s = -2.9$. Therefore, the effect of this pole on the transient response is relatively small. Since no restriction has been imposed on the nondominant pole and no specification has been given concerning the value of the static velocity error coefficient, we conclude that the present design is satisfactory.

Comments. We may place the zero of the compensator at $s = -2$ and the pole at $s = -4$, so that the angle contribution of the lead compensator is 30°. (In this case, the zero of the lead compensator will cancel a pole of the plant, resulting in a second-order system, rather than the third-order system we designed.) It can be seen that K_v in this case is 4 sec^{-1}. Other combinations can be selected that will yield a 30° phase lead. (For different combinations of a zero and pole of the compensator that contribute 30°, the value of α will be different and the value of K_v will also be different.) Although a certain change in the value of K_v can be made by altering the pole–zero location of the lead compensator, if a large increase in the value of K_v is desired, then we must alter the lead compensator to a lag–lead compensator.

Comparison of step responses of the compensated and uncompensated systems. In what follows, we shall examine the unit-step responses of the compensated and uncompensated systems with MATLAB.

The closed-loop transfer function of the compensated system is

$$\frac{C(s)}{R(s)} = \frac{18.7(s + 2.9)}{s(s + 2)(s + 5.4) + 18.7(s + 2.9)}$$
$$= \frac{18.7s + 54.23}{s^3 + 7.4s^2 + 29.5s + 54.23}$$

Hence,

$$\text{numc} = [0 \quad 0 \quad 18.7 \quad 54.23]$$
$$\text{denc} = [1 \quad 7.4 \quad 29.5 \quad 54.23]$$

For the uncompensated system, the closed-loop transfer function is

$$\frac{C(s)}{R(s)} = \frac{4}{s^2 + 2s + 4}$$

Thus,

$$\text{num} = [0 \quad 0 \quad 4]$$
$$\text{den} = [1 \quad 2 \quad 4]$$

MATLAB Program 4–18 produces the unit-step response curves for the two systems. The resulting plot is shown in Figure 4–37. Notice that the compensated system exhibits slightly larger maximum overshoot. The settling time of the compensated system is one-half that of the original system, as expected.

```
MATLAB Program 4–18

>> % ——— Unit-step response ———
>>
>> % ***** Unit-step responses of compensated and uncompensated
>> % systems *****
>>
>> numc = [18.7   54.23];
>> denc = [1   7.4   29.5   54.23];
>> num = [4];
>> den = [1   2   4];
>> t = 0:0.05:5;
>> [c1,x1,t] = step(numc,denc,t);
>> [c2,x2,t] = step(num,den,t);
>> plot(t,c1,t,c1,'o',t,c2,t,c2,'x')
>> grid
>> title('Unit-Step Responses of Compensated and Uncompensated Systems')
>> xlabel('t Sec')
>> ylabel('Outputs c_1 and c_2')
>> text(0.6, 1.32, 'Compensated system')
>> text(1.3,0.68,'Uncompensated system')
```

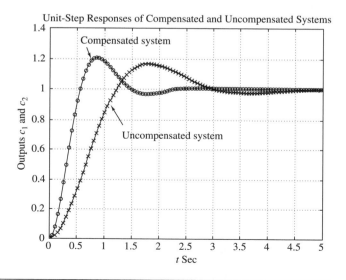

Figure 4–37
Unit-step responses of compensated and uncompensated systems.

Lag compensation techniques. Consider the problem of finding a suitable compensation network for the case where the system exhibits satisfactory transient-response characteristics but unsatisfactory steady-state characteristics. Compensation in this case essentially consists of increasing the open-loop gain without appreciably changing the transient-response characteristics. This means that the root locus in the neighborhood of the dominant closed-loop poles should not be changed appreciably, but the open-loop gain should be increased as much as is needed. This can be accomplished if a lag compensator is put in cascade with the given feedforward transfer function.

To avoid an appreciable change in the root loci, the angle contribution of the lag network should be limited to a small amount, say, between $0°$ and $-5°$. To assure this, we place the pole and zero of the lag network relatively close together and near the origin of the s plane. Then the closed-loop poles of the compensated system will be shifted only slightly from their original locations. Hence, the transient-response characteristics will be changed only slightly.

Consider a lag compensator $G_c(s)$, where

$$G_c(s) = \hat{K}_c \beta \frac{Ts + 1}{\beta Ts + 1} = \hat{K}_c \frac{s + \dfrac{1}{T}}{s + \dfrac{1}{\beta T}} \qquad (\beta > 1) \qquad (4\text{--}12)$$

The pole–zero configuration of the lag network is shown in Figure 4–38.

If we place the zero and pole of the lag compensator very close to each other, then, at $s = s_1$, where s_1 is one of the dominant closed-loop poles, the magnitudes $s_1 + (1/T)$ and $s_1 + [1/(\beta T)]$ are almost equal, or

$$|G_c(s_1)| = \left| \hat{K}_c \frac{s_1 + \dfrac{1}{T}}{s_1 + \dfrac{1}{\beta T}} \right| \doteq \hat{K}_c$$

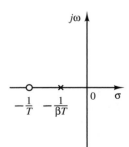

Figure 4–38
Pole–zero configuration of lag network.

To make the angle contribution of the lag compensator small, we require that

$$-5° < \underline{/\dfrac{s_1 + \dfrac{1}{T}}{s_1 + \dfrac{1}{\beta T}}} < 0°$$

This implies that if the gain $\hat{K}_c$ of the lag compensator is set equal to unity, then the transient-response characteristics will not be altered. (This means that the overall gain of the open-loop transfer function can be increased by a factor $\beta > 1$.) If the pole and zero are placed very close to the origin, then the value of β can be made large. (A large value of β may be used, provided that the lag compensator can be physically realized.) Note that the value of T must be large, but its exact value is not critical. However, it should not be too large, in order to avoid difficulties in realizing the phase lag compensator by physical components.

An increase in the gain means an increase in the static error constants. If the open-loop transfer function of the uncompensated system is $G(s)$, then the static velocity error constant K_v of the uncompensated system is

$$K_v = \lim_{s \to 0} sG(s)$$

If the compensator is chosen as given by Equation (4–12), then, for the compensated system with the open-loop transfer function $G_c(s)G(s)$, the static velocity error constant $\hat{K}_v$ becomes

$$\begin{aligned}\hat{K}_v &= \lim_{s \to 0} sG_c(s)G(s) \\ &= \lim_{s \to 0} G_c(s)K_v \\ &= \hat{K}_c \beta K_v\end{aligned}$$

Thus, if the compensator is given by Equation (4–12), then the static velocity error constant is increased by the factor $\hat{K}_c \beta$, where $\hat{K}_c$ is approximately unity.

The main negative effect of lag compensation is that the compensator zero that will be generated near the origin creates a closed-loop pole near the origin. This closed-loop pole and the compensator zero will generate a long tail of small amplitude in the step response, thus increasing the settling time.

Design procedures for lag compensation by the root-locus method. The procedure using the root-locus method to design lag compensators for the system shown in Figure 4–39 will be illustrated in Examples 4–13 and 4–14. (We assume that the uncompensated system meets the transient-response specifications by simple gain adjustment; if this is not the case, refer to lag–lead compensation):

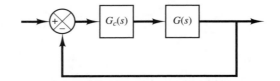

Figure 4–39
Control system.

EXAMPLE 4–13 Consider the system shown in Figure 4–40(a). The feedforward transfer function is

$$G(s) = \frac{1.06}{s(s + 1)(s + 2)}$$

The root-locus plot of the system is shown in Figure 4–40(b). The closed-loop transfer function becomes

$$\frac{C(s)}{R(s)} = \frac{1.06}{s(s + 1)(s + 2) + 1.06}$$

$$= \frac{1.06}{(s + 0.3307 - j0.5864)(s + 0.3307 + j0.5864)(s + 2.3386)}$$

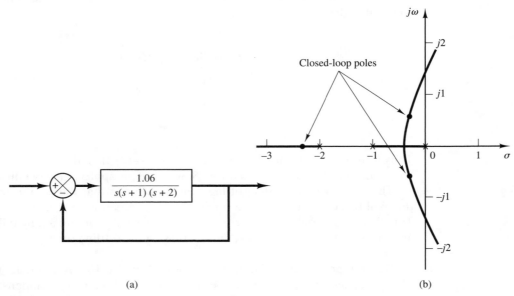

(a) (b)

Figure 4–40
(a) Control system; (b) root-locus plot.

The dominant closed-loop poles are

$$s = -0.3307 \pm j0.5864$$

The damping ratio of the dominant closed-loop poles is $\zeta = 0.491$. The undamped natural frequency of the dominant closed-loop poles is 0.673 rad/sec. The static velocity error constant is 0.53 sec^{-1}.

It is desired to increase the static velocity error constant K_v to about 5 sec^{-1} without appreciably changing the location of the dominant closed-loop poles.

To meet this specification, let us insert a lag compensator as given by Equation (4–12) in cascade with the given feedforward transfer function. To increase the static velocity error constant by a factor of about 10, let us choose $\beta = 10$ and place the zero and pole of the lag compensator at $s = -0.05$ and $s = -0.005$, respectively. The transfer function of the lag compensator becomes

$$G_c(s) = \hat{K}_c \frac{s + 0.05}{s + 0.005}$$

The angle contribution of this lag network near a dominant closed-loop pole is about 4°. Because this angle contribution is not very small, there is a small change in the new root locus near the desired dominant closed-loop poles.

The open-loop transfer function of the compensated system then becomes

$$G_c(s)G(s) = \hat{K}_c \frac{s + 0.05}{s + 0.005} \frac{1.06}{s(s + 1)(s + 2)}$$

$$= \frac{K(s + 0.05)}{s(s + 0.005)(s + 1)(s + 2)}$$

where

$$K = 1.06\hat{K}_c$$

The block diagram of the compensated system is shown in Figure 4–41. The root-locus plot of the compensated system near the dominant closed-loop poles is shown in Figure 4–42(a), together with the original root-locus plot. Figure 4–42(b) shows the root-locus plot of the compensated system near the origin. The MATLAB program to generate the root-locus plots shown in Figures 4–42(a) and (b) is given in MATLAB Program 4–19.

Figure 4–41
Compensated system.

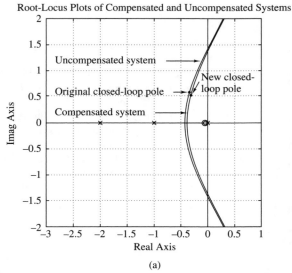

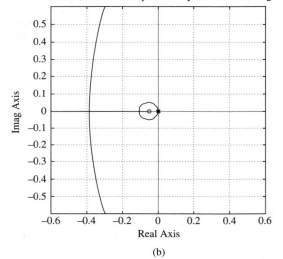

Figure 4–42
(a) Root-locus plots of the compensated system and uncompensated system; (b) root-locus plot of compensated system near the origin.

If the damping ratio of the new dominant closed-loop poles is kept the same, then the poles are obtained from the new root-locus plot as follows:

$$s_1 = -0.31 + j0.55, \qquad s_2 = -0.31 - j0.55$$

The open-loop gain K is

$$K = \left| \frac{s(s + 0.005)(s + 1)(s + 2)}{s + 0.05} \right|_{s=-0.31+j0.55}$$

$$= 1.0235$$

MATLAB Program 4–19

```
>> numc = [1  0.05]; denc = [1  3.005  2.015  0.01  0];
>> num = [1.06]; den = [1  3  2  0];
>> r1 = rlocus (numc,denc);
>> plot(r1); v = [-3  1  -2  2]; axis (v); axis ('square')
>> grid
>> hold
Current plot held
>> r2 = rlocus (num,den);
>> plot (r2);
>> title('Root-Locus Plot of Compensated and Uncompensated Systems')
>> xlabel('Real Axis'); ylabel ('Imag Axis')
>> text(-2.8,0.2,'Compensated System')
>> text(-2.8,1.2,'Uncompensated System')
>> text(-2.8,0.58,'Original closed-loop pole')
>> text(-0.1,0.85,'New closed-')
>> text(-0.1,0.62,'loop pole')
>> hold
Current plot released
>> % Plot root loci of the compensated system near the origin.
>> plot(r1); v = [-0.6  0.6  -0.6  0.6]; axis(v); axis('square')
>> grid
>> title('Root-Locus Plot of the Compensated System near the Origin')
>> xlabel('Real Axis'); ylabel ('Imag Axis')
>>
>> % Manually enter 'o' and 'x' marks on open-loop zero and open-loop
>> % poles, respectively.
```

Then the lag compensator gain $\hat{K}_c$ is determined as

$$\hat{K}_c = \frac{K}{1.06} = \frac{1.0235}{1.06} = 0.9656$$

Thus, the transfer function of the lag compensator designed is

$$G_c(s) = 0.9656 \frac{s + 0.05}{s + 0.005} = 9.656 \frac{20s + 1}{200s + 1} \qquad (4\text{--}13)$$

Then the compensated system has the following open-loop transfer function:

$$G_1(s) = \frac{1.0235(s + 0.05)}{s(s + 0.005)(s + 1)(s + 2)}$$

$$= \frac{5.12(20s + 1)}{s(200s + 1)(s + 1)(0.5s + 1)}$$

The static velocity error constant K_v is

$$K_v = \lim_{s \to 0} sG_1(s) = 5.12 \text{ sec}^{-1}$$

In the compensated system, the static velocity error constant has increased to 5.12 sec^{-1}, or 5.12/0.53 = 9.66 times the original value. (The steady-state error with ramp inputs has decreased to about 10% of that of the original system.) We have essentially accomplished the design objective of increasing the static velocity error constant to about 5 sec^{-1}.

Note that, since the pole and zero of the lag compensator are placed close together and are located very near the origin, their effect on the shape of the original root loci is small. Except for the presence of a small closed root locus near the origin, the root loci of the compensated and the uncompensated systems are similar to each other. However, the value of the static velocity error constant of the compensated system is 9.66 times greater than that of the uncompensated system.

The two other closed-loop poles of the compensated system are as follows:

$$s_3 = -2.326, \quad s_4 = -0.0549$$

The addition of the lag compensator increases the order of the system from 3 to 4, adding another closed-loop pole close to the zero of the lag compensator. (The added closed-loop pole at $s = -0.0549$ is close to the zero at $s = -0.05$.) Such a pair consisting of a zero and a pole creates a long tail of small amplitude in the transient response, as we will see later in the unit-step response. Since the pole at $s = -2.326$ is very far from the $j\omega$-axis compared with the dominant closed-loop poles, the effect of this pole on the transient response is small. Therefore, we may consider the closed-loop poles at $s = -0.31 \pm j0.55$ to be the dominant closed-loop poles.

The undamped natural frequency of the dominant closed-loop poles of the compensated system is 0.631 rad/sec, about 6% less than the original value, 0.673 rad/sec. This implies that the transient response of the compensated system is slower than that of the original system. The response will take a longer time to settle down. The maximum overshoot in the step response will increase in the compensated system. If such adverse effects can be tolerated, the lag compensation as discussed here presents a satisfactory solution to the given design problem.

Next, we shall compare the unit-ramp responses of the compensated system against the uncompensated system and verify that the steady-state performance is much better in the compensated system than the uncompensated system.

To obtain the unit-ramp response with MATLAB, we use the step command for the system $C(s)/[sR(s)]$. Since, for the compensated system,

$$\frac{C(s)}{sR(s)} = \frac{1.0235(s + 0.05)}{s[s(s + 0.005)(s + 1)(s + 2) + 1.0235(s + 0.05)]}$$

$$= \frac{1.0235s + 0.0512}{s^5 + 3.005s^4 + 2.015s^3 + 1.0335s^2 + 0.0512s}$$

we have

$$\text{numc} = [0 \quad 0 \quad 0 \quad 0 \quad 1.0235 \quad 0.0512]$$
$$\text{denc} = [1 \quad 3.005 \quad 2.015 \quad 1.0335 \quad 0.0512 \quad 0]$$

Also, for the uncompensated system,

$$\frac{C(s)}{sR(s)} = \frac{1.06}{s[s(s+1)(s+2)+1.06]}$$

$$= \frac{1.06}{s^4 + 3s^3 + 2s^2 + 1.06s}$$

Hence,

$$\text{num} = [0 \quad 0 \quad 0 \quad 0 \quad 1.06]$$
$$\text{den} = [1 \quad 3 \quad 2 \quad 1.06 \quad 0]$$

MATLAB Program 4–20 produces the plot of the unit-ramp response curves. Figure 4–43 shows the result. Clearly, the compensated system shows much smaller steady-state error (one-tenth of the original steady-state error) in following the unit-ramp input.

MATLAB Program 4–20

```
>> % ——— Unit-ramp response ———
>>
>> % ***** Unit-ramp response will be obtained as the unit-step
>> % response of C(s)/[sR(s)] *****
>> % ***** Enter the numerators and denominators of C1(s)/[sR(s)]
>> % and C2(s)/[sR(s)], where C1(s) and C2(s) are Laplace
>> % transforms of the outputs of the compensated and un-
>> % compensated systems, respectively. *****
>>
>> numc = [1.0235   0.0512];
>> denc = [1   3.005   2.015   1.0335   0.0512   0];
>> num = [1.06];
>> den = [1   3   2   1.06   0];
>>
>> % ***** Specify the time range (such as t= 0:0.1:50) and enter
>> % step command and plot command. *****
>>
>> t = 0:0.1:50;
>> [c1,x1,t] = step(numc,denc,t);
>> [c2,x2,t] = step(num,den,t);
>> plot(t,c1,'-',t,c2,'.',t,t,'--')
>> grid
>> text(2.2,27,'Compensated system');
>> text(26,21.3,'Uncompensated system');
>> title('Unit-Ramp Responses of Compensated and Uncompensated Systems')
>> xlabel('t Sec');
>> ylabel('Outputs c_1 and c_2')
```

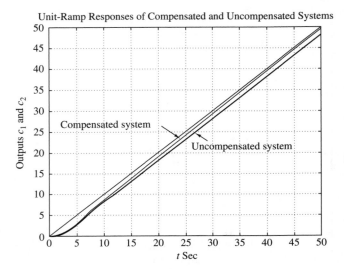

Figure 4–43
Unit-ramp responses of compensated and uncompensated systems. [The compensator is given by Equation (4–13).]

MATLAB Program 4–21 gives the unit-step response curves of the compensated and uncompensated systems. These curves are shown in Figure 4–44. Notice that the lag-compensated system exhibits a larger maximum overshoot and slower response than the

---

MATLAB Program 4–21

```
>> % ——— Unit-step response ———
>>
>> % ***** Enter the numerators and denominators of the
>> % compensated and uncompensated systems *****
>>
>> numc = [1.0235   0.0512];
>> denc = [1   3.005   2.015   1.0335   0.0512];
>> num = [1.06];
>> den = [1   3   2   1.06];
>>
>> % ***** Specify the time range (such as t = 0:0.1:40) and enter
>> % step command and plot command. *****
>>
>> t = 0:0.1:40;
>> [c1,x1,t] = step(numc,denc,t);
>> [c2,x2,t] = step(num,den,t);
>> plot(t,c1,'-',t,c2,'.')
>> grid
>> text(12.8,1.12,'Compensated system')
>> text(13.6,0.88,'Uncompensated system')
>> title('Unit-Step Responses of Compensated and Uncompensated Systems')
>> xlabel('t Sec')
>> ylabel('Outputs c_1 and c_2')
```

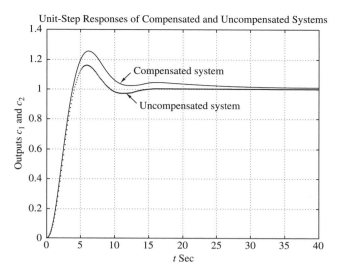

Figure 4–44
Unit-step responses of compensated and uncompensated systems. [The compensator is given by Equation (4–13).]

original uncompensated system. Notice also that a pair consisting of the pole at $s = -0.0549$ and the zero at $s = -0.05$ generates a long tail of small amplitude in the transient response. If we do not desire a larger maximum overshoot and a slower response, we need to use a lag–lead compensator.

Comments. Under certain circumstances, both a lead compensator and a lag compensator may satisfy the given specifications (transient-response specifications and steady-state specifications). Then either type of compensation may be used.

Lag–lead compensation techniques based on the root-locus approach. Lead compensation basically speeds up the response and increases the stability of the system. Lag compensation improves the steady-state accuracy of the system, but reduces the speed of the response.

If improvements in both transient response and steady-state response are desired, then both a lead compensator and a lag compensator may be used simultaneously. Rather than introducing a lead compensator and a lag compensator as separate elements, however, it is economical to use a single lag–lead compensator.

Lag–lead compensation combines the advantages of lag and lead compensations. Since the lag–lead compensator possesses two poles and two zeros, such a compensation increases the order of the system by 2, unless cancellation of pole(s) and zero(s) occurs in the compensated system.

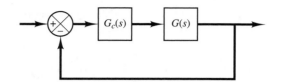

Figure 4–45
Control system.

Consider the system shown in Figure 4–45. Assume that we use the lag–lead compensator

$$G_c(s) = K_c \frac{\beta}{\gamma} \frac{(T_1 s + 1)(T_2 s + 1)}{\left(\dfrac{T_1}{\gamma} s + 1\right)(\beta T_2 s + 1)} = K_c \left(\frac{s + \dfrac{1}{T_1}}{s + \dfrac{\gamma}{T_1}}\right)\left(\frac{s + \dfrac{1}{T_2}}{s + \dfrac{1}{\beta T_2}}\right) \quad (4\text{–}14)$$

where $\beta > 1$ and $\gamma > 1$. (Assume that K_c belongs to the lead portion of the lag–lead compensator.)

In designing lag–lead compensators, we consider two cases: $\gamma \neq \beta$ and $\gamma = \beta$.

Case 1: $\gamma \neq \beta$. In this case, the design process is a combination of the design of the lead compensator and that of the lag compensator. The design procedure for the lag–lead compensator is as follows:

1. From the given performance specifications, determine the desired locations of the dominant closed-loop poles.
2. Using the uncompensated open-loop transfer function $G(s)$, determine the angle deficiency ϕ if the dominant closed-loop poles are to be at the desired locations. The phase-lead portion of the lag–lead compensator must contribute this angle ϕ.
3. Assuming that we later choose T_2 sufficiently large so that the magnitude of the lag portion

$$\left| \frac{s_1 + \dfrac{1}{T_2}}{s_1 + \dfrac{1}{\beta T_2}} \right|$$

is approximately unity, where $s = s_1$ is one of the dominant closed-loop poles, choose the values of T_1 and γ from the requirement that

$$\left| \frac{s_1 + \dfrac{1}{T_1}}{s_1 + \dfrac{\gamma}{T_1}} \right| = \phi$$

The choice of T_1 and γ is not unique. (Infinitely many sets of T_1 and γ are possible.) Then determine the value of K_c from the magnitude condition:

$$\left| K_c \frac{s_1 + \dfrac{1}{T_1}}{s_1 + \dfrac{\gamma}{T_1}} G(s_1) \right| = 1$$

4. If the static velocity error constant K_v is specified, determine the value of β to satisfy the requirement for K_v. The static velocity error constant K_v is given by

$$K_v = \lim_{s \to 0} s G_c(s) G(s)$$

$$= \lim_{s \to 0} s K_c \left(\frac{s + \dfrac{1}{T_1}}{s + \dfrac{\gamma}{T_1}} \right) \left(\frac{s + \dfrac{1}{T_2}}{s + \dfrac{1}{\beta T_2}} \right) G(s)$$

$$= \lim_{s \to 0} s K_c \frac{\beta}{\gamma} G(s)$$

where K_c and γ have been determined in step 3. Hence, given the value of K_v, the value of β can be determined from this last equation. Then, using the value of β thus determined, choose the value of T_2 such that

$$\left| \frac{s_1 + \dfrac{1}{T_2}}{s_1 + \dfrac{1}{\beta T_2}} \right| \doteq 1$$

$$-5° < \left/ \frac{s_1 + \dfrac{1}{T_2}}{s_1 + \dfrac{1}{\beta T_2}} \right. < 0°$$

(The preceding design procedure is illustrated in Example 4–14.)

Case 2. $\gamma = \beta$. If $\gamma = \beta$ is required in Equation (4–14), then the preceeding design procedure for the lag–lead compensator may be modified as follows:

1. From the given performance specifications, determine the desired locations of the dominant closed-loop poles.

2. The lag–lead compensator given by Equation (4–14) is modified to

$$G_c(s) = K_c \frac{(T_1 s + 1)(T_2 s + 1)}{\left(\dfrac{T_1}{\beta} s + 1 \right)(\beta T_2 s + 1)} = K_c \frac{\left(s + \dfrac{1}{T_1} \right)\left(s + \dfrac{1}{T_2} \right)}{\left(s + \dfrac{\beta}{T_1} \right)\left(s + \dfrac{1}{\beta T_2} \right)} \quad (4\text{–}15)$$

where $\beta > 1$. The open-loop transfer function of the compensated system is $G_c(s)G(s)$. If the static velocity error constant K_v is specified, determine the value of the constant K_c from the following equation:

$$K_v = \lim_{s \to 0} sG_c(s)G(s)$$
$$= \lim_{s \to 0} sK_cG(s)$$

3. To have the dominant closed-loop poles at the desired locations, calculate the angle contribution ϕ needed from the phase lead portion of the lag–lead compensator.

4. For the lag–lead compensator, we later choose T_2 sufficiently large so that

$$\left| \frac{s_1 + \dfrac{1}{T_2}}{s_1 + \dfrac{1}{\beta T_2}} \right|$$

is approximately unity, where $s = s_1$ is one of the dominant closed-loop poles. Determine the values of T_1 and β from the magnitude and angle conditions:

$$\left| K_c \left(\frac{s_1 + \dfrac{1}{T_1}}{s_1 + \dfrac{\beta}{T_1}} \right) G(s_1) \right| = 1$$

$$\left/ \frac{s_1 + \dfrac{1}{T_1}}{s_1 + \dfrac{\beta}{T_1}} \right. = \phi$$

5. Using the value of β just determined, choose T_2 so that

$$\left| \frac{s_1 + \dfrac{1}{T_2}}{s_1 + \dfrac{1}{\beta T_2}} \right| \doteq 1$$

$$-5° < \left/ \frac{s_1 + \dfrac{1}{T_2}}{s_1 + \dfrac{1}{\beta T_2}} \right. < 0°$$

The value of βT_2, the largest time constant of the lag–lead compensator, should not be too large to be physically realized. (An example of the design of the lag–lead compensator when $\gamma = \beta$ is given in Example 4–15.)

EXAMPLE 4-14 Consider the control system shown in Figure 4-46. The feedforward transfer function is

$$G(s) = \frac{4}{s(s + 0.5)}$$

This system has closed-loop poles at

$$s = -0.2500 \pm j1.9843$$

The damping ratio is 0.125, the undamped natural frequency is 2 rad/sec, and the static velocity error constant is 8 sec^{-1}.

It is desired to make the damping ratio of the dominant closed-loop poles equal to 0.5 and to increase the undamped natural frequency to 5 rad/sec and the static velocity error constant to 80 sec^{-1}. Design an appropriate compensator to meet all the performance specifications.

Let us assume that we use a lag–lead compensator having the transfer function

$$G_c(s) = K_c \left(\frac{s + \frac{1}{T_1}}{s + \frac{\gamma}{T_1}} \right) \left(\frac{s + \frac{1}{T_2}}{s + \frac{1}{\beta T_2}} \right) \quad (\gamma > 1, \beta > 1)$$

where $\gamma \neq \beta$. Then the compensated system will have the open-loop transfer function

$$G_c(s)G(s) = K_c \left(\frac{s + \frac{1}{T_1}}{s + \frac{\gamma}{T_1}} \right) \left(\frac{s + \frac{1}{T_2}}{s + \frac{1}{\beta T_2}} \right) G(s)$$

From the performance specifications, the dominant closed-loop poles must be at

$$s = -2.50 \pm j4.33$$

Since

$$\left/ \frac{4}{s(s + 0.5)} \right|_{s = -2.50 + j4.33} = -235°$$

the phase lead portion of the lag–lead compensator must contribute 55° so that the root locus passes through the desired location of the dominant closed-loop poles.

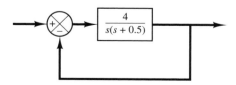

Figure 4-46
Control system.

To design the phase lead portion of the compensator, we first determine the locations of the zero and pole that will give 55° contribution. There are many possible choices, but we shall here choose the zero at $s = -0.5$ so that this zero will cancel the pole at $s = -0.5$ of the plant. Once the zero is chosen, the pole can be located such that the angle contribution is 55°. By simple calculation or graphical analysis, the pole must be located at $s = -5.021$. Thus, the phase lead portion of the lag–lead compensator becomes

$$K_c \frac{s + \dfrac{1}{T_1}}{s + \dfrac{\gamma}{T_1}} = K_c \frac{s + 0.5}{s + 5.021}$$

Thus,

$$T_1 = 2, \qquad \gamma = \frac{5.021}{0.5} = 10.04$$

Next, we determine the value of K_c from the magnitude condition:

$$\left| K_c \frac{s + 0.5}{s + 5.021} \frac{4}{s(s + 0.5)} \right|_{s=-2.5+j4.33} = 1$$

Hence,

$$K_c = \left| \frac{(s + 5.021)s}{4} \right|_{s=-2.5+j4.33} = 6.26$$

The phase lag portion of the compensator can be designed as follows: First the value of β is determined to satisfy the requirement on the static velocity error constant:

$$K_v = \lim_{s \to 0} sG_c(s)G(s) = \lim_{s \to 0} sK_c \frac{\beta}{\gamma} G(s)$$

$$= \lim_{s \to 0} s(6.26) \frac{\beta}{10.04} \frac{4}{s(s + 0.5)} = 4.988\beta = 80$$

Hence,

$$\beta = 16.04$$

Finally, we choose the value of T_2 large enough so that

$$\left| \frac{s + \dfrac{1}{T_2}}{s + \dfrac{1}{16.04 T_2}} \right|_{s=-2.5+j4.33} \doteq 1$$

and

$$-5° < \left|\frac{s + \dfrac{1}{T_2}}{s + \dfrac{1}{16.04T_2}}\right|_{s=-2.5+j4.33} < 0°$$

Since $T_2 \doteq 5$ (or any number greater than 5) satisfies these two requirements, we may choose

$$T_2 = 5$$

Now, the transfer function of the designed lag–lead compensator is given by

$$G_c(s) = (6.26)\left(\frac{s + \dfrac{1}{2}}{s + \dfrac{10.04}{2}}\right)\left(\frac{s + \dfrac{1}{5}}{s + \dfrac{1}{16.04 \times 5}}\right)$$

$$= 6.26\left(\frac{s + 0.5}{s + 5.02}\right)\left(\frac{s + 0.2}{s + 0.01247}\right)$$

$$= \frac{10(2s + 1)(5s + 1)}{(0.1992s + 1)(80.19s + 1)}$$

The compensated system will have the open-loop transfer function

$$G_c(s)G(s) = \frac{25.04(s + 0.2)}{s(s + 5.02)(s + 0.01247)}$$

Because of the cancellation of the $(s + 0.5)$ terms, the compensated system is a third-order system. (Mathematically, this cancellation is exact, but practically, such cancellation will not be exact because some approximations are usually involved in deriving the mathematical model of the system and, as a result, the time constants are not precise.) The root-locus plot of the compensated system is shown in Figure 4–47(a). An enlarged view of the root-locus plot near the origin is shown in Figure 4–47(b). Because the angle contribution of the phase lag portion of the lag–lead compensator is quite small, there is only a small change in the location of the dominant closed-loop poles from the desired location, $s = -2.5 \pm j4.33$. In fact, the new closed-loop poles are located at $s = -2.4123 \pm j4.2756$. (The new damping ratio is $\zeta = 0.491$.) Therefore, the compensated system meets all the required performance specifications. The third closed-loop pole of the compensated system is located at $s = -0.2078$. Since this closed-loop pole is very close to the zero at $s = -0.2$, the effect of the pole on the response is small. (Note that, in general, if a pole and a zero lie close to each other on the negative real axis near the origin, then such a pole–zero combination will yield a long tail of small amplitude in the transient response.)

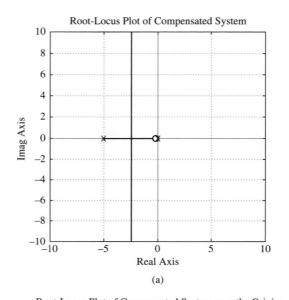

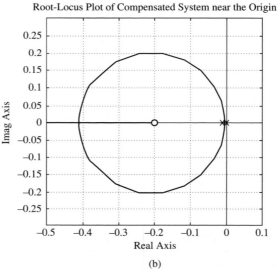

Figure 4-47
(a) Root-locus plot of the compensated system; (b) root-locus plot near the origin.

The unit-step response curves and unit-ramp response curves before and after compensation are shown in Figure 4–48.

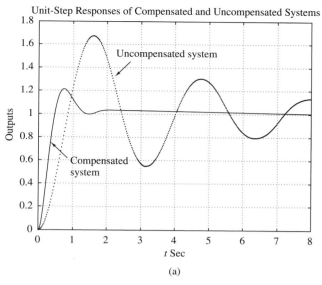

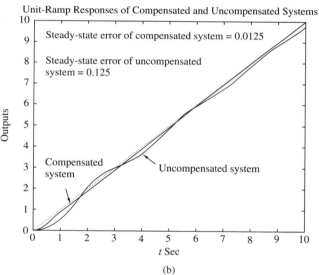

Figure 4–48
Transient response curves for the compensated system and uncompensated system.
(a) Unit-step response curves; (b) unit-ramp response curves.

EXAMPLE 4–15 Consider the control system of Example 4–14. Suppose that we use a lag–lead compensator of the form given by Equation (4–15), or

$$G_c(s) = K_c \frac{\left(s + \dfrac{1}{T_1}\right)\left(s + \dfrac{1}{T_2}\right)}{\left(s + \dfrac{\beta}{T_1}\right)\left(s + \dfrac{1}{\beta T_2}\right)} \quad (\beta > 1)$$

Assuming that the specifications are the same as those given in Example 4–14, design a compensator $G_c(s)$.

The desired locations for the dominant closed-loop poles are at

$$s = -2.50 \pm j4.33$$

The open-loop transfer function of the compensated system is

$$G_c(s)G(s) = K_c \frac{\left(s + \dfrac{1}{T_1}\right)\left(s + \dfrac{1}{T_2}\right)}{\left(s + \dfrac{\beta}{T_1}\right)\left(s + \dfrac{1}{\beta T_2}\right)} \cdot \frac{4}{s(s + 0.5)}$$

Since the requirement on the static velocity error constant K_v is 80 sec^{-1}, we have

$$K_v = \lim_{s \to 0} sG_c(s)G(s) = \lim_{s \to 0} K_c \frac{4}{0.5} = 8K_c = 80$$

Thus,

$$K_c = 10$$

The time constant T_1 and the value of β are determined from

$$\left|\frac{s + \dfrac{1}{T_1}}{s + \dfrac{\beta}{T_1}}\right| \left|\frac{40}{s(s + 0.5)}\right|_{s=-2.5+j4.33} = \left|\frac{s + \dfrac{1}{T_1}}{s + \dfrac{\beta}{T_1}}\right| \frac{8}{4.77} = 1$$

$$\left/\frac{s + \dfrac{1}{T_1}}{s + \dfrac{\beta}{T_1}}\right|_{s=-2.5+j4.33} = 55°$$

(The angle deficiency of 55° was obtained in Example 4–14.) Referring to Figure 4–49, we can easily locate points A and B such that

$$\angle APB = 55°, \quad \frac{\overline{PA}}{\overline{PB}} = \frac{4.77}{8}$$

(Use a graphical approach or a trigonometric approach.) The result is

$$\overline{AO} = 2.38, \quad \overline{BO} = 8.34$$

or

$$T_1 = \frac{1}{2.38} = 0.420, \quad \beta = 8.34T_1 = 3.503$$

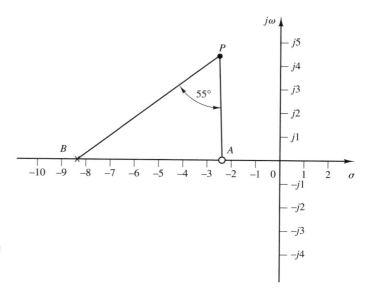

Figure 4–49
Determination of the desired pole–zero location.

The phase lead portion of the lag–lead network thus becomes

$$10\left(\frac{s + 2.38}{s + 8.34}\right)$$

For the phase lag portion, we may choose

$$T_2 = 10$$

Then

$$\frac{1}{\beta T_2} = \frac{1}{3.503 \times 10} = 0.0285$$

Thus, the lag–lead compensator becomes

$$G_c(s) = (10)\left(\frac{s + 2.38}{s + 8.34}\right)\left(\frac{s + 0.1}{s + 0.0285}\right)$$

The compensated system will have the open-loop transfer function

$$G_c(s)G(s) = \frac{40(s + 2.38)(s + 0.1)}{(s + 8.34)(s + 0.0285)s(s + 0.5)}$$

No cancellation occurs in this case, and the compensated system is of fourth order. Because the angle contribution of the phase lag portion of the lag–lead network is small, the dominant closed-loop poles are located very near the desired location. In fact, the dominant closed-loop poles are located at $s = -2.4539 \pm j4.3099$. The other two closed-loop poles are located at

$$s = -0.1003; \quad s = -3.8604$$

Since the closed-loop pole at $s = -0.1003$ is very close to a zero at $s = -0.1$, they almost cancel each other. Thus, the effect of this closed-loop pole is very small. The remaining closed-loop pole ($s = -3.8604$) does not quite cancel the zero at $s = -2.38$. The effect of this zero is to cause a larger overshoot in the step response than is present a similar system without such a zero. The unit-step response curves of the compensated and uncompensated systems are shown in Figure 4–50(a). The unit-ramp response curves for both systems are depicted in Figure 4–50(b).

The maximum overshoot in the step response of the compensated system is approximately 38%. (This is much larger than the maximum overshoot of 21% in the design presented in Example 4–14.) It is possible to decrease the maximum overshoot by a small amount from 38%, but not to 20% if $\gamma = \beta$ is required, as in this example. Note that by not requiring that $\gamma = \beta$, we have an additional parameter to play with and thus can reduce the maximum overshoot.

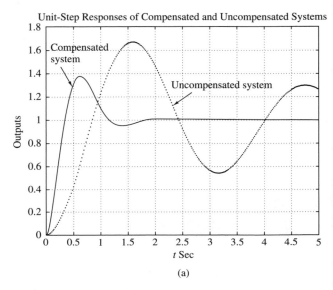

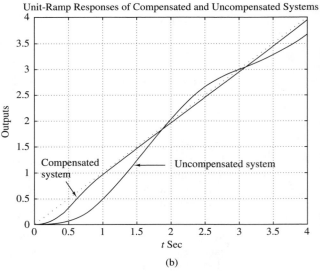

Figure 4–50
(a) Unit-step response curves for the compensated and uncompensated systems; (b) unit-ramp response curves for both systems.

Basic principle for designing a parallel compensated system. Thus far, we have presented series compensation techniques that use lead, lag, or lag–lead compensators. In this section, we discuss a parallel compensation technique. Because, in parallel compensation design, the controller (or compensator) is in a minor loop, the design may seem to be more complicated than in the series compensation case. It is, however, not complicated if we rewrite the characteristic equation to be of the same form as the characteristic equation for the series compensated system. In what follows, we present a simple design problem involving parallel compensation. From Figure 4–51(a), the closed-loop transfer function for the system with series compensation is

$$\frac{C}{R} = \frac{G_c G}{1 + G_c GH}$$

The characteristic equation is

$$1 + G_c GH = 0$$

Given G and H, the design problem becomes that of determining the compensator G_c that satisfies the given specification.

The closed-loop transfer function for the system with parallel compensation [Figure 4–51(b)] is

$$\frac{C}{R} = \frac{G_1 G_2}{1 + G_2 G_c + G_1 G_2 H}$$

The characteristic equation is

$$1 + G_1 G_2 H + G_2 G_c = 0$$

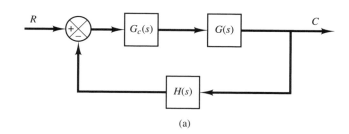

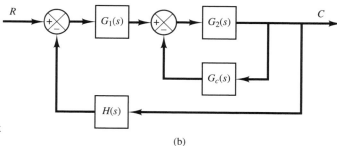

Figure 4–51
(a) Series compensation;
(b) parallel or feedback compensation.

Dividing this characteristic equation by the sum of the terms that do not involve G_c, we obtain

$$1 + \frac{G_c G_2}{1 + G_1 G_2 H} = 0 \qquad (4\text{--}16)$$

If we define

$$G_f = \frac{G_2}{1 + G_1 G_2 H}$$

then Equation (4–16) becomes

$$1 + G_c G_f = 0$$

Since G_f is a fixed transfer function, the design of G_c becomes the same as the case of series compensation. Hence, the same design approach applies to the parallel compensated system.

Velocity-feedback systems. A velocity-feedback system (tachometer feedback system) is an example of parallel compensated systems. The controller (or compensator) in such a system is a gain element. The gain of the feedback element in a minor loop must be determined properly so that the entire system satisfies the given design specifications. The characteristic of such a velocity-feedback system is that the variable parameter does not appear as a multiplying factor in the open-loop transfer function, so direct application of the root-locus design technique is not possible. However, by rewriting the characteristic equation such that the variable parameter appears as a multiplying factor, the root-locus approach to the design is possible.

An example of control system design using a parallel compensation technique is presented in Example 4–16.

EXAMPLE 4–16 Consider the system shown in Figure 4–52. Draw a root-locus diagram. Then determine the value of k such that the damping ratio of the dominant closed-loop poles is 0.4.

Here, the system involves velocity feedback. The open-loop transfer function is

$$\text{Open-loop transfer function} = \frac{20}{s(s+1)(s+4) + 20ks}$$

Notice that the adjustable variable k does not appear as a multiplicative factor. The characteristic equation for the system is

$$s^3 + 5s^2 + 4s + 20ks + 20 = 0 \qquad (4\text{--}17)$$

We define

$$20k = K$$

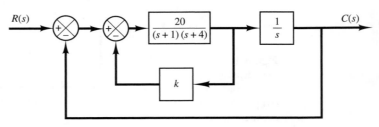

Figure 4–52
Control system.

Then Equation (4–17) becomes

$$s^3 + 5s^2 + 4s + Ks + 20 = 0 \tag{4-18}$$

Dividing both sides of Equation (4–18) by the sum of the terms that do not contain K, we get

$$1 + \frac{Ks}{s^3 + 5s^2 + 4s + 20} = 0$$

or

$$1 + \frac{Ks}{(s + j2)(s - j2)(s + 5)} = 0 \tag{4-19}$$

Equation (4–19) is of the form of Equation (4–2).

We shall now sketch the root loci of the system given by Equation (4–19). Notice that the open-loop poles are located at $s = j2$, $s = -j2$, and $s = -5$, and the open-loop zero is located at $s = 0$. The root locus exists on the real axis between 0 and -5. Since

$$\lim_{s \to \infty} \frac{Ks}{(s + j2)(s - j2)(s + 5)} = \lim_{s \to \infty} \frac{K}{s^2}$$

we have

$$\text{Angles of asymptote} = \frac{\pm 180°(2k + 1)}{2} = \pm 90°$$

The intersection of the asymptotes with the real axis can be found from

$$\lim_{s \to \infty} \frac{Ks}{s^3 + 5s^2 + 4s + 20} = \lim_{s \to \infty} \frac{K}{s^2 + 5s + \cdots} = \lim_{s \to \infty} \frac{K}{(s + 2.5)^2}$$

as

$$s = -2.5$$

The angle of departure (angle θ) from the pole at $s = j2$ is obtained as follows:

$$\theta = 180° - 90° - 21.8° + 90° = 158.2°$$

Thus, the angle of departure from the pole $s = j2$ is $158.2°$. Figure 4–53 shows a root-locus plot of the system. Notice that two branches of the root locus originate from the poles at $s = \pm j2$ and terminate on the zeros at infinity. The remaining branch originates from the pole at $s = -5$ and terminates on the zero at $s = 0$.

Note that the closed-loop poles with $\zeta = 0.4$ must lie on straight lines passing through the origin and making the angles $\pm 66.42°$ with the negative real axis. In the present case, there are two intersections of the root-locus branch in the upper-half s plane and the straight line of angle $66.42°$. Thus, two values of K will give the damping ratio ζ of the closed-loop poles equal to 0.4. At point P, the value of K is

$$K = \left| \frac{(s + j2)(s - j2)(s + 5)}{s} \right|_{s=-1.0490+j2.4065} = 8.9801$$

Hence,

$$k = \frac{K}{20} = 0.4490 \qquad \text{at point } P$$

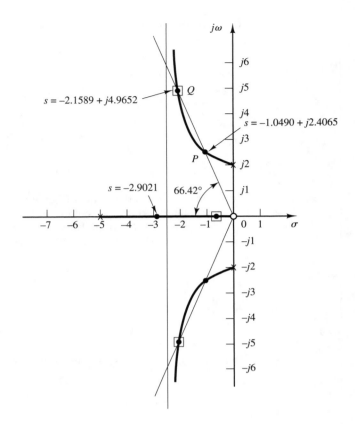

Figure 4–53
Root-locus plot for the system shown in Figure 4–52.

At point Q, the value of K is

$$K = \left| \frac{(s + j2)(s - j2)(s + 5)}{s} \right|_{s=-2.1589+j4.9652} = 28.260$$

Consequently,

$$k = \frac{K}{20} = 1.4130 \quad \text{at point } Q$$

Thus, we have two solutions for this problem. For $k = 0.4490$, the three closed-loop poles are located at

$$s = -1.0490 + j2.4065, \quad s = -1.0490 - j2.4065, \quad s = -2.9021$$

For $k = 1.4130$, the three closed-loop poles are located at

$$s = -2.1589 + j4.9652, \quad s = -2.1589 - j4.9652, \quad s = -0.6823$$

It is important to point out that the zero at the origin is the open-loop zero, not the closed-loop zero. This is evident because the original system shown in Figure 4–52 does not have a closed-loop zero, since

$$\frac{G(s)}{R(s)} = \frac{20}{s(s+1)(s+4) + 20(1+ks)}$$

The open-loop zero at $s = 0$ was introduced in the process of modifying the characteristic equation such that the adjustable variable $K = 20k$ was to appear as a multiplicative factor.

We have obtained two different values of k to satisfy the requirement that the damping ratio of the dominant closed-loop poles be equal to 0.4. The closed-loop transfer function with $k = 0.4490$ is given by

$$\frac{C(s)}{R(s)} = \frac{20}{s^3 + 5s^2 + 12.98s + 20}$$

$$= \frac{20}{(s + 1.0490 + j2.4065)(s + 1.0490 - j2.4065)(s + 2.9021)}$$

The closed-loop transfer function with $k = 1.4130$ is given by

$$\frac{C(s)}{R(s)} = \frac{20}{s^3 + 5s^2 + 32.26s + 20}$$

$$= \frac{20}{(s + 2.1589 + j4.9652)(s + 2.1589 - j4.9652)(s + 0.6823)}$$

Notice that the system with $k = 0.4490$ has a pair of dominant complex-conjugate closed-loop poles, while in the system with $k = 1.4130$, the real closed-loop pole at $s = -0.6823$ is dominant and the complex-conjugate closed-loop poles are not dominant. In this case, the response characteristic is determined primarily by the real closed-loop pole.

Let us compare the unit-step responses of both systems. MATLAB Program 4–22 plots the unit-step response curves in one diagram. The resulting curves [$c_1(t)$ for $k = 0.4490$ and $c_2(t)$ for $k = 1.4130$] are shown in Figure 4–54.

From Figure 4–54, we notice that the response of the system with $k = 0.4490$ is oscillatory. (The effect of the closed-loop pole at $s = -2.9021$ on the unit-step response is small.) For the system with $k = 1.4130$, the oscillations due to the closed-loop poles at $s = -2.1589 \pm j4.9652$ damp out much faster than the purely exponential response due to the closed-loop pole at $s = -0.6823$.

The system with $k = 0.4490$ (which exhibits a faster response with relatively small overshoot) has a much better response characteristic than the system with $k = 1.4130$ (which exhibits a slow, overdamped response). Therefore, we should choose $k = 0.4490$ for the current system.

MATLAB Program 4–22

```
>> % ——— Unit-step response ———
>>
>> % ***** Enter numerators and denominators of systems with
>> % k = 0.4490 and k = 1.4130, respectively. *****
>>
>> num1 = [20];
>> den1 = [1   5   12.98   20];
>> num2 = [20];
>> den2 = [1   5   32.26   20];
>> t = 0:0.1:10;
>> [c1,x1,t] = step(num1,den1,t);
>> [c2,x2,t] = step(num2,den2,t);
>> plot(t,c1,t,c2)
>> text(2.5,1.12,'k = 0.4490')
>> text(3.7,0.85,'k = 1.4130')
>> grid
>> title('Unit-step Responses of Two Systems')
>> xlabel('t Sec')
>> ylabel('Outputs c_1 and c_2')
```

Figure 4–54
Unit-step response curves for the system shown in Figure 4–52 when the damping ratio ζ of the dominant closed-loop poles is set equal to 0.4. (Two possible values of k give the damping ratio $\zeta = 0.4$.)

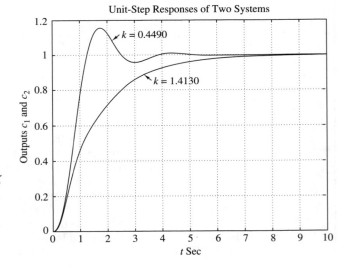

Frequency-Response Analysis

5-1 PLOTTING BODE DIAGRAMS WITH MATLAB

The command bode computes magnitudes and phase angles of the frequency response of continuous-time, linear, time-invariant systems.

When the command bode (without left-hand arguments) is entered into the computer, MATLAB produces a Bode plot on the screen, where the magnitude is given in decibels. The most commonly used bode commands are

> bode(num,den)
> bode(num,den,w)
> bode(A,B,C,D)
> bode(A,B,C,D,w)
> bode(A,B,C,D,iu,w)
> bode(sys)

When invoked with left-hand arguments, such as

> [mag,phase,w] = bode(num,den,w)

bode returns the frequency response of the system in matrices mag, phase, and w. No plot is drawn on the screen. The matrices mag and phase contain, respectively, magnitudes and phase angles of the frequency response of the system, evaluated at user-specified frequency points. The phase angle is returned in degrees. The magnitude can be converted to decibels with the statement

> magdB = 20*log10(mag)

Commonly used bode commands with left-hand arguments are

[mag,phase,w] = bode(num,den)
[mag,phase,w] = bode(num,den,w)
[mag,phase,w] = bode(A,B,C,D)
[mag,phase,w] = bode(A,B,C,D,w)
[mag,phase,w] = bode(A,B,C,D,iu,w)
[mag,phase,w] = bode(sys)

To specify the frequency range, use the command logspace(d1,d2) or logspace (d1,d2,n). logspace(d1,d2) generates a vector of 50 points equally spaced logarithmically between decades 10^{d1} and 10^{d2}. (The 50 points include both endpoints; there are 48 points between the endpoints.) To generate 50 points between 0.1 rad/sec and 100 rad/sec, enter the command

$$w = \text{logspace}(-1,2)$$

logspace(d1,d2,n) generates n points equally spaced logarithmically between decades 10^{d1} and 10^{d2}. (The n points include both endpoints.) For example, to generate 100 points between 1 rad/sec and 1000 rad/sec, enter the command

$$w = \text{logspace}(0,3,100)$$

To incorporate the user-specified frequency points when plotting Bode diagrams, the bode command must include the frequency vector w as in the commands bode(num,den,w) and [mag,phase,w] = bode(A,B,C,D,w).

EXAMPLE 5–1 Consider the following transfer function:

$$G(s) = \frac{25}{s^2 + 4s + 25}$$

Plot a Bode diagram of this transfer function.

When the system is defined in the form

$$G(s) = \frac{\text{num}(s)}{\text{den}(s)}$$

use the command bode(num,den) to draw the Bode diagram. [When the numerator and denominator contain the polynomial coefficients in descending powers of s, bode(num,den) draws the Bode diagram.] MATLAB Program 5–1 shows a program to plot the Bode diagram for this system. The resulting diagram is shown in Figure 5–1.

MATLAB Program 5–1
>> num = [25]; >> den = [1   4   25]; >> bode(num,den) >> grid >> title('Bode Diagram of G(s) = 25/(s^2 + 4s + 25)')

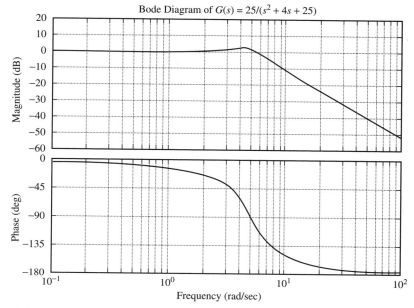

Figure 5-1
Bode diagram of $G(s) = \dfrac{25}{s^2 + 4s + 25}$.

EXAMPLE 5-2 Consider the system shown in Figure 5-2. The open-loop transfer function is

$$G(s) = \frac{9(s^2 + 0.2s + 1)}{s(s^2 + 1.2s + 9)}$$

Plot a bode diagram.

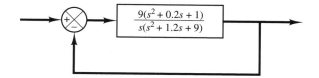

Figure 5-2
Control system.

MATLAB Program 5-2 plots a Bode diagram of this system. The resulting plot is shown in Figure 5-3. The frequency range in this case is automatically determined to be from 0.1 to 100 rad/sec.

MATLAB Program 5-2

```
>> num = [9   1.8   9];
>> den = [1   1.2   9   0];
>> bode(num,den)
>> grid
>> title('Bode Diagram of G(s) = 9(s^2 + 0.2s + 1)/[s(s^2 + 1.2s + 9)]')
```

Section 5-1 / Plotting Bode Diagrams with MATLAB

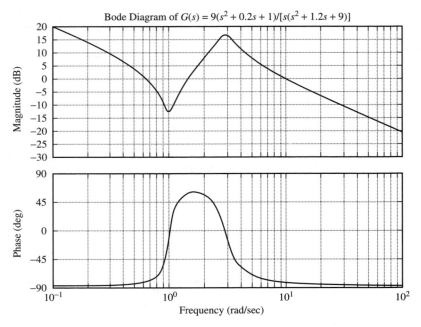

Figure 5-3
Bode diagram of $G(s) = \dfrac{9(s^2 + 0.2s + 1)}{s(s^2 + 1.2s + 9)}$.

If it is desired to plot the Bode diagram from 0.01 to 1000 rad/sec, enter the following command:

$$w = \text{logspace}(-2,3,100)$$

This command generates 100 points equally spaced logarithmically between 0.01 and 100 rad/sec. (Note that such a vector w specifies the frequencies in radians per second at which the frequency response will be calculated.)

If we use the command

$$\text{bode}(\text{num},\text{den},w)$$

then the frequency range is as the user specified, but the magnitude range and phase-angle range will be automatically determined. (See MATLAB Program 5-3 and the resulting plot in Figure 5-4.)

MATLAB Program 5-3

```
>> num = [9   1.8   9];
>> den = [1   1.2   9   0];
>> w = logspace(-2,3,100);
>> bode(num,den,w)
>> grid
>> title('Bode Diagram of G(s) = 9(s^2 + 0.2s + 1)/[s(s^2 + 1.2s + 9)]')
```

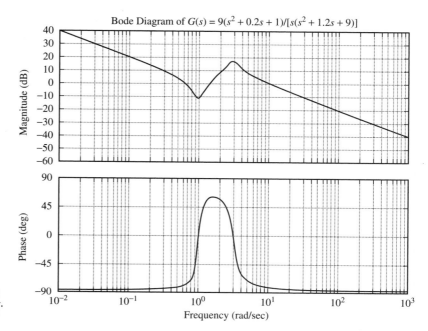

Figure 5–4
Bode diagram of
$$G(s) = \frac{9(s^2 + 0.2s + 1)}{s(s^2 + 1.2s + 9)}.$$

To specify the magnitude range and phase-angle range, use the following command:

[mag,phase,w] = bode(num,den,w)

The matrices mag and phase contain the magnitudes and phase angles, respectively, of the frequency response, evaluated at the user-specified frequency points. The phase angle is returned in degrees. The magnitude can be converted to decibels with the statement

magdb = 20*log10(mag)

If we wish to specify the magnitude range to be, for example, between −50 dB and +50 dB, then we enter lines at −50 dB and +50 dB in the plot by specifying dBmax (maximum magnitude) and dBmin (minimum magnitude) as follows:

dBmax = 50*ones(1,100);
dBmin = −50*ones(1,100);

Then we enter the following semilog plot command:

semilogx(w,magdB,'o',w,magdB,'-',w,dBmax,'--',w,dBmin,':')

(Note that the number of dBmax points and the number of dBmin points must be equal to the number of frequency points in w. In this example, all numbers are 100.) Then the screen will show the magnitude curve magdB with 'o' marks.

The range for the magnitude is normally a multiple of 5, 10, 20, or 50 dB. (There are exceptions.) For the present case, the range for the magnitude will be from −50 dB to +50 dB.

For the phase angle, if we wish to specify the range to be, for example, between −150° and +150°, we enter lines at −150° and +150° in the program with the commands pmax (maximum phase angle) and pmin (minimum phase angle) as follows:

pmax = 150*ones(1,100);
pmin = −150*ones(1,100);

Then we enter the semilog plot command:

semilogx(w,phase,'o',w,phase,'-',w,pmax,'--',w,pmin,':')

(The number of pmax points and the number of pmin points must be equal to the number of frequency points in w.) The screen will show the phase curve.

The range for the phase angle is normally a multiple of 5°, 10°, 50°, or 100°. (There are exceptions.) For the present case, the range for the phase angle will be from −150° to +150°.

MATLAB Program 5–4 produces the Bode diagram for the system such that the frequency range is from 0.01 to 1000 rad/sec, the magnitude range is from −50 to +50 dB (the magnitude range is a multiple of 50 dB), and the phase-angle range is from −150° to +150° (the phase-angle range is a multiple of 50°). Figure 5–5 shows the Bode diagram obtained with MATLAB Program 5–4.

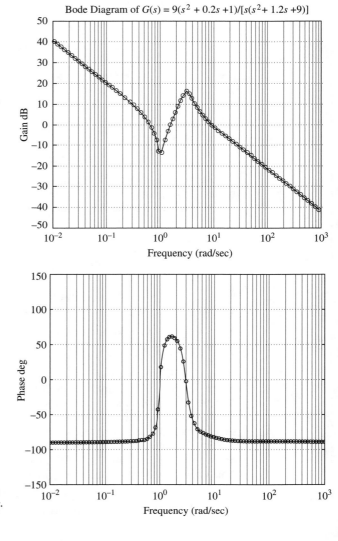

Figure 5–5
Bode diagram of
$$G(s) = \frac{9(s^2 + 0.2s + 1)}{s(s^2 + 1.2s + 9)}.$$

MATLAB Program 5–4

```
>> % ——— Bode diagram ———
>>
>> % ***** Enter the numerator and denominator of the transfer
>> % function *****
>>
>> num = [9   1.8   9];
>> den = [1   1.2   9   0];
>>
>> % ***** Specify the frequency range and enter the command
>> % [mag,phase,w] = bode(num,den,w) *****
>>
>> w = logspace(-2,3,100);
>> [mag,phase,w] = bode(num,den,w);
>>
>> % Convert mag to decibels *****
>>
>> magdB = 20*log10(mag);
>>
>> % ***** Specify the range for the magnitude. For the system
>> % considered, the magnitude range should include -50 dB
>> % and +50 dB. Enter dBmax and dBmin into the program and
>> % draw dBmax line and dBmin line. To plot the magdB curve
>> % enter the following dBmax, dBmin, and semilogx commands: *****
>>
>> dBmax = 50*ones(1,100);
>> dBmin = -50*ones(1,100);
>> semilogx(w,magdB,'o',w,magdB,'-',w,dBmax,'-',w,dBmin,'-')
>>
>> % ***** Enter grid, title, xlabel, and ylabel *****
>>
>> grid
>> title('Bode Diagram of G(s) = 9(s^2+0.2s+1)/[s(s^2+1.2s+9)]')
>> xlabel('Frequency (rad/sec)')
>> ylabel('Gain dB')
>>
>> % ***** Next, we shall plot the phase-angle curve *****
>> % ***** Specify the range for the phase angle. For the system
>> % considered, the phase-angle range should include -150 degrees
>> % and +150 degrees. Enter pmax and pmin into the program and
>> % draw pmax line and pmin line. To plot the phase curve, enter
>> % the following pmax, pmin, and semilog commands: *****
>>
>> pmax = 150*ones(1,100);
>> pmin = -150*ones(1,100);
>> semilogx(w,phase,'o',w,phase,'-',w,pmax,'-',w,pmin,'-')
>>
>> % ***** Enter grid, xlabel, and ylabel *****
>>
>> grid
>> xlabel('Frequency (rad/sec)')
>> ylabel('Phase deg')
```

Obtaining Bode diagrams of systems defined in state space. Consider the system defined by

$$\dot{\mathbf{x}} = \mathbf{Ax} + \mathbf{Bu}$$
$$\mathbf{y} = \mathbf{Cx} + \mathbf{Du}$$

where $\mathbf{x}$ = state vector (n-vector)
$\mathbf{y}$ = output vector (m-vector)
$\mathbf{u}$ = control vector (r-vector)
$\mathbf{A}$ = state matrix ($n \times n$ matrix)
$\mathbf{B}$ = control matrix ($n \times r$ matrix)
$\mathbf{C}$ = output matrix ($m \times n$ matrix)
$\mathbf{D}$ = direct transmission matrix ($m \times r$ matrix)

A Bode diagram for this system may be obtained by entering the command

bode(A,B,C,D)

or others listed at the beginning of this section.

The command bode(A,B,C,D) produces a series of Bode plots, one for each input of the system, with the frequency range automatically determined. (More points are used when the response is changing rapidly.)

The command bode(A,B,C,D,iu), where iu is the ith input of the system, produces the Bode diagrams from the input iu to all the outputs $(y_1, y_2, \ldots, y_m)$ of the system, with a frequency range automatically determined. (The scalar iu is an index into the inputs of the system and specifies which input is to be used for plotting Bode diagrams). If the control vector $\mathbf{u}$ has three inputs such that

$$\mathbf{u} = \begin{bmatrix} u_1 \\ u_2 \\ u_3 \end{bmatrix}$$

then iu must be set to either 1, 2, or 3.

If the system has only one input u, then either of the following commands may be used:

bode(A,B,C,D)

or

bode(A,B,C,D,1)

EXAMPLE 5-3 Consider the following system:

$$\begin{bmatrix} \dot{x}_1 \\ \dot{x}_2 \end{bmatrix} = \begin{bmatrix} 0 & 1 \\ -25 & -4 \end{bmatrix} \begin{bmatrix} x_1 \\ x_2 \end{bmatrix} + \begin{bmatrix} 0 \\ 25 \end{bmatrix} u$$

$$y = \begin{bmatrix} 1 & 0 \end{bmatrix} \begin{bmatrix} x_1 \\ x_2 \end{bmatrix}$$

This system has one input u and one output y. By using the command

bode(A,B,C,D)

and entering MATLAB Program 5–5 into the computer, we obtain the Bode diagram shown in Figure 5–6.

MATLAB Program 5–5

```
>> A = [0   1;-25   -4];
>> B = [0;25];
>> C = [1   0];
>> D = [0];
>> bode(A,B,C,D)
>> grid
>> title('Bode Diagram')
```

If we replace the command bode(A,B,C,D) in MATLAB Program 5–5 with

bode(A,B,C,D,1)

then MATLAB will produce the Bode diagram identical to that shown in Figure 5–6.

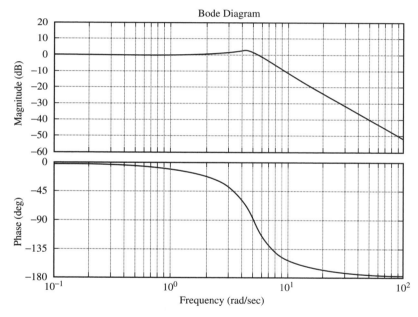

Figure 5–6
Bode diagram of the system considered in Example 5–3.

EXAMPLE 5–4 Consider the system defined by

$$\begin{bmatrix} \dot{x}_1 \\ \dot{x}_2 \end{bmatrix} = \begin{bmatrix} 0 & 1 \\ -25 & -4 \end{bmatrix} \begin{bmatrix} x_1 \\ x_2 \end{bmatrix} + \begin{bmatrix} 1 & 1 \\ 0 & 1 \end{bmatrix} \begin{bmatrix} u_1 \\ u_2 \end{bmatrix}$$

$$\begin{bmatrix} y_1 \\ y_2 \end{bmatrix} = \begin{bmatrix} 1 & 0 \\ 0 & 1 \end{bmatrix} \begin{bmatrix} x_1 \\ x_2 \end{bmatrix}$$

Obtain the sinusoidal transfer functions $Y_1(j\omega)/U_1(j\omega)$, $Y_2(j\omega)/U_1(j\omega)$, $Y_1(j\omega)/U_2(j\omega)$, and $Y_2(j\omega)/U_2(j\omega)$. In deriving $Y_1(j\omega)/U_1(j\omega)$ and $Y_2(j\omega)/U_1(j\omega)$, assume that $U_2(j\omega) = 0$. Similarly, in obtaining $Y_1(j\omega)/U_2(j\omega)$ and $Y_2(j\omega)/U_2(j\omega)$, assume that $U_1(j\omega) = 0$.

The transfer matrix expression for the system defined by

$$\dot{\mathbf{x}} = \mathbf{Ax} + \mathbf{Bu}$$
$$\dot{\mathbf{y}} = \mathbf{Cx} + \mathbf{Du}$$

is given by

$$\mathbf{Y}(s) = \mathbf{G}(s)\mathbf{U}(s)$$

where

$$\mathbf{G}(s) = \mathbf{C}(s\mathbf{I} - \mathbf{A})^{-1}\mathbf{B} + \mathbf{D}$$

is the transfer matrix. For the system considered here, the transfer matrix becomes

$$\mathbf{C}(s\mathbf{I} - \mathbf{A})^{-1}\mathbf{B} + \mathbf{D} = \begin{bmatrix} 1 & 0 \\ 0 & 1 \end{bmatrix} \begin{bmatrix} s & -1 \\ 25 & s+4 \end{bmatrix}^{-1} \begin{bmatrix} 1 & 1 \\ 0 & 1 \end{bmatrix}$$

$$= \frac{1}{s^2 + 4s + 25} \begin{bmatrix} s+4 & 1 \\ -25 & s \end{bmatrix} \begin{bmatrix} 1 & 1 \\ 0 & 1 \end{bmatrix}$$

$$= \begin{bmatrix} \dfrac{s+4}{s^2 + 4s + 25} & \dfrac{s+5}{s^2 + 4s + 25} \\ \dfrac{-25}{s^2 + 4s + 25} & \dfrac{s-25}{s^2 + 4s + 25} \end{bmatrix}$$

Hence,

$$\begin{bmatrix} Y_1(s) \\ Y_2(s) \end{bmatrix} = \begin{bmatrix} \dfrac{s+4}{s^2 + 4s + 25} & \dfrac{s+5}{s^2 + 4s + 25} \\ \dfrac{-25}{s^2 + 4s + 25} & \dfrac{s-25}{s^2 + 4s + 25} \end{bmatrix} \begin{bmatrix} U_1(s) \\ U_2(s) \end{bmatrix}$$

Assuming that $U_2(j\omega) = 0$, we find $Y_1(j\omega)/U_1(j\omega)$ and $Y_2(j\omega)/U_1(j\omega)$ as follows:

$$\frac{Y_1(j\omega)}{U_1(j\omega)} = \frac{j\omega + 4}{(j\omega)^2 + 4j\omega + 25}$$

$$\frac{Y_2(j\omega)}{U_1(j\omega)} = \frac{-25}{(j\omega)^2 + 4j\omega + 25}$$

Similarly, assuming that $U_1(j\omega) = 0$, we find $Y_1(j\omega)/U_2(j\omega)$ and $Y_2(j\omega)/U_2(j\omega)$ as follows:

$$\frac{Y_1(j\omega)}{U_2(j\omega)} = \frac{j\omega + 5}{(j\omega)^2 + 4j\omega + 25}$$

$$\frac{Y_2(j\omega)}{U_2(j\omega)} = \frac{j\omega - 25}{(j\omega)^2 + 4j\omega + 25}$$

Notice that $Y_2(j\omega)/U_2(j\omega)$ is a non-minimum-phase transfer function.

EXAMPLE 5–5 Referring to Example 5–4, plot Bode diagrams for the system, using MATLAB.

MATLAB Program 5–6 produces Bode diagrams for the system. There are four sets of diagrams: two for input 1 and two for input 2. All of the diagrams are shown in Figure 5–7.

MATLAB Program 5–6
```
>> A = [0    1;-25   -4];
>> B = [1    1;0    1];
>> C = [1    0;0    1];
>> D = [0    0;0    0];
>> bode(A,B,C,D)
>> grid
```

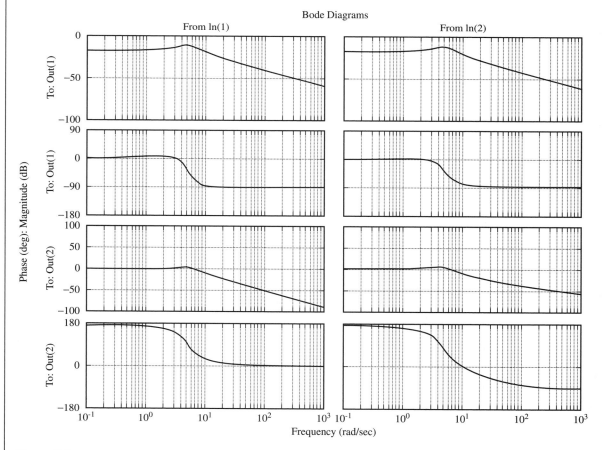

Figure 5–7
Bode diagrams.

EXAMPLE 5-6 Using MATLAB, plot Bode diagrams for the closed-loop system shown in Figure 5–8 for $K = 1$, $K = 10$, and $K = 20$. Plot three magnitude curves in one diagram and three phase-angle curves in another diagram.

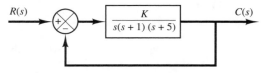

Figure 5–8
Closed-loop system.

The closed-loop transfer function of the system is given by

$$\frac{C(s)}{R(s)} = \frac{K}{s(s + 1)(s + 5) + K}$$

$$= \frac{K}{s^3 + 6s^2 + 5s + K}$$

Hence, the numerator and denominator of $C(s)/R(s)$ are

$$\text{num} = [K]$$
$$\text{den} = [1 \quad 6 \quad 5 \quad K]$$

A possible MATLAB program is shown in MATLAB Program 5–7. The resulting Bode diagrams are shown in Figures 5–9(a) and (b).

MATLAB Program 5–7

```
>> w = logspace(-1,2,200);
>>
>> for i = 1:3;
      if i = 1; K = 1;[mag,phase,w] = bode([K],[1   6   5   K],w);
        mag1dB = 20*log10(mag); phase1 = phase; end;
      if i = 2; K = 10;[mag,phase,w] = bode([K],[1   6   5   K],w);
        mag2dB = 20*log10(mag); phase2 = phase; end;
      if i = 3; K = 20;[mag,phase,w] = bode([K],[1   6   5   K],w);
        mag3dB = 20*log10(mag); phase3 = phase; end;
      end
>> semilogx(w,mag1dB, '-',w,mag2dB,'-',w,mag3dB,'-')
>> grid
>> title('Bode Diagrams of G(s) = K/[s(s + 1)(s + 5)], where K = 1, K = 10, and K = 20')
>> xlabel('Frequency (rad/sec)')
>> ylabel('Gain (dB)')
>> text(1.2,-31,'K = 1')
>> text(1.1,-8,'K = 10')
>> text(11,-31,'K = 20')
>> semilogx(w,phase1,'-',w,phase2,'-'w,phase3,'-')
>> grid
>> xlabel('Frequency (rad/sec)')
>> ylabel('Phase (deg)')
>> text(0.2,-90,'K = 1')
>> text(0.2,-20,'K = 10')
>> text(1.6,-20,'K = 20')
```

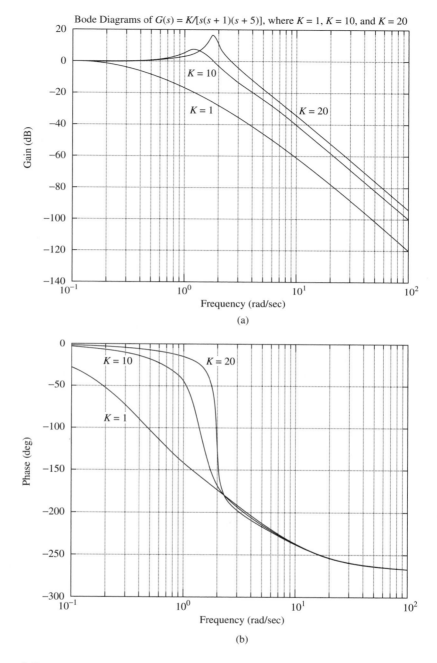

Figure 5–9
Bode diagrams: (a) Magnitude-versus-frequency curves;
(b) phase-angle-versus-frequency curves.

5-2 PLOTTING NYQUIST DIAGRAMS WITH MATLAB

Nyquist plots, just like Bode diagrams, are commonly used in the frequency-response representation of linear time-invariant feedback control systems. Nyquist plots, also called Nyquist diagrams, are polar plots, while Bode diagrams are rectangular plots. One plot or the other may be more convenient for a particular operation, but a given operation can always be carried out in either plot.

The polar plot of a sinusoidal transfer function $G(j\omega)$ is a plot of the magnitude of $G(j\omega)$ versus the phase angle of $G(j\omega)$ on polar coordinates as ω is varied from zero to infinity. Thus, the polar plot is the locus of vectors $|G(j\omega)|\underline{/G(j\omega)}$ as ω is varied from zero to infinity. Note that in polar plots a positive (negative) phase angle is measured counter-clockwise (clockwise) from the positive real axis.

Nyquist stability analysis. *Nyquist stability criterion:* In the system shown in Figure 5–10, if the open-loop transfer function $G(s)H(s)$ has k poles in the right-half s plane, then, for stability, the $G(s)H(s)$ locus, as a representative point s traces on the modified Nyquist path in the clockwise direction, must encircle the $-1 + j0$ point k times in the counterclockwise direction. [The Nyquist path is a closed contour that consists of the entire $j\omega$-axis from $\omega = -\infty$ to $+\infty$ and a semicircular path of infinite radius in the right-half s plane. Thus, the Nyquist path encloses the entire right-half s plane. The direction of the path is clockwise. The modified Nyquist path is a Nyquist path that avoids any pole(s) and zero(s) that lie on the $j\omega$-axis by going around using a semicircular path of infinitesimal radius at each such pole or zero.]

Remarks on the nyquist stability criterion

1. The Nyquist stability criterion can be expressed as

$$Z = N + P$$

where Z = number of zeros of $1 + G(s)H(s)$ in the right-half s plane
N = number of clockwise encirclements of the $-1 + j0$ point
P = number of poles of $G(s)H(s)$ in the right-half s plane

If P is not zero, then, for a stable control system, we must have $Z = 0$, or $N = -P$, which means that we must have P counterclockwise encirclements of the $-1 + j0$ point.

If $G(s)H(s)$ does not have any poles in the right-half s plane, then $Z = N$. Thus, for stability, there must be no encirclement of the $-1 + j0$

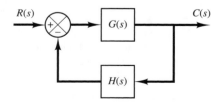

Figure 5–10
Closed-loop system.

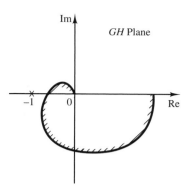

Figure 5–11
Region enclosed by a Nyquist plot.

point by the $G(j\omega)H(j\omega)$ locus. In this case, it is not necessary to consider the locus for the entire $j\omega$-axis—only for the positive-frequency portion. The stability of such a system can be determined by seeing if the $-1 + j0$ point is enclosed by the Nyquist plot of $G(j\omega)H(j\omega)$. The region enclosed by the Nyquist plot is shown in Figure 5–11. For stability, the $-1 + j0$ point must lie outside the shaded region.

2. We must be careful when testing the stability of multiple-loop systems, since they may include poles in the right-half s plane. (Note that although an inner loop may be unstable, the entire closed-loop system can be made stable by proper design.) Simple inspection of encirclements of the $-1 + j0$ point by the $G(j\omega)H(j\omega)$ locus is not sufficient to detect instability in multiple-loop systems. In such cases, however, whether any pole of $1 + G(s)H(s)$ is in the right-half s plane can be determined easily by applying the Routh stability criterion to the denominator of $G(s)H(s)$.

If transcendental functions, such as transport lag e^{-Ts}, are included in $G(s)H(s)$, they must be approximated by a series expansion before the Routh stability criterion can be applied. A few forms of series expansion of e^{-Ts} were presented in Section 4–6.

3. If the locus of $G(j\omega)H(j\omega)$ passes through the $-1 + j0$ point, then zeros of the characteristic equation, or closed-loop poles, are located on the $j\omega$-axis. This is not desirable for practical control systems. For a well-designed closed-loop system, none of the roots of the characteristic equation should lie on the $j\omega$-axis.

Stability analysis. In what follows, we shall present several examples illustrating the stability analysis of control systems by means of the Nyquist stability criterion.

If the Nyquist path in the s plane encircles Z zeros and P poles of $1 + G(s)H(s)$ and does not pass through any poles or zeros of $1 + G(s)H(s)$ as a representative point s moves in the clockwise direction along the path, then the corresponding contour in the $G(s)H(s)$ plane encircles the $-1 + j0$ point $N = Z - P$ times in the clockwise direction. (Negative values of N imply counterclockwise encirclements.)

In examining the stability of linear control systems with the use of the Nyquist stability criterion, we see that three possibilities can occur:

1. There is no encirclement of the $-1 + j0$ point. This implies that the system is stable if there are no poles of $G(s)H(s)$ in the right-half s plane; otherwise, the system is unstable.
2. There are one or more counterclockwise encirclements of the $-1 + j0$ point. In this case, the system is stable if the number of counterclockwise encirclements is the same as the number of poles of $G(s)H(s)$ in the right-half s plane; otherwise, the system is unstable.
3. There are one or more clockwise encirclements of the $-1 + j0$ point. In this case, the system is unstable.

The MATLAB command nyquist computes the frequency response for continuous-time, linear, time-invariant systems. When invoked without left-hand arguments, nyquist produces a Nyquist plot or a Nyquist diagram on the screen.

The command

$$\text{nyquist(num,den)}$$

draws the Nyquist plot of the transfer function

$$G(s) = \frac{\text{num}(s)}{\text{den}(s)}$$

where num and den contain the polynomial coefficients in descending powers of s. Other commonly used nyquist commands are

$$\text{nyquist(num,den,w)}$$
$$\text{nyquist(A,B,C,D)}$$
$$\text{nyquist(A,B,C,D,w)}$$
$$\text{nyquist(A,B,C,D,iu,w)}$$
$$\text{nyquist(sys)}$$

The command involving the user-specified frequency vector w, such as

$$\text{nyquist(num,den,w)}$$

calculates the frequency response at the specified frequency points in radians per second. Note that if any of the commands listed above is used, the subsequent use of the grid command will not produce x–y grid lines (horizontal and vertical grid lines). If x–y grid lines are desired, use any one of the commands with left-hand arguments listed below, together with a grid command. See, for instance, Examples 5–7, 5–8 and 5–10 among others.

When invoked with left-hand arguments such as

$$[\text{re,im,w}] = \text{nyquist(num,den)}$$
$$[\text{re,im,w}] = \text{nyquist(num,den,w)}$$
$$[\text{re,im,w}] = \text{nyquist(A,B,C,D)}$$
$$[\text{re,im,w}] = \text{nyquist(A,B,C,D,w)}$$
$$[\text{re,im,w}] = \text{nyquist(A,B,C,D,iu,w)}$$
$$[\text{re,im,w}] = \text{nyquist(sys)}$$

MATLAB returns the frequency response of the system in the matrices re, im, and w. No plot is drawn on the screen. The matrices re and im contain, respectively, the real and imaginary parts of the frequency response of the system, evaluated at the frequency points specified in the vector w. Note that re and im have as many columns as outputs and one row for each element in w. To draw a Nyquist plot, use a plot command.

EXAMPLE 5-7 Consider the following open-loop transfer function:

$$G(s) = \frac{1}{s^2 + 0.8s + 1}$$

Draw a Nyquist plot of this transfer function with MATLAB.

Since the system is given in the form of the transfer function, the command

nyquist(num,den)

may be used to draw a Nyquist plot. MATLAB Program 5-8 produces the Nyquist plot shown in Figure 5-12. In this plot, the ranges for the real axis and imaginary axis are automatically determined.

MATLAB Program 5-8

```
>> num = [1];
>> den = [1   0.8   1];
>> nyquist(num,den)
>> title('Nyquist Plot of G(s) = 1/(s^2 + 0.8s + 1)')
```

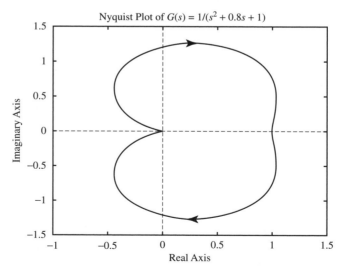

Figure 5-12
Nyquist plot of $G(s) = \dfrac{1}{s^2 + 0.8s + 1}$.

Section 5-2 / Plotting Nyquist Diagrams with MATLAB

If we wish to draw the Nyquist plot with manually determined ranges—for example, from −2 to 2 on the real axis and from −2 to 2 on the imaginary axis—enter the following command into the computer:

$$v = [-2 \quad 2 \quad -2 \quad 2];$$
$$\text{axis}(v);$$

Or you may combine the preceding two lines into one:

$$\text{axis}([-2 \quad 2 \quad -2 \quad 2]);$$

(See MATLAB Program 5–9 and the resulting Nyquist plot shown in Figure 5–13.)

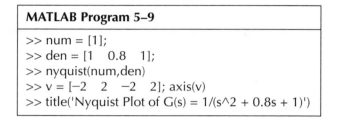

MATLAB Program 5–9

```
>> num = [1];
>> den = [1   0.8   1];
>> nyquist(num,den)
>> v = [-2   2   -2   2]; axis(v)
>> title('Nyquist Plot of G(s) = 1/(s^2 + 0.8s + 1)')
```

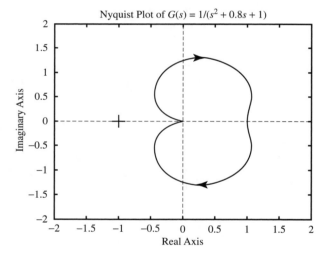

Figure 5–13
Nyquist plot of $G(s) = \dfrac{1}{s^2 + 0.8s + 1}$.

Caution. In drawing a Nyquist plot, where a MATLAB operation involves "Divide by zero," the resulting Nyquist plot may be erroneous or undesirable for any further analysis. For example, if the transfer function $G(s)$ is given by

$$G(s) = \frac{1}{s(s + 1)}$$

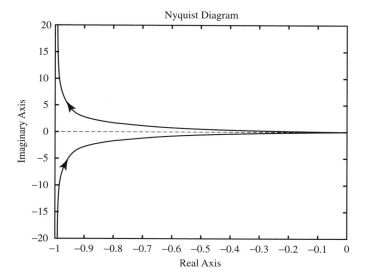

Figure 5-14
Erroneous Nyquist plot.

then the MATLAB command

$$\text{num} = [1];$$
$$\text{den} = [1 \quad 1 \quad 0];$$
$$\text{nyquist(num,den)}$$

produces an undesirable Nyquist plot. An example of an undesirable Nyquist plot is shown in Figure 5-14. If such a plot appears on the computer, then it can be corrected if we specify axis(v). For example, if we enter the axis command

$$v = [-2 \quad 2 \quad -5 \quad 5]; \text{axis(v)}$$

in the program, then a correct Nyquist plot can be obtained. (See Example 5-8.)

EXAMPLE 5-8 Draw a Nyquist plot for the transfer function

$$G(s) = \frac{1}{s(s+1)}$$

MATLAB Program 5-10 will produce a correct Nyquist plot on the computer, even though a warning message "Divide by zero" may appear on the screen. The resulting Nyquist plot is shown in Figure 5-15.

MATLAB Program 5-10

```
>> num = [1];
>> den = [1  1  0];
>> nyquist(num,den)
>> v = [-2  2  -5  5]; axis(v)
>> title('Nyquist Plot of G(s) = 1/[s(s + 1)]')
```

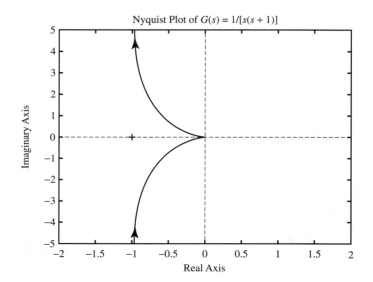

Figure 5-15
Nyquist plot of $G(s) = \dfrac{1}{s(s+1)}$.

Notice that the Nyquist plot shown in Figure 5–15 includes the loci for both $\omega > 0$ and $\omega < 0$. If we wish to draw the plot for only the positive frequency region ($\omega > 0$), then we need to use the command

$$[\text{re,im,w}] = \text{nyquist(num,den,w)}$$

A MATLAB program using this nyquist command is shown in MATLAB Program 5–11. The resulting Nyquist plot is presented in Figure 5–16.

MATLAB Program 5-11

```
>> num = [1];
>> den = [1   1   0];
>> w = 0.1:0.1:100;
>> [re,im,w] = nyquist(num,den,w);
>> plot(re,im)
>> v = [-2   2   -5   5]; axis(v)
>> grid
>> title('Nyquist Plot of G(s) = 1/[s(s + 1)]')
>> xlabel('Real Axis')
>> ylabel('Imag Axis')
```

---

240   Chapter 5 / Frequency-Response Analysis

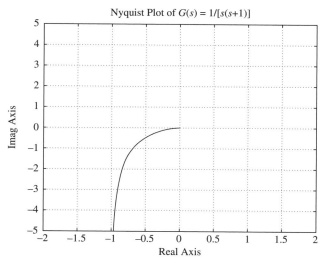

Figure 5–16
Nyquist plot of $G(s) = \dfrac{1}{s(s+1)}$ for $\omega > 0$.

Drawing Nyquist plots of a system defined in state space. Consider the system defined by

$$\dot{\mathbf{x}} = \mathbf{Ax} + \mathbf{Bu}$$
$$\mathbf{y} = \mathbf{Cx} + \mathbf{Du}$$

where $\mathbf{x}$ = state vector (n-vector)
$\mathbf{y}$ = output vector (m-vector)
$\mathbf{u}$ = control vector (r-vector)
$\mathbf{A}$ = state matrix ($n \times n$ matrix)
$\mathbf{B}$ = control matrix ($n \times r$ matrix)
$\mathbf{C}$ = output matrix ($m \times n$ matrix)
$\mathbf{D}$ = direct transmission matrix ($m \times r$ matrix)

Nyquist plots for this system may be obtained by the use of the command

nyquist(A,B,C,D)

This command produces a series of Nyquist plots, one for each input–output combination of the system. The frequency range is automatically determined.

The command

nyquist(A,B,C,D,iu)

produces Nyquist plots from the single input iu to all the outputs of the system, with the frequency range determined automatically. The scalar iu is an index into the inputs of the system and specifies which input to use for the frequency response.

The command

nyquist(A,B,C,D,iu,w)

utilizes the user-supplied frequency vector w, which specifies the frequencies in radians per second at which the frequency response should be calculated.

EXAMPLE 5-9 Consider the system defined by

$$\begin{bmatrix} \dot{x}_1 \\ \dot{x}_2 \end{bmatrix} = \begin{bmatrix} 0 & 1 \\ -25 & -4 \end{bmatrix} \begin{bmatrix} x_1 \\ x_2 \end{bmatrix} + \begin{bmatrix} 0 \\ 25 \end{bmatrix} u$$

$$y = \begin{bmatrix} 1 & 0 \end{bmatrix} \begin{bmatrix} x_1 \\ x_2 \end{bmatrix} + [0]u$$

Draw a Nyquist plot of this system.

The system has a single input u and a single output y. A Nyquist plot may be obtained by entering the command

nyquist(A,B,C,D)

or

nyquist(A,B,C,D,1)

MATLAB Program 5–12 will provide the Nyquist plot, shown in Figure 5–17. (Note that we obtain the identical result by using either of these two commands.)

MATLAB Program 5–12

```
>> A = [0   1;-25   -4];
>> B = [0;25];
>> C = [1   0];
>> D = [0];
>> nyquist(A,B,C,D)
>> v = [-0.6   1.2   -1.5   1.5]; axis(v)
```

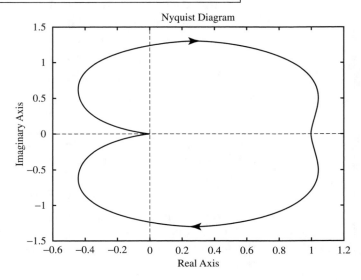

Figure 5–17
Nyquist plot of system considered in Example 5–9.

EXAMPLE 5-10 Consider the system defined by

$$\begin{bmatrix} \dot{x}_1 \\ \dot{x}_2 \end{bmatrix} = \begin{bmatrix} -1 & -1 \\ 6.5 & 0 \end{bmatrix} \begin{bmatrix} x_1 \\ x_2 \end{bmatrix} + \begin{bmatrix} 1 & 1 \\ 1 & 0 \end{bmatrix} \begin{bmatrix} u_1 \\ u_2 \end{bmatrix}$$

$$\begin{bmatrix} y_1 \\ y_2 \end{bmatrix} = \begin{bmatrix} 1 & 0 \\ 0 & 1 \end{bmatrix} \begin{bmatrix} x_1 \\ x_2 \end{bmatrix} + \begin{bmatrix} 0 & 0 \\ 0 & 0 \end{bmatrix} \begin{bmatrix} u_1 \\ u_2 \end{bmatrix}$$

This system involves two inputs and two outputs. There are four sinusoidal output–input relationships: $Y_1(j\omega)/U_1(j\omega)$, $Y_2(j\omega)/U_1(j\omega)$, $Y_1(j\omega)/U_2(j\omega)$, and $Y_2(j\omega)/U_2(j\omega)$. Draw Nyquist plots for the system. (When considering input u_1, we assume that input u_2 is zero, and vice versa.)

The four individual Nyquist plots can be obtained by the use of the command

nyquist(A,B,C,D)

MATLAB Program 5–13 produces the four Nyquist plots, shown in Figure 5–18.

MATLAB Program 5–13

```
>> A = [-1  -1;6.5   0];
>> B = [1   1;1   0];
>> C = [1   0;0   1];
>> D = [0   0;0   0];
>> nyquist(A,B,C,D)
```

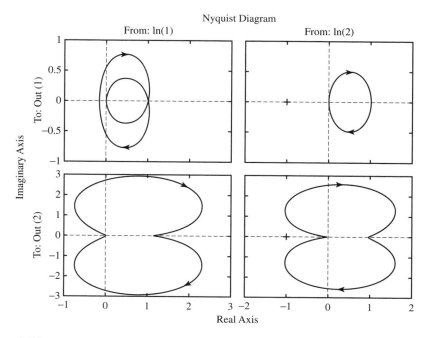

Figure 5–18
Nyquist plot of system considered in Example 5–10.

EXAMPLE 5–11 Consider a unity-feedback system with the following open-loop transfer function:

$$G(s) = \frac{20(s^2 + s + 0.5)}{s(s + 1)(s + 10)}$$

Draw a Nyquist plot of this transfer function with MATLAB, and examine the stability of the closed-loop system.

MATLAB Program 5–14 produces the Nyquist diagram shown in Figure 5–19. From the figure, we see that the Nyquist plot does not encircle the $-1 + j0$ point. Hence, $N = 0$ in the Nyquist stability criterion. Since no open-loop poles lie in the right-half s plane, $P = 0$. Therefore, $Z = N + P = 0$. The closed-loop system is stable.

```
MATLAB Program 5–14

>> num = [20   20   10];
>> den = [1   11   10   0];
>> nyquist(num,den)
>> v = [-2   3   -3   3]; axis(v)
```

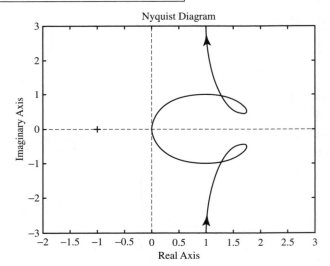

Figure 5–19
Nyquist plot of
$G(s) = \dfrac{20(s^2 + s + 0.5)}{s(s + 1)(s + 10)}$.

EXAMPLE 5–12 Consider the system discussed in Example 5–11. Draw the Nyquist plot for only the positive frequency region.

Drawing a Nyquist plot for only the positive frequency region can be done by the use of the following command:

$$[re,im,w] = nyquist(num,den,w)$$

The frequency region may be divided into several subregions by the use of different increments. For example, the frequency region of interest may be divided into three subregions as follows:

w1 = 0.1:0.1:10;
w2 = 10:2:100;
w3 = 100:10:500;
w = [w1   w2   w3]

MATLAB Program 5–15 uses this frequency region and produces the Nyquist plot shown in Figure 5–20.

MATLAB Program 5–15

```
>> num = [20   20   10];
>> den = [1   11   10   0];
>> w1 = 0.1:0.1:10; w2 = 10:2:100; w3 = 100:10:500;
>> w = [w1   w2   w3];
>> [re,im,w] = nyquist(num,den,w);
>> plot(re,im)
>> v = [-3   3   -5   1]; axis(v)
>> grid
>> title('Nyquist Plot of G(s) = 20(s^2 + s + 0.5)/[s(s + 1)(s + 10)]')
>> xlabel('Real Axis')
>> ylabel('Imaginary Axis')
```

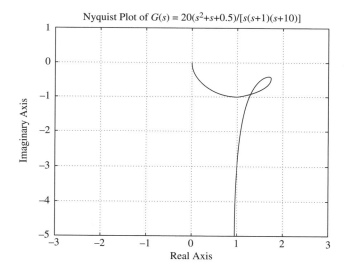

Figure 5–20
Nyquist plot for the positive frequency region.

EXAMPLE 5–13 Referring to Example 5–12, draw a Nyquist plot of

$$G(s) = \frac{20(s^2 + s + 0.5)}{s(s + 1)(s + 10)}$$

On the plot, locate frequency points where $\omega = 0.2, 0.3, 0.5, 1, 2, 6, 10,$ and 20 rad/sec. Also, find the magnitudes and phase angles of $G(j\omega)$ at the specified frequency points.

In MATLAB Program 5–15, we used the frequency vector w, which consists of three frequency subvectors: w1, w2, and w3. Instead of such a w, we may simply use the frequency vector ww = logspace(d_1,d_2,n). MATLAB Program 5–16 uses the following two frequency vectors:

ww = logspace(-1,2,100)
w = [0.2   0.3   0.5   1   2   6   10   20]

This MATLAB program produces the Nyquist plot and locates the specified frequency points on the polar locus, as shown in Figure 5–21.

MATLAB Program 5–16

```
>> num = [20   20   10];
>> den = [1   11   10   0];
>> ww = logspace(-1,2,100);
>> [re,im,ww] = nyquist (num,den,ww);
>> plot(re,im)
>> v = [-2   3   -5   0]; axis(v)
>> grid
>> title('Nyquist Diagram')
>> xlabel('Real Axis')
>> ylabel('Imaginary Axis')
>> hold
```
Current plot held
```
>> w = [0.2   0.3   0.5   1   2   6   10   20];
>> [re,im,w] = nyquist(num,den,w);
>> plot(re,im,'o')
>> text(1.1,-4.8,'w = 0.2')
>> text(1.1,-3.1,'0.3')
>> text(1.25,-1.7,'0.5')
>> text(1.37,-0.4,'1')
>> text(1.8,-0.3,'2')
>> text(1.4,-1.1,'6')
>> text(0.77,-0.8,'10')
>> text(0.037,-0.8,'20')
>>
>> % ——— To get the values of magnitude and phase (in degrees) of G(jw)
>> % at the specified w values, enter the command [mag,phase,w]
>> % = bode(num,den,w) ———
>>
>> [mag,phase,w] = bode(num,den,w);
>>
>> % ——— The following table shows the specified frequency values w and
>> % the corresponding values of magnitude and phase (in degrees):———
>>
>> [w   mag   phase]

ans =

    0.2000    4.9176   -78.9571
    0.3000    3.2426   -72.2244
    0.5000    1.9975   -55.9925
    1.0000    1.5733   -24.1455
    2.0000    1.7678   -14.4898
    6.0000    1.6918   -31.0946
   10.0000    1.4072   -45.0285
   20.0000    0.8933   -63.4385
```

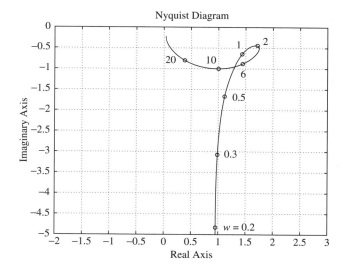

Figure 5–21
Polar plot of $G(j\omega)$ given in Example 5–13.

EXAMPLE 5–14 Consider a unity-feedback, positive-feedback system with the following open-loop transfer function:

$$G(s) = \frac{s^2 + 4s + 6}{s^2 + 5s + 4}$$

Draw a Nyquist plot of this transfer function.

Solution. The Nyquist plot of the positive-feedback system can be obtained by defining num and den as

$$\text{num} = [-1 \quad -4 \quad -6]$$
$$\text{den} = [1 \quad 5 \quad 4]$$

and using the command nyquist(num,den). MATLAB Program 5–17 produces the Nyquist plot, as shown in Figure 5–22.

This system is unstable, because the $-1 + j0$ point is encircled once clockwise. Note that this is a special case where the Nyquist plot passes through the $-1 + j0$ point and also encircles that point once clockwise. This means that the closed-loop system is degenerate: The system behaves as if it is an unstable first-order system. Following is the closed-loop transfer function of the positive-feedback system:

$$\frac{C(s)}{R(s)} = \frac{s^2 + 4s + 6}{s^2 + 5s + 4 - (s^2 + 4s + 6)}$$

$$= \frac{s^2 + 4s + 6}{s - 2}$$

MATLAB Program 5–17

```
>> num = [-1  -4  -6];
>> den = [1  5  4];
>> nyquist(num,den);
>> grid
>> title('Nyquist Plot of G(s) = -(s^2 + 4s + 6)/(s^2 + 5s + 4)')
```

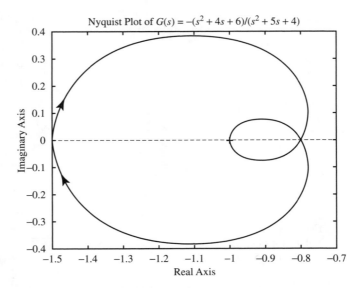

Figure 5-22
Nyquist plot for positive-feedback system.

Note that the Nyquist plot for the positive-feedback case is a mirror image about the imaginary axis of the Nyquist plot for the negative-feedback case. This may be seen from Figure 5-23, which was obtained with the use of MATLAB Program 5-18. (Note that the positive-feedback case is unstable, but the negative-feedback case is stable.)

MATLAB Program 5-18
| ```
>> num1 = [1    4    6];
>> den1 = [1    5    4];
>> num2 = [-1   -4   -6];
>> den2 = [1    5    4];
>> nyquist(num1,den1);
>> hold on
>> nyquist(num2,den2);
>> v = [-2    2    -1    1];
>> axis(v);
>> title('Nyquist Plots of G(s) and -G(s)')
>> text(1.0,0.5,'G(s)')
>> text(0.65,-0.53,'Use this Nyquist')
>> text(0.62,-0.65,'plot for negative-')
>> text(0.65,-0.77,'feedback system')
>> text(-1.3,0.5,'-G(s)')
>> text(-1.60,-0.53,'Use this Nyquist')
>> text(-1.62,-0.65,'plot for positive-')
>> text(-1.60,-0.77,'feedback system')
``` |
```

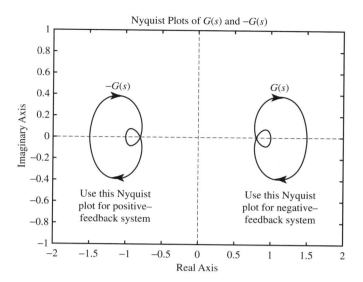

Figure 5–23
Nyquist plots for positive-feedback system and negative-feedback system.

EXAMPLE 5–15 Consider the control system shown in Figure 5–24. The open-loop transfer function $G(s)$ is

$$G(s) = \frac{10K}{s(s+5)(0.1s+1)}$$

Plot Nyquist diagrams of $G(s)$ for $K = 1, 7.5,$ and 20.

A possible MATLAB program is shown in MATLAB Program 5–19. The resulting Nyquist diagrams are shown in Figure 5–25. From the diagrams, we see that the system is stable when $K = 1$. It is critically stable when $K = 7.5$. The system is unstable when $K = 20$.

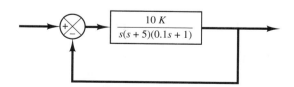

Figure 5–24
Control system.

MATLAB Program 5-19

```
>> den = [0.1   1.5   5   0];
>> for i = 1:3;
     if i = 1; K = 1; [re1,im1] = nyquist([10*K],den); end;
     if i = 2; K = 7.5; [re2,im2] = nyquist([10*K],den); end;
     if i = 3; K = 20; [re3,im3] = nyquist([10*K],den); end;
   end
>> plot(re1,im1,'-',re2,im2,'o',re2,im2,'-',re3,im3,'x',re3,im3,'-')
>> v = [-5   1   -5   1]; axis(v)
>> grid
>> title('Nyquist Diagrams of G(s) = 10K/[s(s + 5)(0.1s + 1)] for K = 1, 7.5, and 20')
>> xlabel('Real Axis')
>> ylabel('Imaginary Axis')
>> text(-0.4,-3.7,'K = 1')
>> text(-2.75,-2.7,'K = 7.5')
>> text(-4.35,-1.7,'K = 20')
```

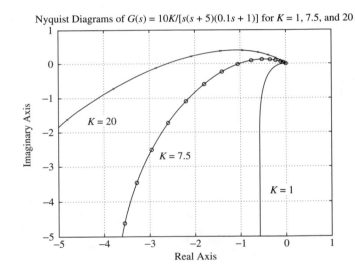

Figure 5-25
Nyquist diagrams.

5-3 LOG-MAGNITUDE-VERSUS-PHASE PLOTS

Another approach to graphically portraying a system's frequency-response characteristics is to use the log-magnitude-versus-phase plot, which is a plot of the logarithmic magnitude in decibels versus the phase angle or phase margin for a frequency range of interest. [The phase margin is the difference between the actual phase angle ϕ and $-180°$; that is, $\phi - (-180°) = 180° + \phi$.] The curve is graduated in terms of the frequency ω. Such log-magnitude-versus-phase plots are commonly called Nichols plots.

In the Bode diagram, the frequency-response characteristics of $G(j\omega)$ are shown on semilog paper by two separate curves—the log-magnitude curve and the phase-angle curve—while in the log-magnitude-versus-phase plot, the two curves in the Bode diagram are combined into one. In the manual approach, the log-magnitude-versus-phase plot can easily be constructed by reading values of the log magnitude and phase angle from the Bode diagram. Notice that in the log-magnitude-versus-phase plot a change in the gain constant of $G(j\omega)$ merely shifts the curve upward (for increasing gain) or downward (for decreasing gain), but the shape of the curve remains the same.

Advantages of the log-magnitude-versus-phase plot are that the relative stability of the closed-loop system can be determined quickly and that compensation can be worked out easily.

The log-magnitude-versus-phase plot for the sinusoidal transfer function $G(j\omega)$ and that for $1/G(j\omega)$ are skew symmetrical about the origin, since

$$\left|\frac{1}{G(j\omega)}\right| \text{ in dB} = -|G(j\omega)| \text{ in dB}$$

and

$$\angle\frac{1}{G(j\omega)} = -\angle G(j\omega)$$

Figure 5–26 compares frequency-response curves of

$$G(j\omega) = \frac{1}{1 + 2\zeta\left(j\dfrac{\omega}{\omega_n}\right) + \left(j\dfrac{\omega}{\omega_n}\right)^2}$$

in three different representations. In the log-magnitude-versus-phase plot, the vertical distance between the points $\omega = 0$ and $\omega = \omega_r$, where ω_r is the resonant frequency, is the peak value of $G(j\omega)$ in decibels.

M circles and N circles. Consider the unity-feedback control system with the open-loop transfer function $G(s)$. Define the magnitude of the closed-loop frequency response as M and the phase angle as α, or

$$\frac{C(j\omega)}{R(j\omega)} = \frac{G(j\omega)}{1 + G(j\omega)} = Me^{j\alpha}$$

In what follows, we shall find that the constant-magnitude loci and constant phase-angle loci are convenient in determining the closed-loop frequency response from the open-loop frequency response.

Constant-magnitude loci (M circles). To obtain the constant-magnitude loci, let us first note that $G(j\omega)$ is a complex quantity and can be written as

$$G(j\omega) = X + jY$$

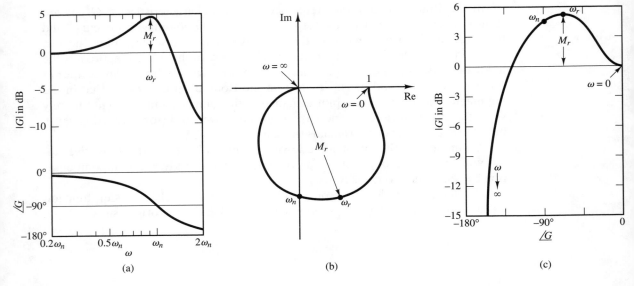

Figure 5–26
Three representations of the frequency response of $\dfrac{1}{1 + 2\zeta\left(j\dfrac{\omega}{\omega_n}\right) + \left(j\dfrac{\omega}{\omega_n}\right)^2}$ for $\zeta > 0$.

(a) Bode diagram; (b) polar plot; (c) log-magnitude-versus-phase plot.

where X and Y are real quantities. Then M is given by

$$M = \frac{|X + jY|}{|1 + X + jY|}$$

and M^2 is

$$M^2 = \frac{X^2 + Y^2}{(1 + X)^2 + Y^2}$$

Hence,

$$X^2(1 - M^2) - 2M^2X - M^2 + (1 - M^2)Y^2 = 0 \tag{5-1}$$

If $M = 1$, then, from Equation (5–1), we obtain $X = -\tfrac{1}{2}$. This is the equation of a straight line parallel to the Y axis and passing through the point $\left(-\tfrac{1}{2}, 0\right)$.

If $M \neq 1$, Equation (5–1) can be written

$$X^2 + \frac{2M^2}{M^2 - 1}X + \frac{M^2}{M^2 - 1} + Y^2 = 0$$

If the term $M^2/(M^2 - 1)^2$ is added to both sides of this last equation, we obtain

$$\left(X + \frac{M^2}{M^2 - 1}\right)^2 + Y^2 = \frac{M^2}{(M^2 - 1)^2} \tag{5-2}$$

Equation (5–2) is the equation of a circle with center at $X = -M^2/(M^2 - 1)$, $Y = 0$ and radius $|M/(M^2 - 1)|$.

The constant M loci on the $G(s)$ plane are thus a family of circles. The center and radius of the circle for a given value of M can be easily calculated. For example, for $M = 1.3$, the center is at $(-2.45, 0)$ and the radius is 1.88. A family of constant M circles is shown in Figure 5–27. Note that as M becomes larger compared with 1, the M circles become smaller and converge to the $-1 + j0$ point. For $M > 1$, the centers of the M circles lie to the left of the $-1 + j0$ point. Similarly, as M becomes smaller compared with 1, the M circle becomes smaller and converges to the origin. For $0 < M < 1$, the centers of the M circles lie to the right of the origin. $M = 1$ corresponds to the locus of points equidistant from the origin and from the $-1 + j0$ point. As stated earlier, it is a straight line passing through the point $\left(-\frac{1}{2}, 0\right)$ and parallel to the imaginary axis. (The constant M circles corresponding to $M > 1$ lie to the left of the $M = 1$ line, and those corresponding to $0 < M < 1$ lie to the right of the $M = 1$ line.) The M circles are symmetrical with respect to the straight line corresponding to $M = 1$ and with respect to the real axis.

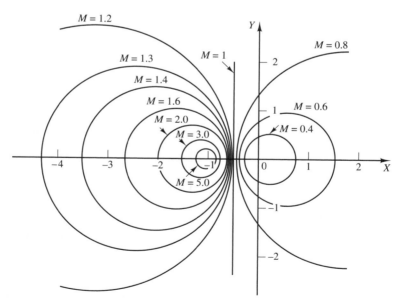

Figure 5–27
A family of constant M circles.

Constant phase-angle loci (N circles). We shall obtain the phase angle in terms of X and Y. Since

$$\angle e^{j\alpha} = \left/ \frac{X + jY}{1 + X + jY} \right.$$

the phase angle α is

$$\alpha = \tan^{-1}\left(\frac{Y}{X}\right) - \tan^{-1}\left(\frac{Y}{1+X}\right)$$

If we define

$$\tan \alpha = N$$

then

$$N = \tan\left[\tan^{-1}\left(\frac{Y}{X}\right) - \tan^{-1}\left(\frac{Y}{1+X}\right)\right]$$

Since

$$\tan(A - B) = \frac{\tan A - \tan B}{1 + \tan A \tan B}$$

we obtain

$$N = \frac{\dfrac{Y}{X} - \dfrac{Y}{1+X}}{1 + \dfrac{Y}{X}\left(\dfrac{Y}{1+X}\right)} = \frac{Y}{X^2 + X + Y^2}$$

or

$$X^2 + X + Y^2 - \frac{1}{N}Y = 0$$

The addition of $\left(\frac{1}{4}\right) + 1/(2N)^2$ to both sides of this last equation yields

$$\left(X + \frac{1}{2}\right)^2 + \left(Y - \frac{1}{2N}\right)^2 = \frac{1}{4} + \left(\frac{1}{2N}\right)^2 \tag{5-3}$$

This is an equation of a circle with center at $X = -\frac{1}{2}, Y = 1/(2N)$ and with radius $\sqrt{\frac{1}{4} + 1/(2N)^2}$. For example, if $\alpha = 30°$, then $N = \tan \alpha = 0.577$, and the center and the radius of the circle corresponding to $\alpha = 30°$ are found to be $(-0.5, 0.866)$ and unity, respectively. Since Equation (5–3) is satisfied when $X = Y = 0$ and $X = -1, Y = 0$, regardless of the value of N, each circle passes through the origin and the $-1 + j0$ point. The constant α loci can be drawn easily once the value of N is given. A family of constant N circles is shown in Figure 5–28 with α as a parameter.

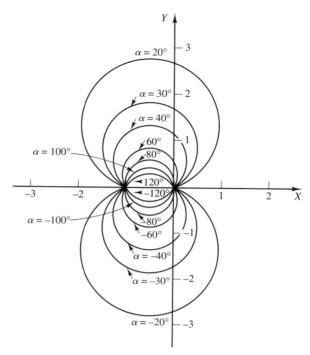

Figure 5–28
A family of constant N circles.

Note that the constant N locus for a given value of α is actually not the entire circle, but only an arc. In other words, the $\alpha = 30°$ and $\alpha = -150°$ arcs are parts of the same circle. This is so because the tangent of an angle remains the same if $\pm 180°$ (or multiples thereof) is added to the angle.

The use of M and N circles enables us to find the entire closed-loop frequency response from the open-loop frequency response $G(j\omega)$ without calculating the magnitude and phase of the closed-loop transfer function at each frequency. The intersections of the $G(j\omega)$ locus with the M circles and N circles give the values of M and N at frequency points on the $G(j\omega)$ locus.

The N circles are multivalued in the sense that the circle for $\alpha = \alpha_1$ and that for $\alpha = \alpha_1 \pm 180°n (n = 1, 2, \ldots)$ are the same. In using the N circles for the determination of the phase angle of closed-loop systems, we must interpret the proper value of α. To avoid any error, we start at zero frequency, which corresponds to $\alpha = 0°$, and proceed to higher frequencies. The phase-angle curve must be continuous.

Graphically, the intersections of the $G(j\omega)$ locus and the M circles give the values of M at the frequencies denoted on the $G(j\omega)$ locus. Thus, the constant M circle with the smallest radius that is tangent to the $G(j\omega)$ locus gives the value of the resonant peak magnitude M_r. If it is desired to keep the resonant peak value less than a certain value, then the system should not enclose the critical point (the $-1 + j0$ point), and at the same time, there should be no intersections with the particular M circle and the $G(j\omega)$ locus.

Nichols chart. In dealing with design problems, we find it convenient to construct the M and N loci in the log-magnitude-versus-phase plane. The chart consisting of the M and N loci in the log-magnitude-versus-phase diagram is called the Nichols chart. The $G(j\omega)$ locus drawn on the Nichols chart gives both the gain characteristics and phase characteristics of the closed-loop transfer function at the same time. The Nichols chart is shown in Figure 5–29 for phase angles between 0° and −240°.

Note that the critical point (the $-1 + j0$ point) is mapped to the Nichols chart as the point (0 dB, −180°). The Nichols chart contains curves of constant closed-loop magnitude and phase angle. The designer can graphically determine the phase margin, gain margin, resonant peak magnitude, resonant frequency, and bandwidth of the closed-loop system from the plot of the open-loop locus, $G(j\omega)$.

The Nichols chart is symmetric about the −180° axis. The M and N loci repeat for every 360°, and there is symmetry at every 180° interval. The M loci are

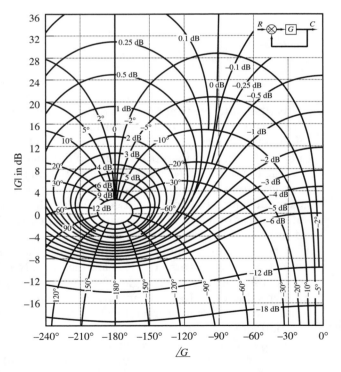

Figure 5–29
Nichols chart.

centered about the critical point (0 dB, $-180°$). The Nichols chart is useful for determining the frequency response of the closed loop from that of the open loop. If the open-loop frequency-response curve is superimposed on the Nichols chart, the intersections of the open-loop frequency-response curve $G(j\omega)$ with the M and N loci give the values of the magnitude M and phase angle α of the closed-loop frequency response at each frequency point. If the $G(j\omega)$ locus does not intersect the $M = M_r$ locus, but is tangent to it, then the resonant peak value of M of the closed-loop frequency response is given by M_r. The resonant frequency is given by the frequency at the point of tangency.

Getting a Nichols plot with MATLAB. MATLAB commands for the Nichols plot are

```
nichols(num,den)
nichols(num,den,w)
nichols(A,B,C,D)
nichols(A,B,C,D,w)
nichols(A,B,C,D,iu,w)
nichols(sys)
[mag,phase,w] = nichols(num,den)
[mag,phase,w] = nichols(num,den,w)
[mag,phase,w] = nichols(A,B,C,D)
[mag,phase,w] = nichols(A,B,C,D,w)
[mag,phase,w] = nichols(A,B,C,D,iu,w)
[mag,phase,w] = nichols(sys)
```

A Nichols plot can be drawn with any of the Nichols commands. If the Nichols command used does not involve any left-hand argument, the Nichols plot is automatically generated. If a left-hand argument is present, the Nichols command does not plot the Nichols diagram, but returns the magnitude and phase (in degrees). To plot the curve, we need to use a plot command. The command ngrid is used for drawing a Nichols chart grid.

EXAMPLE 5–16 Consider the system shown in Figure 5–30. Draw a Nichols plot of $G(s)$.
MATLAB Program 5–20 produces a Nichols plot, which is shown in Figure 5–31.

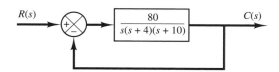

Figure 5–30
Control system.

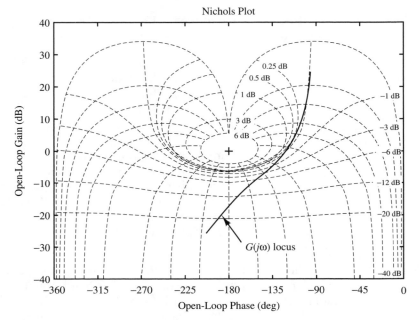

MATLAB Program 5–20

```
>> num = [80];
>> den = [1   14   40   0];
>> w = logspace(-1,1,100);
>> nichols(num,den,w);
>> ngrid;
>> title('Nichols Plot');
>> text(-160, -30, 'G(j\omega) locus')
```

Figure 5–31
Nichols plot.

EXAMPLE 5–17 Consider a unity-feedback control system with the open-loop transfer function

$$G(s) = \frac{100}{s(s + 7.5)}$$

Draw a Nichols plot of $G(s)$.

MATLAB Program 5–21 produces a Nichols plot for this $G(s)$. The frequency range is arbitrarily chosen to be $1 \leq \omega \leq 100$. Figure 5–32 shows the Nichols plot for the system. Notice that the $G(j\omega)$ locus is tangent to the 3-dB oval locus at a point near $\omega = 8$ rad/sec.

MATLAB Program 5–21

```
>> num = [100];
>> den = [1   7.5   0];
>> w = [1   4   8   10   20   40   100];
>> [mag,phase,w] = nichols(num,den,w);
>> magdB = 20*log10(mag);
>> plot(phase,magdB,phase,magdB,'o');
>> v = [-250   0   -60   40]; axis(v);
>> ngrid
>> title('Nichols Plot');
>> xlabel('Open-Loop Phase (deg)');
>> ylabel('Open-Loop Gain (dB)');
>> text(-90,22,'\omega = 1');
>> text(-115,9,'\omega = 4');
>> text(-132,0,'\omega = 8')
>> text(-165,-26,'\omega = 40');
>> text(-170,-41,'\omega = 100');
```

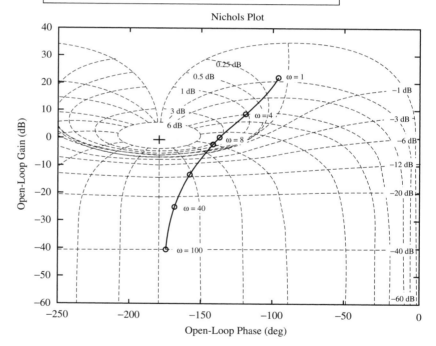

Figure 5–32
Nichols plot for
$G(s) = \dfrac{100}{s(s+7.5)}$.

This means that the closed-loop transfer function

$$\frac{C(j\omega)}{R(j\omega)} = \frac{G(j\omega)}{1+G(j\omega)}$$

has the maximum magnitude of about 3 dB near $\omega = 8$ rad/sec, a fact that can be checked with the Bode diagram of $G(j\omega)/[1 + G(j\omega)]$ shown in Figure 5–33. The MATLAB Program for getting the Bode diagram is given in MATLAB Program 5–22.

MATLAB Program 5–22

```
>> num = [100];
>> den = [1   7.5   0];
>> sys1 = tf(num,den);
>> sys = feedback(sys1,[1]);
>> w = logspace(0,2,100);
>> bode(sys,w)
>> grid
>> title('Bode Diagram of Closed-Loop System G/(1 + G)')
```

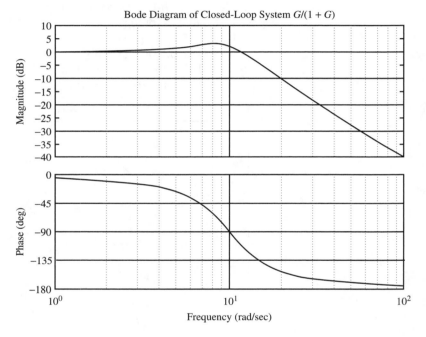

Figure 5–33
Bode diagram of $G(j\omega)/[1 + G(j\omega)]$.

EXAMPLE 5–18 Consider a unity-feedback control system with the open-loop transfer function

$$G(s) = \frac{45s^2 + 20s + 1}{s^5 + 10s^4 + 20s^3 + 30s^2 + 5s}$$

MATLAB Program 5–23 produces the Nichols plot shown in Figure 5–34. From the plot, we see that the locus is tangent to the $M = 5.5$-dB locus near $\omega = 2.8$ rad/sec. This can be checked with the Bode diagram of the closed-loop transfer function $G(s)/[1 + G(s)]$, which shows the maximum magnitude of 5.5 dB near $\omega = 2.8$ rad/sec. (See Figure 5–35.)

MATLAB Program 5-23

```
>> num = [45   20   1];
>> den = [1   10   20   30   5   0];
>> w = logspace(-3,2,300);
>> nichols(num,den,w);
>> ngrid
>> hold
Current plot held
>> ww = [1   3   10   100];
>> [mag,phase,ww] = nichols(num,den,ww);
>> magdB = 20*log10(mag);
>> plot(phase,magdB,'o')
>> gtext('1')
>> gtext('3')
>> gtext('10')
>> gtext('100')
>> hold
Current plot released
>> sys 1 = tf(num,den);
>> sys = feedback(sys1,[1])
Transfer function:
            45 s^2 + 20 s + 1
   ───────────────────────────────────────
   s^5 + 10 s^4 + 20 s^3 + 75 s^2 + 25 s + 1
>> bode(sys,w)
>> grid
```

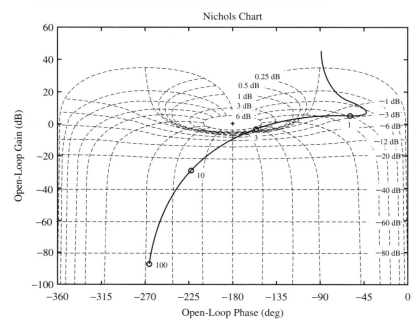

Figure 5-34
Nichols plot of $G(s)$ given in Example 5-18.

Section 5-3 / Log-Magnitude-Versus-Phase Plots

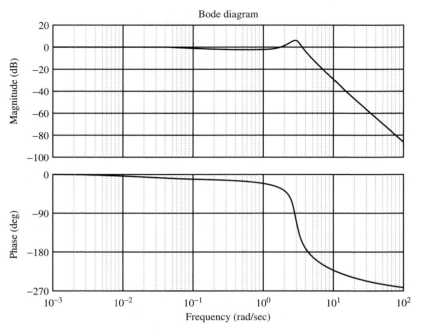

Figure 5–35
Bode diagram for the closed-loop transfer function $G(s)/[1 + G(s)]$.

5–4 PHASE MARGIN AND GAIN MARGIN

Figure 5–36 shows polar plots of $G(j\omega)$ for three different values of the open-loop gain K. For a large value of the gain K, the system is unstable. As the gain is decreased to a certain value, the $G(j\omega)$ locus passes through the $-1 + j0$ point. This means that, with this gain value, the system is on the verge of instability and will exhibit sustained oscillations. For a small value of the gain K, the system is stable.

In general, the closer the $G(j\omega)$ locus comes to encircling the $-1 + j0$ point, the more oscillatory is the system response. The closeness of the $G(j\omega)$ locus to the $-1 + j0$ point can be used as a measure of the margin of stability. (This does not apply, however, to conditionally stable systems.) It is common practice to represent the closeness in terms of phase margin and gain margin.

Phase margin: The phase margin is that amount of additional phase lag at the gain crossover frequency required to bring the system to the verge of instability. The gain crossover frequency is the frequency at which $|G(j\omega)|$, the magnitude of the open-loop transfer function, is unity. The phase margin γ is $180°$ plus the phase angle ϕ of the open-loop transfer function at the gain crossover frequency, or

$$\gamma = 180° + \phi$$

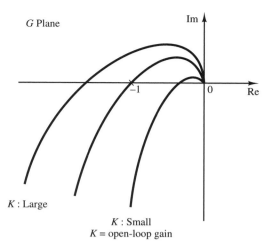

Figure 5–36
Polar plots of
$$\frac{K(1 + j\omega T_a)(1 + j\omega T_b)\cdots}{(j\omega)(1 + j\omega T_1)(1 + j\omega T_2)\cdots}.$$

Figures 5–37(a), (b), and (c) illustrate the phase margin of both a stable system and an unstable system in Bode diagrams, polar plots, and log-magnitude-versus-phase plots. In the polar plot, a line may be drawn from the origin to the point at which the unit circle crosses the $G(j\omega)$ locus. The angle from the negative real axis to this line is the phase margin, which is positive for $\gamma > 0$ and negative for $\gamma < 0$. For a minimum-phase system to be stable, the phase margin must be positive. In the log-magnitude-versus-phase plots, the critical point in the complex plane corresponds to the intersection of the 0-dB and $-180°$ lines.

Gain margin: The gain margin is the reciprocal of the magnitude $|G(j\omega)|$ at the frequency at which the phase angle is $-180°$. Defining the phase crossover frequency ω_1 to be the frequency at which the phase angle of the open-loop transfer function equals $-180°$ gives the gain margin K_g:

$$K_g = \frac{1}{|G(j\omega_1)|}$$

In terms of decibels,

$$K_g \text{ dB} = 20 \log K_g = -20 \log|G(j\omega_1)|$$

The gain margin expressed in decibels is positive if K_g is greater than unity and negative if K_g is smaller than unity. Thus, a positive gain margin (in decibels) means that the system is stable, and a negative gain margin (in decibels) means that the system is unstable. The gain margin is shown in Figures 5–37(a), (b), and (c).

For a stable minimum-phase system, the gain margin indicates how much the gain can be increased before the system becomes unstable. For an unstable system, the gain margin is indicative of how much the gain must be decreased to make the system stable.

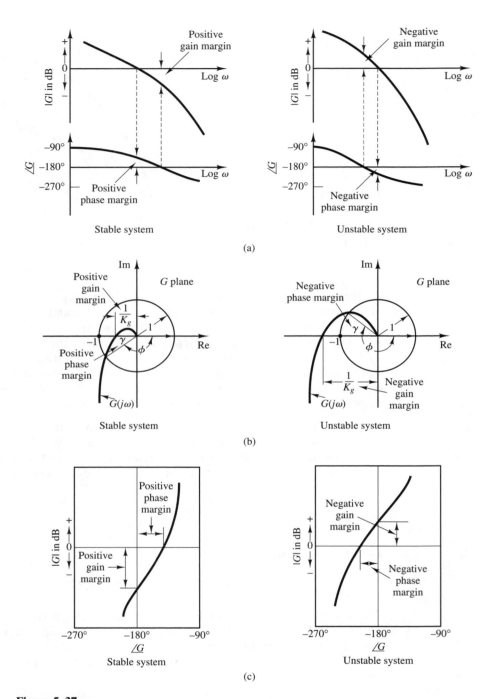

Figure 5-37
Phase and gain margins of stable and unstable systems. (a) Bode diagrams; (b) polar plots; (c) log-magnitude-versus-phase plots.

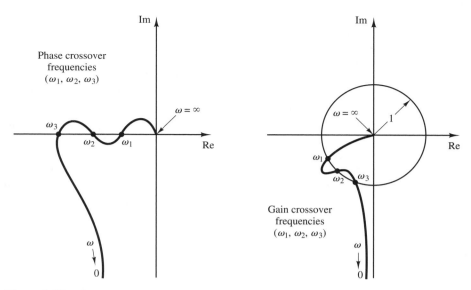

Figure 5–38
Polar plots showing more than two phase or gain crossover frequencies.

The gain margin of a first- or second-order system is infinite, since the polar plots for such systems do not cross the negative real axis. Thus, theoretically, first- or second-order systems cannot be unstable. (Note, however, that so-called first- or second-order systems are only approximations, in the sense that small time lags are neglected in deriving the system equations. If these small lags are accounted for, the so-called first- or second-order systems may become unstable.)

Note that, for a non-minimum-phase system with an unstable open loop, the stability condition will not be satisfied unless the $G(j\omega)$ plot encircles the $-1 + j0$ point. Hence, such a stable non-minimum-phase system will have negative phase and gain margins.

Note also that conditionally stable systems will have two or more phase crossover frequencies, and some higher-order systems with complicated numerator dynamics may have two or more gain crossover frequencies as well, as is shown in Figure 5–38. For stable systems having two or more gain crossover frequencies, the phase margin is measured at the highest gain crossover frequency.

A few comments on phase and gain margins. The phase and gain margins of a control system are a measure of the closeness of the polar plot to the $-1 + j0$ point. Therefore, these margins may be used as design criteria.

Observe that neither the gain margin alone nor the phase margin alone gives a sufficient indication of relative stability. Both should be given in any determination of relative stability.

For a minimum-phase system, both the phase and gain margins must be positive for the system to be stable. Negative margins indicate instability.

Proper phase and gain margins ensure us against variations in the system components and are specified for definite positive values. The two values bound the behavior of the closed-loop system near the resonant frequency. For satisfactory performance, the phase margin should be between 30° and 60° and the gain margin should be greater than 6 dB. With these values, a minimum-phase system has guaranteed stability, even if the open-loop gain and time constants of the components vary to a certain extent. Although the phase and gain margins give only rough estimates of the effective damping ratio of the closed-loop system, they do offer a convenient means for designing control systems or adjusting the gain constants of systems.

For minimum-phase systems, the magnitude and phase characteristics of the open-loop transfer function are definitely related. The requirement that the phase margin be between 30° and 60° means that, in a Bode diagram, the slope of the log-magnitude curve at the gain crossover frequency should be more gradual than -40 dB/decade. In most practical cases a slope of -20 dB/decade is desirable at the gain crossover frequency for stability. If the slope is -40 dB/decade, the systems could be either stable or unstable. (Even if the system is stable, however, the phase margin is small.) If the slope at the gain crossover frequency is -60 dB/decade or steeper, the system is most likely unstable.

For non-minimum-phase systems, the correct interpretation of stability margins requires careful study. The best way to determine the stability of non-minimum-phase systems is to use the Nyquist diagram rather than the Bode diagram.

EXAMPLE 5–19 Obtain the phase and gain margins of the system shown in Figure 5–39 for the two cases where $K = 10$ and $K = 100$.

The phase and gain margins can easily be obtained from the Bode diagram. A Bode diagram of the given open-loop transfer function with $K = 10$ is shown in Figure 5–40(a). For $K = 10$, we have

$$\text{Phase margin} = 21°, \quad \text{Gain margin} = 8 \text{ dB}$$

Therefore, the system gain may be increased by 8 dB before the instability occurs.

Increasing the gain from $K = 10$ to $K = 100$ shifts the 0-dB axis down by 20 dB, as shown in Figure 5–40(b). In that case,

$$\text{Phase margin} = -30°, \quad \text{Gain margin} = -12 \text{ dB}$$

Thus, the system is stable for $K = 10$, but unstable for $K = 100$.

Notice that one of the very convenient aspects of the Bode diagram is the ease with which the effects of gain changes can be evaluated. To obtain satisfactory performance, we must increase the phase margin to $30° \sim 60°$. This can be done by decreasing the gain K. Decreasing K is not desirable, however, since a small value of K will yield a large error for the ramp input. This suggests that reshaping of the open-loop frequency-response curve by adding compensation may be necessary. Compensation techniques are discussed in detail in Section 5–5.

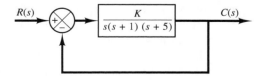

Figure 5–39
Control system.

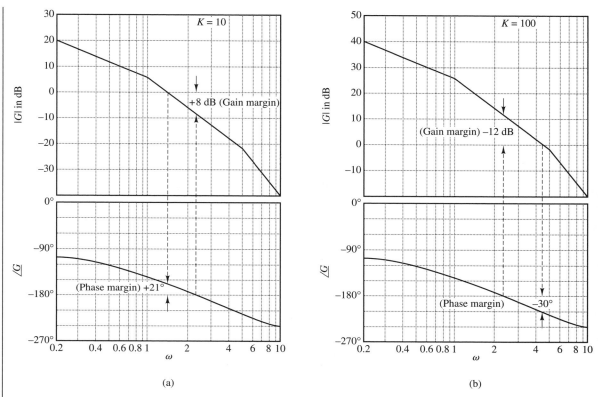

Figure 5–40
Bode diagrams of the system shown in Figure 5–39; (a) with $K = 10$ and (b) with $K = 100$.

Obtaining gain margin, phase margin, phase crossover frequency, and gain crossover frequency with MATLAB. The gain margin, phase margin, phase crossover frequency, and gain crossover frequency can be obtained easily with MATLAB. The command used is

[Gm,Pm,wcp,wcg] = margin(sys)

where Gm is the gain margin, Pm is the phase margin, wcp is the phase crossover frequency, and wcg is the gain crossover frequency. Example 5–20 gives details on how to use this command.

EXAMPLE 5–20 Draw a Bode diagram of the open-loop transfer function $G(s)$ of the closed-loop system shown in Figure 5–41. Determine the gain margin, phase margin, phase crossover frequency, and gain crossover frequency with MATLAB.

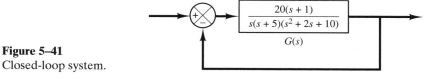

Figure 5–41
Closed-loop system.

Section 5–4 / Phase Margin and Gain Margin

A MATLAB program to plot a Bode diagram and to obtain the gain margin, phase margin, phase crossover frequency, and gain crossover frequency is shown in MATLAB Program 5–24. The Bode diagram of $G(s)$ is shown in Figure 5–42.

MATLAB Program 5–24

```
>> num = [20   20];
>> den = conv([1   5   0],[1   2   10]);
>> sys = tf(num,den);
>> w = logspace(-1,2,100);
>> bode(sys,w)
>> [Gm,Pm,wcp,wcg] = margin(sys);
>> GmdB = 20*log10(Gm);
>> [GmdB   Pm   wcp   wcg]

ans =

    9.9301   103.6573   4.0132   0.4426
```

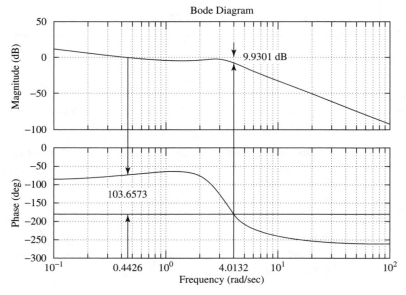

Figure 5–42
Bode diagram of $G(s)$ shown in Figure 5–41.

MATLAB approach to getting resonant peak, resonant frequency, and bandwidth. The resonant peak is the value of the maximum magnitude (in decibels) of the closed-loop frequency response. The resonant frequency is the frequency that yields the maximum magnitude. MATLAB commands used to obtain the resonant peak and resonant frequency are as follows:

```
[mag,phase,w] = bode(num,den,w);   or   [mag,phase,w] = bode(sys,w);
[Mp,k] = max(mag);
resonant_peak = 20*log10(Mp);
resonant_frequency = w(k)
```

The bandwidth can be obtained by entering the following lines into the program:

```
n = 1;
while 20*log10(mag(n)) > = −3; n = n + 1;
end
bandwidth = w(n)
```

Example 5–21 presents a detailed MATLAB program.

EXAMPLE 5–21 Consider the system shown in Figure 5–43. Using MATLAB, obtain a Bode diagram for the closed-loop transfer function. Also, obtain the resonant peak, resonant frequency, and bandwidth.

MATLAB Program 5–25 produces a Bode diagram for the closed-loop system, as well as the resonant peak, resonant frequency, and bandwidth. The resulting Bode diagram is

---

MATLAB Program 5–25

```
>> num = [1];
>> den = [0.5   1.5   1   0];
>> sys1 = tf(num,den);
>> sys = feedback(sys1,1);
>> w = logspace(−1,1);
>> bode(sys,w)
>> grid
>> [mag,phase,w] = bode(sys,w);
>> [Mp,k] = max(mag);
>> resonant_peak = 20*log10(Mp)

resonant_peak =
    5.2388

>> resonant_frequency = w(k)

resonant_frequency =
    0.7906

>> n = 1;
>> while 20*log(mag(n)) > = −3; n = n + 1;
    end
>> bandwidth = w(n)

bandwidth =
    1.2649
```

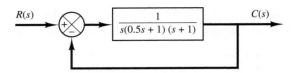

Figure 5–43
Closed-loop system.

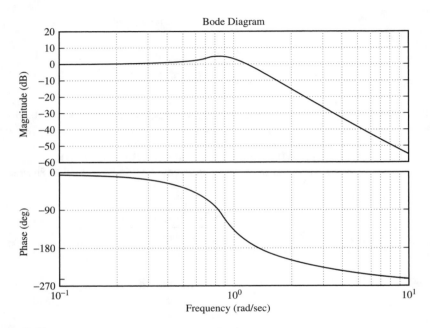

Figure 5–44
Bode diagram of the closed-loop transfer function of the system shown in Figure 5–43.

shown in Figure 5–44. The resonant peak is 5.2388 dB, the resonant frequency is 0.7906 rad/sec, and the bandwidth is 1.2649 rad/sec. These values can be verified from the figure.

5-5 FREQUENCY-RESPONSE APPROACH TO CONTROL SYSTEMS COMPENSATION

In this section, we consider lead compensation, lag compensation, and lag–lead compensation of control systems via a Bode diagram approach.

Lead compensation. We illustrate a lead compensation technique with the use of a few examples.

EXAMPLE 5-22 Consider the control system shown in Figure 5–45. Determine the value of the gain K such that the phase margin is 60°. What is the gain margin with this value of K?

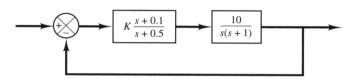

Figure 5–45
Control system.

The open-loop transfer function is

$$G(s) = K\frac{s + 0.1}{s + 0.5} \frac{10}{s(s + 1)}$$

$$= \frac{K(10s + 1)}{s^3 + 1.5s^2 + 0.5s}$$

MATLAB Program 5–26 plots the Bode diagram of $G(s)$ when $K = 1$. Figure 5–46 shows the Bode diagram produced by that program. From the diagram, the required phase margin of 60° occurs at the frequency $\omega = 1.15$ rad/sec. The magnitude of $G(j\omega)$ at this frequency is found to be 14.5 dB. Then the gain K must satisfy the following equation:

$$20 \log K = -14.5 \text{ dB}$$

or

$$K = 0.188$$

MATLAB Program 5–26

```
>> num = [10  1];
>> den = [1   1.5   0.5   0];
>> bode(num,den)
>> grid
>> title('Bode Diagram of G(s) = (10s + 1)/[s(s + 0.5)(s + 1)]')
```

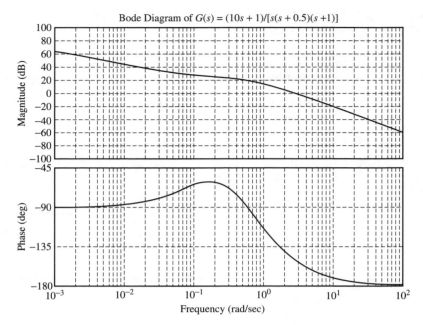

Figure 5–46
Bode diagram of $G(s) = \dfrac{10s + 1}{s(s + 0.5)(s + 1)}$.

Thus, we have determined the value of the gain K. Since the angle curve does not cross the $-180°$ line, the gain margin is $+\infty$ dB.

To verify the results, let us draw a Nyquist plot of G for the frequency range

$$w = 0.5:0.01:1.15$$

The endpoint of the locus ($\omega = 1.15$ rad/sec) will be on a unit circle in the Nyquist plane. To check the phase margin, it is convenient to draw the Nyquist plot on a polar diagram, using polar grids.

To draw the Nyquist plot on a polar diagram, we first define a complex vector z by

$$z = re + i*im = re^{i\theta}$$

where r and θ (theta) are given by

$$r = abs(z)$$
$$theta = angle(z)$$

The abs function finds the square root of the sum of the real part squared and imaginary part squared: angle means $\tan^{-1}$ (imaginary part divided by real part).

If we use the command

$$\text{polar(theta,r)}$$

MATLAB will produce a plot in polar coordinates.

MATLAB Program 5–27 produces the Nyquist plot of $G(j\omega)$, where ω is between 0.5 and 1.15 rad/sec. The resulting plot is shown in Figure 5–47. Notice that the point $G(j1.15)$ lies on the unit circle and the phase angle of this point is $-120°$. Hence, the phase margin is 60°. The fact that $G(j1.15)$ is on the unit circle verifies that at $\omega = 1.15$ rad/sec the magnitude is equal to 1, or 0 dB. (Thus, $\omega = 1.15$ is the gain crossover frequency.) Hence, $K = 0.188$ gives the desired phase margin of 60°.

MATLAB Program 5–27

```
>> % *****Nyquist plot in rectangular coordinates*****
>>
>> num = [1.88   0.188];
>> den = [1   1.5   0.5   0];
>> w = 0.5:0.01:1.15;
>> [re,im,w] = nyquist(num,den,w);
>>
>> % *****Convert rectangular coordinates into polar coordinates
>> % by defining z, r, theta as follows*****
>>
>> z = re + i*im;
>> r = abs(z);
>> theta = angle(z);
>>
>> % *****To draw polar plot, enter command 'polar(theta,r)'*****
>>
>> polar(theta,r)
>> text(-1,3,'Check of Phase Margin')
>> text(0.3,-1.7,'Nyquist plot')
>> text(-2.2,-0.75,'Phase margin')
>> text(-2.2,-1.1,'is 60 degrees')
>> text(1.45,-0.7,'Unit circle')
```

EXAMPLE 5–23 Consider the system shown in Figure 5–48. Design a lead compensator $G_c(s)$ such that the closed-loop system will have a phase margin of 50° and a gain margin of not less than 10 dB. Assume that

$$G_c(s) = K_c \alpha \left(\frac{Ts + 1}{\alpha Ts + 1} \right) \quad (0 < \alpha < 1)$$

It is desired that the bandwidth of the closed-loop system be $1 \sim 2$ rad/sec. What are the values of M_r and ω_r of the compensated system?

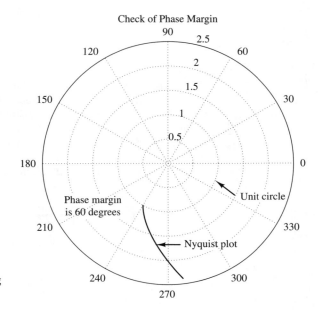

Figure 5–47
Nyquist plot of $G(j\omega)$ showing that the phase margin is 60°.

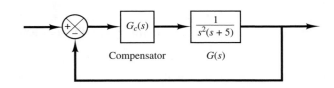

Figure 5–48
Closed-loop system.

Note that the lead angle that $G_c(j\omega)$ provides is

$$\phi = \angle G_c(j\omega) = \tan^{-1}(\omega T) - \tan^{-1}(\alpha\omega T)$$

The frequency ω_m that produces the maximum lead angle ϕ_m can be obtained by solving $\partial\phi/\partial\omega = 0$. The result is $\omega_m = 1/(\sqrt{\alpha}T)$. The maximum phase lead angle ϕ_m is then obtained as

$$\phi_m = \tan^{-1}(\omega_m T) - \tan^{-1}(\alpha\omega_m T) = \tan^{-1}(1/\sqrt{\alpha}) - \tan^{-1}(\sqrt{\alpha})$$
$$= \tan^{-1}\left(\frac{1-\alpha}{2\sqrt{\alpha}}\right) = \sin^{-1}\frac{1-\alpha}{1+\alpha}$$

or

$$\sin\phi_m = \frac{1-\alpha}{1+\alpha}$$

Since the bandwidth of the closed-loop system is close to the gain crossover frequency, we choose the gain crossover frequency to be 1 rad/sec. At $\omega = 1$, the phase angle of $G(j\omega)$ is $-191.31°$. Hence, the lead network needs to supply $50° + 11.31° = 61.31°$ at $\omega = 1$. Thus, α can be determined from

$$\sin \phi_m = \sin 61.31° = \frac{1-\alpha}{1+\alpha} = 0.8772$$

as follows:

$$\alpha = 0.06541$$

Since we chose $\omega_m = 1$, we get

$$\omega_m = 1/(\sqrt{\alpha}T) = 1/(\sqrt{0.06541}\,T) = 1$$

from which we obtain

$$T = 3.910, \qquad \alpha T = 0.2558$$

Then we get

$$\frac{G_c(s)G(s)}{0.06541K_c} = \frac{3.910s + 1}{0.2558s + 1} \frac{0.2}{s^2(0.2s + 1)}$$

A Bode diagram of $G_c(s)G(s)/(0.06541K_c)$ can be obtained with MATLAB Program 5–28. The Bode diagram is shown in Figure 5–49.

MATLAB Program 5–28

```
>> num = [0.782   0.2];
>> den = [0.05116   0.4558   1   0   0];
>> w = logspace(-1,1,100);
>> bode(num,den,w);
>> grid
>> ww = [0.5   1   2];
>> [mag,phase,ww] = bode(num,den,ww);
>> magdB = 20*log10(mag);
>> [ww   magdB]

ans =

   0.5000    4.7804
   1.0000   -2.3063
   2.0000   -9.7403
```

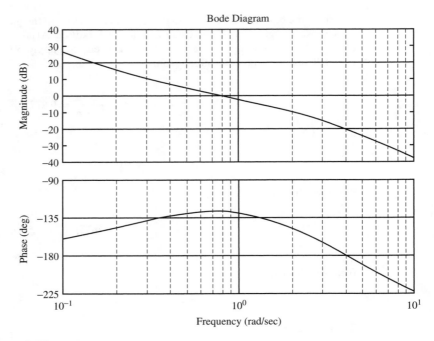

Figure 5–49
Bode diagram for $G_c(s)G(s)/0.06541K_c$.

From MATLAB Program 5–28, we find that the magnitude of $G_c(j\omega)G(j\omega)/(0.06541K_c)$ with $\omega = 1$ is -2.3063 dB. Thus, the magnitude curve must be raised by 2.3063 dB so that the magnitude of $G_c(j\omega)G(j\omega)$ equals 0 dB at $\omega = 1$ rad/sec. Hence, we set

$$20 \log 0.06541 K_c = 2.3063$$

or

$$0.06541 K_c = 1.3041$$

which yields

$$K_c = 19.94$$

Next, we obtain a Bode diagram of $G_c(s)G(s)$ with $K_c = 19.94$. We have

$$G_c(s)G(s) = 0.06541 K_c \frac{3.910s + 1}{0.2558s + 1} \cdot \frac{0.2}{s^2(0.2s + 1)}$$

$$= \frac{1.01994s + 0.26086}{0.05116s^4 + 0.4558s^3 + s^2}$$

MATLAB Program 5–29 generates the Bode diagram of the open-loop system (Figure 5–50) and the Bode diagram of the closed-loop system (Figure 5–51).

MATLAB Program 5–29

```
>> numc = [1.01994   0.26086];
>> denc = [0.05116   0.4558   1   0   0];
>> w = logspace(-1,2,100);
>> bode(numc,denc,w);
>> grid
>> [Gm,Pm,wcp,wcg] = margin(numc,denc);
>> GmdB = 20*log10(Gm);
>> [GmdB   Pm   wcp   wcg]

ans =

   17.7408   49.9939   4.1535   1.0001

>> gtext('Gain margin = 17.74 dB')
>> gtext('Phase margin = 49.99 degrees')
>>
>> % Figure 5–50 shows the Bode diagram of the open-loop system (sys1),
>> % and Figure 5–51 shows that of the closed-loop system (sys).
>>
>> sys1 = tf(numc,denc);
>> sys = feedback(sys1,[1])
```

Transfer function:

$$\frac{1.02\ s\ +\ 0.2609}{0.05116\ s^4\ +\ 0.4558\ s^3\ +\ s^2\ +\ 1.02\ s\ +\ 0.2609}$$

```
>> [mag,phase,w] = bode(sys,w);
>> [Mr,r] = max (mag);
>> MrdB = 20*log10 (Mr);
>> wr = w(r);
>> [MrdB   wr]

ans =

   2.1116   0.6136

>> gtext('Resonant peak = 2.1116 dB')
>> gtext('Resonant peak frequency = 0.6136 rad/sec')
>>
>> n = 1;
>> while 20*log10(mag(n)) >-3; n = n+1;
   end
>> Bandwidth = w(n)

Bandwidth =

   1.8738

>> gtext('Bandwidth = 1.8738 rad/sec')
```

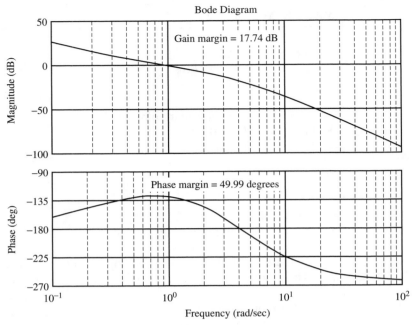

Figure 5–50
Bode diagram of the open-loop system (sys1).

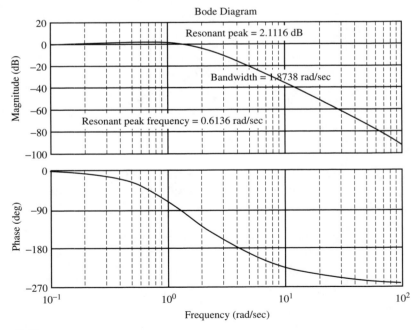

Figure 5–51
Bode diagram of the closed-loop system (sys).

EXAMPLE 5–24 Consider the system shown in Figure 5–52(a). Design a compensator such that the closed-loop system will satisfy the requirement that the static velocity error constant = 20 sec^{-1}, the phase margin = 50°, and the gain margin $\geq$ 10 dB.

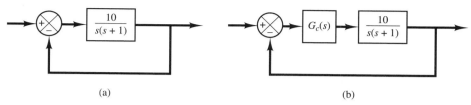

Figure 5–52
(a) Control system; (b) compensated system.

To satisfy this requirement, we shall try a lead compensator of the form

$$G_c(s) = K_c \alpha \frac{Ts + 1}{\alpha Ts + 1}$$

$$= K_c \frac{s + \dfrac{1}{T}}{s + \dfrac{1}{\alpha T}}$$

(If the lead compensator does not work, then we shall employ a compensator of a different form.) The compensated system is shown in Figure 5–52(b).

We define

$$G_1(s) = KG(s) = \frac{10K}{s(s+1)}$$

where $K = K_c \alpha$. The first step in the design is to adjust the gain K to meet the steady-state performance specification or to provide the required static velocity error constant K_v. Since the latter is given as 20 sec^{-1}, we have

$$K_v = \lim_{s \to 0} sG_c(s)G(s)$$

$$= \lim_{s \to 0} s \frac{Ts+1}{\alpha Ts+1} G_1(s)$$

$$= \lim_{s \to 0} \frac{s10K}{s(s+1)}$$

$$= 10K = 20$$

or

$$K = 2$$

With $K = 2$, the compensated system will satisfy the steady-state requirement.
Next, we plot the Bode diagram of

$$G_1(s) = \frac{20}{s(s+1)}$$

Section 5–5 / Frequency-Response Approach to Control Systems Compensation   279

MATLAB Program 5–30 produces the Bode diagram shown in Figure 5–53. From this plot, the phase margin is found to be 12.8°. The gain margin is $+\infty$ dB.

MATLAB Program 5–30

```
>> num = [20];
>> den = [1   1   0];
>> w = logspace(-1,2,100);
>> bode(num,den,w)
>> grid
>> title('Bode Diagram of G1(s) = 20/[s(s+1)]')
>> [Gm,Pm,wcp,wcg] = margin(num,den);
>> GmdB = 20*log10(Gm);
>> [GmdB,Pm,wcp,wcg]

ans =

     Inf   12.7580   Inf   4.4165

>> gtext('Phase margin = 12.8 degrees')
>> gtext('Gain margin = Inf dB')
```

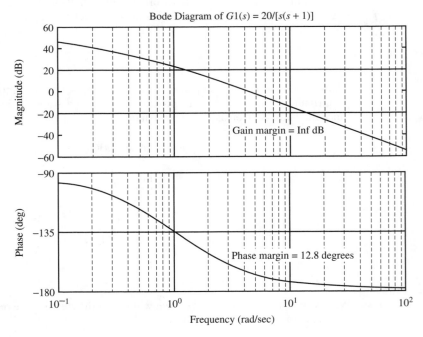

Figure 5–53
Bode diagram of $G_1(s) = 20/[s(s+1)]$.

Since the specification calls for a phase margin of 50°, the additional phase lead necessary to satisfy the phase-margin requirement is 37.2°. A lead compensator can contribute this amount.

Noting that the addition of a lead compensator modifies the magnitude curve in the Bode diagram, we realize that the gain crossover frequency will be shifted to the right. We must offset the increased phase lag of $G_1(j\omega)$ due to this increase in the gain crossover frequency. Taking the shift of the gain crossover frequency into consideration, we may assume that ϕ_m, the maximum phase lead required, is approximately 41°. (This means that approximately 4° has been added to compensate for the shift in the gain crossover frequency.) Since

$$\sin \phi_m = \frac{1-\alpha}{1+\alpha}$$

$\phi_m = 41°$ corresponds to $\alpha = 0.2077$. Note that $\alpha = 0.21$ corresponds to $\phi_m = 40.76°$. Whether we choose $\phi_m = 41°$ or $\phi_m = 40.76°$ does not make much difference in the final solution. Hence, let us choose $\alpha = 0.21$.

Once the attenuation factor α has been determined on the basis of the required phase-lead angle, the next step is to determine the corner frequencies $\omega = 1/T$ and $\omega = 1/(\alpha T)$ of the lead compensator. Notice that the maximum phase-lead angle ϕ_m occurs at the geometric mean of the two corner frequencies, or $\omega = 1/(\sqrt{\alpha}T)$.

The amount of the modification in the magnitude curve at $\omega = 1/(\sqrt{\alpha}T)$ due to the inclusion of the term $(Ts+1)/(\alpha Ts+1)$ is

$$\left|\frac{1+j\omega T}{1+j\omega\alpha T}\right|_{\omega=\frac{1}{\sqrt{\alpha}T}} = \left|\frac{1+j\frac{1}{\sqrt{\alpha}}}{1+j\alpha\frac{1}{\sqrt{\alpha}}}\right| = \frac{1}{\sqrt{\alpha}}$$

Note that

$$\frac{1}{\sqrt{\alpha}} = \frac{1}{\sqrt{0.21}} = 6.7778 \text{ dB}$$

We need to find the frequency point at which the total magnitude becomes 0 dB when the lead compensator is added.

From Figure 5–53, we see that the frequency point where the magnitude of $G_1(j\omega)$ is -6.7778 dB occurs between $\omega = 1$ and 10 rad/sec. Hence, we plot a new Bode diagram of $G_1(j\omega)$ in the frequency range between $\omega = 1$ and 10 to locate the exact point where $G_1(j\omega) = -6.7778$ dB. MATLAB Program 5–31 produces the Bode diagram in this frequency range; the diagram is shown in Figure 5–54. From this diagram, we learn that the

---

MATLAB Program 5–31

```
>> num = [20];
>> den = [1   1   0];
>> w = logspace(0,1,100);
>> bode(num,den,w)
>> grid
>> title('Bode Diagram of G1(s) = 20/[s(s + 1)]')
```

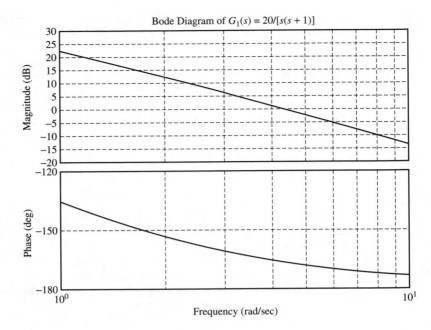

Figure 5-54
Bode diagram of $G_1(s)$.

frequency point where $|G_1(j\omega)| = -6.7778$ dB occurs near $\omega = 6.5$ rad/sec. Let us select this frequency to be the new gain crossover frequency, or $\omega_c = 6.5$ rad/sec. Noting that this frequency corresponds to $1/(\sqrt{\alpha}T)$, or

$$\omega_c = \frac{1}{\sqrt{\alpha}T}$$

we obtain

$$\frac{1}{T} = \omega_c\sqrt{\alpha} = 6.5\sqrt{0.21} = 2.9787$$

and

$$\frac{1}{\alpha T} = \frac{\omega_c}{\sqrt{\alpha}} = \frac{6.5}{\sqrt{0.21}} = 14.1842$$

The lead compensator thus determined is

$$G_c(s) = K_c \frac{s + 2.9787}{s + 14.1842} = K_c \alpha \frac{0.3357s + 1}{0.07050s + 1}$$

where K_c is

$$K_c = \frac{K}{\alpha} = \frac{2}{0.21} = 9.5238$$

282 Chapter 5 / Frequency-Response Analysis

Thus, the transfer function of the compensator becomes

$$G_c(s) = 9.5238 \frac{s + 2.9787}{s + 14.1842} = 2\frac{0.3357s + 1}{0.07050s + 1}$$

MATLAB Program 5–32 produces the Bode diagram of this lead compensator, the diagram is shown in Figure 5–55.

MATLAB Program 5–32

```
>> numc = [9.5238   28.3685];
>> denc = [1   14.1842];
>> w = logspace(-1,3,100);
>> bode(numc,denc,w)
>> grid
>> title('Bode Diagram of Gc(s) = 9.5238(s + 2.9787)/(s + 14.1842)')
```

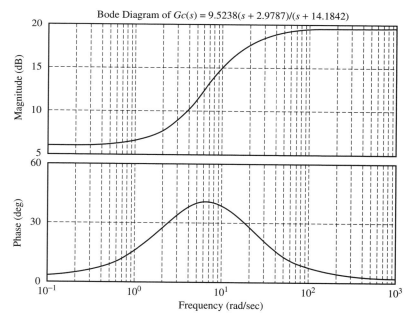

Figure 5–55
Bode diagram of $G_c(s)$.

The open-loop transfer function of the designed system is

$$G_c(s)G(s) = 9.5238 \frac{s + 2.9787}{s + 14.1842} \frac{10}{s(s + 1)}$$

$$= \frac{95.238s + 283.6854}{s^3 + 15.1842s^2 + 14.1842s}$$

Section 5–5 / Frequency-Response Approach to Control Systems Compensation

MATLAB Program 5–33 will produce the Bode diagram of $G_c(s)G(s)$; the diagram itself is shown in Figure 5–56.

MATLAB Program 5–33

```
>> num = [95.238   283.6854];
>> den = [1    15.1842   14.1842   0];
>> w = logspace(-1,3,100);
>> bode(num,den,w)
>> grid
>> title('Bode Diagram of Gc(s)G(s)')
>> [Gm,Pm,wcp,wcg] = margin(num,den);
>> GmdB = 20*log10 (Gm);
>> [GmdB,Pm,wcp,wcg]

ans =
       Inf    49.3687       Inf    6.6027

>> gtext('Gain margin = Inf dB')
>> gtext('Phase margin = 49.4 degrees')
```

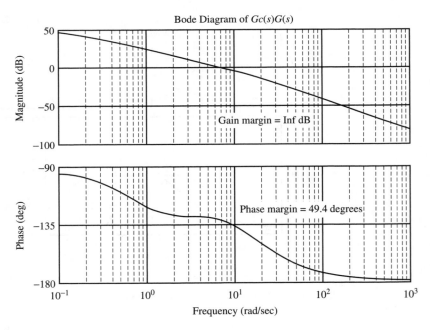

Figure 5–56
Bode diagram of $G_c(s)G(s)$.

Since the phase margin is approximately 50°, the gain margin is $+\infty$ dB, and the static velocity error constant K_v is 20 sec^{-1}, all the specifications are met. Before we conclude this problem, however, we need to check the transient-response characteristics.

Unit-Step Response: We shall compare the unit-step response of the compensated system with that of the original, uncompensated system.

The closed-loop transfer function of the original, uncompensated system is

$$\frac{C(s)}{R(s)} = \frac{10}{s^2 + s + 10}$$

The closed-loop transfer function of the compensated system is

$$\frac{C(s)}{R(s)} = \frac{95.238s + 283.6854}{s^3 + 15.1842s^2 + 109.4222s + 283.6854}$$

MATLAB Program 5–34 produces the unit-step responses of the uncompensated and compensated systems. The resulting response curves are shown in Figure 5–57. Clearly, the compensated system exhibits a satisfactory response. Note that the closed-loop zero and poles are located as follows:

$$\text{Zero at } s = -2.9787$$

$$\text{Poles at } s = -5.2270 \pm j5.7141, \quad s = -4.7303$$

Unit-Ramp Response: It is worthwhile to check the unit-ramp response of the compensated system. Since $K_v = 20 \text{ sec}^{-1}$, the steady-state error following the unit-ramp input will be $1/K_v = 0.05$. The static velocity error constant of the uncompensated system is 10 sec^{-1}. Hence, the original uncompensated system will have twice as large a steady-state error in following the unit-ramp input.

MATLAB Program 5–34

```
>> %*****Unit-step responses*****
>>
>> num1 = [10];
>> den1 = [1   1   10];
>> num2 = [95.238   283.6854];
>> den2 = [1   15.1842   109.4222   283.6854];
>> t = 0:0.01:6;
>> [c1,x1,t] = step(num1,den1,t);
>> [c2,x2,t] = step(num2,den2,t);
>> plot(t,c1,'.',t,c2,'-')
>> grid
>> title('Unit-Step Responses of Compensated and Uncompensated Systems')
>> xlabel('t Sec')
>> ylabel('Outputs')
>> text(1.1,0.5,'Compensated system')
>> text(1.7,1.46,'Uncompensated system')
```

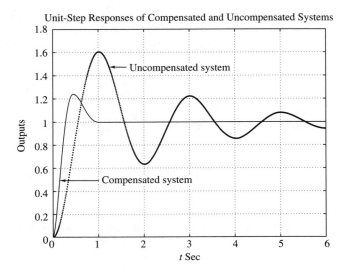

Figure 5-57
Unit-step responses of the compensated and uncompensated systems.

MATLAB Program 5–35 produces the unit-ramp response curves, shown in Figure 5–58. [Note that the unit-ramp response is obtained as the unit-step response of $C(s)/sR(s)$.] The compensated system has a steady-state error equal to one-half that of the original, uncompensated system.

MATLAB Program 5–35

```
>> %*****Unit-ramp responses*****
>>
>> num1 = [10];
>> den1 = [1   1   10   0];
>> num2 = [95.238   283.6854];
>> den2 = [1   15.1842   109.4222   283.6854   0];
>> t = 0:0.01:3;
>> [c1,x1,t] = step(num1,den1,t);
>> [c2,x2,t] = step(num2,den2,t);
>> plot(t,c1,'.',t,c2,'-',t,t,'--')
>> grid
>> title('Unit-Ramp Responses of Compensated and Uncompensated Systems')
>> xlabel('t Sec')
>> ylabel('Outputs')
>> text(0.1,1.3,'Compensated system')
>> text(1.2,0.65,'Uncompensated system')
```

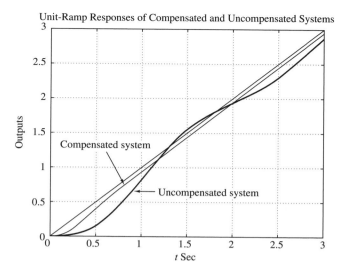

Figure 5–58
Unit-ramp responses of the compensated and uncompensated systems.

Lag compensation. We next illustrate a lag compensation technique with the use of two examples.

EXAMPLE 5–25 Consider the system shown in Figure 5–59. The open-loop transfer function is given by

$$G(s) = \frac{1}{s(s+1)(0.5s+1)}$$

It is desired to compensate the system so that the static velocity error constant K_v is 5 sec^{-1}, the phase margin is at least 40°, and the gain margin is at least 10 dB.

We shall use a lag compensator of the form

$$G_c(s) = K_c \beta \frac{Ts+1}{\beta Ts+1} = K_c \frac{s + \dfrac{1}{T}}{s + \dfrac{1}{\beta T}} \quad (\beta > 1)$$

We define

$$K_c \beta = K$$

and

$$G_1(s) = KG(s) = \frac{K}{s(s+1)(0.5s+1)}$$

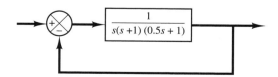

Figure 5–59
Control system.

The first step in the design is to adjust the gain K to meet the required static velocity error constant. Thus,

$$K_v = \lim_{s \to 0} sG_c(s)G(s) = \lim_{s \to 0} s\frac{Ts + 1}{\beta Ts + 1}G_1(s) = \lim_{s \to 0} sG_1(s)$$

$$= \lim_{s \to 0} \frac{sK}{s(s + 1)(0.5s + 1)} = K = 5$$

or

$$K = 5$$

With $K = 5$, the compensated system satisfies the steady-state performance requirement. We next plot the Bode diagram of

$$G_1(j\omega) = \frac{5}{j\omega(j\omega + 1)(0.5j\omega + 1)}$$

The magnitude curve and phase-angle curve of $G_1(j\omega)$ are shown in Figure 5–60. From this plot, the phase margin is found to be $-20°$, which means that the gain-adjusted, but uncompensated, system is unstable.

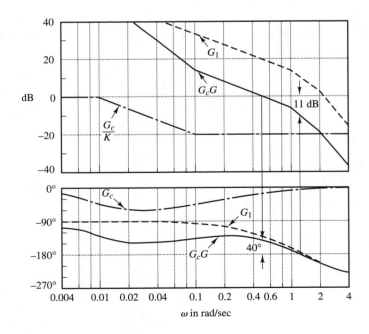

Figure 5–60
Bode diagrams for G_1 (gain-adjusted but uncompensated open-loop transfer function), G_c (compensator), and G_cG (compensated open-loop transfer function).

Noting that the addition of a lag compensator modifies the phase curve of the Bode diagram, we must allow 5° to 12° to the specified phase margin to compensate for the modification of the phase curve. Since the frequency corresponding to a phase margin of 40° is 0.7 rad/sec, the new gain crossover frequency (of the compensated system) must be chosen near this value. To avoid overly large time constants for the lag compensator, we shall choose the corner frequency $\omega = 1/T$ (which corresponds to the zero of the lag compensator) to be 0.1 rad/sec. Since this corner frequency is not too far below the new gain crossover frequency, the modification in the phase curve may not be small. Hence, we add about 12° to the given phase margin as an allowance to account for the lag angle introduced by the lag compensator. The required phase margin is now 52°. The phase angle of the uncompensated open-loop transfer function is −128° at about $\omega = 0.5$ rad/sec. So we choose the new gain crossover frequency to be 0.5 rad/sec. To bring the magnitude curve down to 0 dB at this new gain crossover frequency, the lag compensator must give the necessary attenuation, which in this case is −20 dB. Hence,

$$20 \log \frac{1}{\beta} = -20$$

or

$$\beta = 10$$

The other corner frequency $\omega = 1(\beta T)$, which corresponds to the pole of the lag compensator, is then determined to be

$$\frac{1}{\beta T} = 0.01 \text{ rad/sec}$$

Thus, the transfer function of the lag compensator is

$$G_c(s) = K_c(10)\frac{10s + 1}{100s + 1} = K_c \frac{s + \frac{1}{10}}{s + \frac{1}{100}}$$

Since the gain K was determined to be 5 and β was determined to be 10, we have

$$K_c = \frac{K}{\beta} = \frac{5}{10} = 0.5$$

The open-loop transfer function of the compensated system is

$$G_c(s)G(s) = \frac{5(10s + 1)}{s(100s + 1)(s + 1)(0.5s + 1)}$$

The magnitude and phase-angle curves of $G_c(j\omega)G(j\omega)$ are also shown in Figure 5–60.

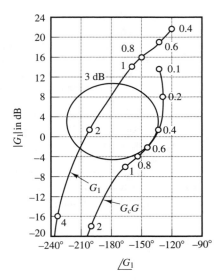

Figure 5–61
Log-magnitude-versus-phase plots of G_1 (gain-adjusted, but uncompensated, open-loop transfer function) and G_cG (compensated open-loop transfer function).

The phase margin of the compensated system is about 40°, which is the required value. The gain margin is about 11 dB, which is quite acceptable. The static velocity error constant is 5 sec^{-1} as required. The compensated system, therefore, satisfies both the steady-state and relative stability requirements.

Note that the new gain crossover frequency is decreased from approximately 1 to 0.5 rad/sec. This means that the bandwidth of the system is reduced.

To further show the effects of lag compensation, the log-magnitude-versus-phase plots of the gain-adjusted, but uncompensated, system $G_1(j\omega)$ and of the compensated system $G_c(j\omega)G(j\omega)$ are shown in Figure 5–61. The plot of $G_1(j\omega)$ clearly shows that the gain-adjusted, but uncompensated, system is unstable. The addition of the lag compensator stabilizes the system. The plot of $G_c(j\omega)G(j\omega)$ is tangent to the $M = 3$ dB locus. Thus, the resonant peak value is 3 dB, or 1.4, and this peak occurs at $\omega = 0.5$ rad/sec.

Compensators designed by different methods or by different designers (even using the same approach) may look sufficiently different. Any of the well-designed systems, however, will give similar transient and steady-state performance. The best among many alternatives may be chosen from the economic consideration that the time constants of the lag compensator should not be too large.

Finally, we shall examine the unit-step response and unit-ramp response of the compensated system and the original, uncompensated system without gain adjustment. The closed-loop transfer functions of the compensated and uncompensated systems are

$$\frac{C(s)}{R(s)} = \frac{50s + 5}{50s^4 + 150.5s^3 + 101.5s^2 + 51s + 5}$$

and

$$\frac{C(s)}{R(s)} = \frac{1}{0.5s^3 + 1.5s^2 + s + 1}$$

respectively. MATLAB Program 5–36 will produce the unit-step and unit-ramp responses of the compensated and uncompensated systems. The resulting unit-step response curves and unit-ramp response curves are shown in Figures 5–62 and 5–63, respectively. From the response curves, we find that the designed system satisfies the given specifications and is satisfactory.

MATLAB Program 5–36

```
>> %*****Unit-step response*****
>>
>> num = [1];
>> den = [0.5   1.5   1   1];
>> numc = [50   5];
>> denc = [50   150.5   101.5   51   5];
>> t = 0:0.1:40;
>> [c1,x1,t] = step(num,den,t);
>> [c2,x2,t] = step(numc,denc,t);
>> plot(t,c1,'.',t,c2,'-')
>> grid
>> title('Unit-Step Responses of Compensated and Uncompensated Systems')
>> xlabel('t Sec')
>> ylabel('Outputs')
>> text(12.7,1.27, 'Compensated system')
>> text(12.2,0.7,'Uncompensated system')
>>
>> %*****Unit-ramp response*****
>>
>> num1 = [1];
>> den1 = [0.5   1.5   1   1   0];
>> num1c = [50   5];
>> den1c = [50   150.5   101.5   51   5   0];
>> t = 0:0.1:20;
>> [y1,z1,t] = step(num1,den1,t);
>> [y2,z2,t] = step(num1c,den1c,t);
>> plot(t,y1,'.',t,y2,'-',t,t,'--');
>> grid
>> title('Unit-Ramp Responses of Compensated and Uncompensated Systems')
>> xlabel('t Sec')
>> ylabel('Outputs')
>> text(8.3,3,'Compensated system')
>> text(8.3,5,'Uncompensated system')
```

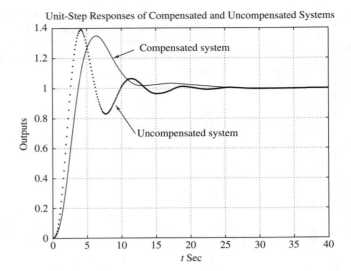

Figure 5–62
Unit-step response curves for the compensated and uncompensated systems (Example 5–25.)

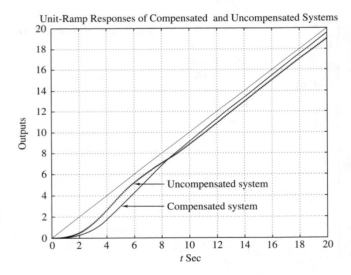

Figure 5–63
Unit-ramp response curves for the compensated and uncompensated systems (Example 5–25.)

Note that the zero and poles of the designed closed-loop system are as follows:

$$\text{Zero at } s = -0.1$$
$$\text{Poles at } s = -0.2859 \pm j0.5196, \quad s = -0.1228, \quad s = -2.3155$$

The dominant closed-loop poles are very close to the $j\omega$-axis, with the result that the response is slow. Also, a pair consisting of the closed-loop pole at $s = -0.1228$ and the zero at $s = -0.1$ produces a slowly decreasing tail of small amplitude.

EXAMPLE 5–26 Consider the system shown in Figure 5–64. Design a compensator such that the static velocity error constant is 4 sec^{-1}, the phase margin is 55°, and the gain margin is 10 dB or more. Obtain the unit-step and unit-ramp response curves of the compensated system with MATLAB.

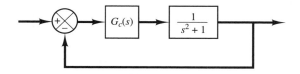

Figure 5–64
Control system.

Since the plant does not have an integrator, it is necessary to add an integrator to the compensator. Let us choose the compensator to be

$$G_c(s) = \frac{K}{s}\hat{G}_c(s)$$

where

$$\lim_{s \to 0}\hat{G}_c(s) = 1$$

[$\hat{G}_c(s)$ is to be determined later.] Since the static velocity error constant K_v is specified as 4 sec^{-1}, we have

$$K_v = \lim_{s \to 0} sG_c(s)\frac{1}{s^2 + 1} = \lim_{s \to 0} s\frac{K}{s}\hat{G}_c(s)\frac{1}{s^2 + 1}$$
$$= K = 4$$

Thus,

$$G_c(s) = \frac{4}{s}\hat{G}_c(s)$$

Next, we plot a Bode diagram of

$$G(s) = \frac{4}{s(s^2 + 1)}$$

MATLAB Program 5–37 produces a Bode diagram of $G(s)$. The resulting Bode diagram is shown in Figure 5–65.

MATLAB Program 5–37

```
>> num = [4];
>> den = [1   0.00000000001   1   0];
>> w = logspace(-1,1,200);
>> bode(num,den,w);
>> grid
>> title('Bode Diagram of 4/[s(s^2+1)]')
```

We need a phase margin of 55° and gain margin of 10 dB or more. From the Bode diagram of Figure 5–65, we notice that the gain crossover frequency is approximately $\omega = 1.8$ rad/sec. Let us assume that the gain crossover frequency of the compensated system is somewhere between $\omega = 1$ and $\omega = 10$ rad/sec.

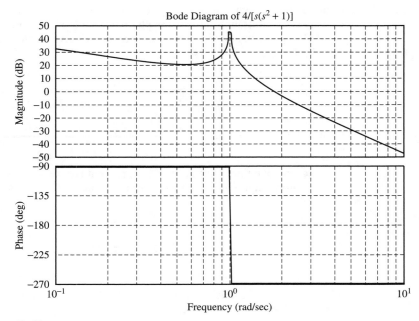

Figure 5-65
Bode diagram of $4/[s(s^2 + 1)]$.

We choose $\hat{G}_c(s)$ to be

$$\hat{G}_c(s) = (as + 1)(bs + 1)$$

and select $a = 5$. Then $(as + 1)$ will contribute up to 90° phase lead in the high-frequency region. MATLAB Program 5-38 produces the Bode diagram of

$$\frac{4(5s + 1)}{s(s^2 + 1)}$$

The resulting Bode diagram is shown in Figure 5-66.

MATLAB Program 5-38

```
>> num = [20   4];
>> den = [1   0.00000000001   1   0];
>> w = logspace(-2,1,101);
>> bode(num,den,w);
>> grid
>> title('Bode Diagram of G(s) = 4(5s+1)/[s(s^2+1)]')
```

On the basis of the Bode diagram of Figure 5-66, we choose the value of b. The term $(bs + 1)$ needs to give a phase margin of 55°. By simple MATLAB trials, we find that $b = 0.25$ gives the phase margin of 55° and a gain margin of $+\infty$ dB. Therefore, choosing $b = 0.25$, we have

$$\hat{G}_c(s) = (5s + 1)(0.25s + 1)$$

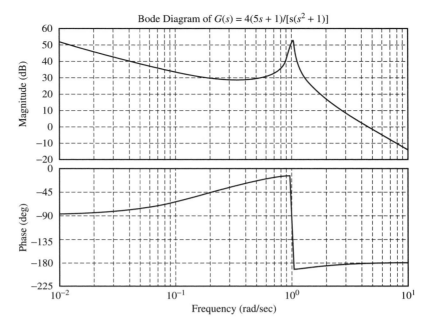

Figure 5–66
Bode diagram of $G(s) = 4(5s + 1)/[s(s^2 + 1)]$.

and the open-loop transfer function of the designed system becomes

$$\text{Open-loop transfer function} = \frac{4(5s + 1)(0.25s + 1)}{s} \frac{1}{s^2 + 1}$$

$$= \frac{5s^2 + 21s + 4}{s^3 + s}$$

Note that the controller designed here is a PID controller. MATLAB Program 5–39 produces the Bode diagram of the open-loop transfer function. The resulting Bode diagram is shown in Figure 5–67. From the designed open-loop transfer function, we see that the static velocity error constant is 4 sec^{-1}. From the Bode diagram we see that the phase margin is 55° and the gain margin is $+\infty$ dB. Therefore, the designed system satisfies all the requirements. Thus, the designed system is acceptable. (Note that there exist infinitely many systems that satisfy all the requirements; the system we have designed is just one of them.)

MATLAB Program 5–39

```
>> num = [5   21   4];
>> den = [1   0   1   0];
>> w = logspace(-2,2,100);
>> bode(num,den,w);
>> grid
>> title('Bode Diagram of 4(5s+1)(0.25s+1)/[s(s^2+1)]')
```

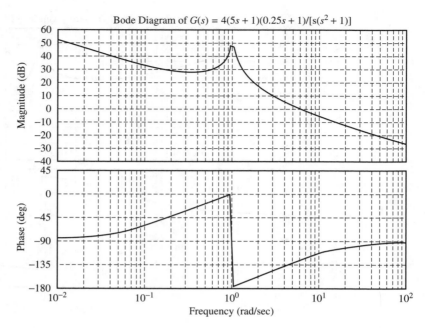

Figure 5–67
Bode diagram of $4(5s + 1)(0.25s + 1)/[s(s^2 + 1)]$.

Next, we shall obtain the unit-step response and the unit-ramp response of the designed system. The closed-loop transfer function is

$$\frac{C(s)}{R(s)} = \frac{5s^2 + 21s + 4}{s^3 + 5s^2 + 22s + 4}$$

Note that the closed-loop zeros are located at

$$s = -4, \quad s = -0.2$$

The closed-loop poles are located at

$$s = -2.4052 + j3.9119$$
$$s = -2.4052 - j3.9119$$
$$s = -0.1897$$

Notice that the complex-conjugate closed-loop poles have a damping ratio of 0.5237. MATLAB Program 5–40 produces the unit-step response and the unit-ramp response. The resulting unit-step response curve is shown in Figure 5–68 and the unit-ramp response curve in Figure 5–69. Notice that the closed-loop pole at $s = -0.1897$ and the zero at $s = -0.2$ produce a long tail of small amplitude in the unit-step response.

MATLAB Program 5–40

```
>> %*****Unit-step response*****
>>
>> num = [5   21   4];
>> den = [1   5   22   4];
>> t = 0:0.01:14;
>> c = step(num,den,t);
>> plot(t,c)
>> grid
>> title('Unit-Step Response of Compensated System')
>> xlabel('t (sec)')
>> ylabel('Output c(t)')
>>
>> %*****Unit-ramp response*****
>>
>> num1 = [5   21   4];
>> den1 = [1   5   22   4   0];
>> t = 0:0.02:20;
>> c = step(num1,den1,t);
>> plot(t,c,'-',t,t,'--'); grid
>> title('Unit-Ramp Response of Compensated System')
>> xlabel('t (sec)')
>> ylabel('Unit-Ramp Input and Output c(t)')
>> text(10.7,8,'Compensated System')
```

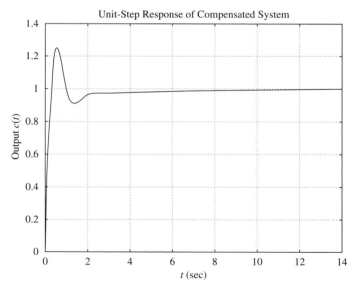

Figure 5–68
Unit-step response curve.

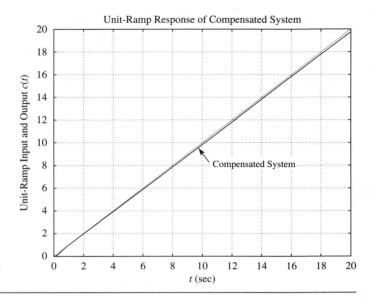

Figure 5–69
Unit-ramp input and the output curve.

Lag–lead compensation. We present a lag–lead compensation technique with the use of an example.

EXAMPLE 5-27 Consider the unity-feedback system whose open-loop transfer function is

$$G(s) = \frac{K}{s(s+1)(s+4)}$$

Design a lag–lead compensator $G_c(s)$ such that the static velocity error constant is 10 sec^{-1}, the phase margin is 50°, and the gain margin is 10 dB or more.

We shall design a lag–lead compensator of the form

$$G_c(s) = K_c \frac{\left(s + \dfrac{1}{T_1}\right)\left(s + \dfrac{1}{T_2}\right)}{\left(s + \dfrac{\beta}{T_1}\right)\left(s + \dfrac{1}{\beta T_2}\right)}$$

Then the open-loop transfer function of the compensated system is $G_c(s)G(s)$. Since the gain K of the plant is adjustable, let us assume that $K_c = 1$. Then $\lim_{s \to 0} G_c(s) = 1$. From the requirement on the static velocity error constant, we obtain

$$K_v = \lim_{s \to 0} sG_c(s)G(s) = \lim_{s \to 0} sG_c(s)\frac{K}{s(s+1)(s+4)}$$

$$= \frac{K}{4} = 10$$

Hence,

$$K = 40$$

We first plot a Bode diagram of the uncompensated system with $K = 40$. MATLAB Program 5–41 plots this Bode diagram. The diagram obtained is shown in Figure 5–70.

MATLAB Program 5–41

```
>> num = [40];
>> den = [1   5   4   0];
>> w = logspace (−1,1,100);
>> bode (num,den,w)
>> grid
>> title('Bode Diagram of G(s) = 40/[s(s+1)(s+4)]')
>> [Gm,Pm,wcp,wcg] = margin (num,den);
Warning: The closed-loop system is unstable.
>> GmdB = 20*log10(Gm);
>> [GmdB, Pm, wcp, wcg]

ans =

    −6.0206   −15.0110    2.0000    2.7797

>> gtext ('Gain margin = −6.02 dB')
>> gtext ('Phase margin = −15.01 degrees')
```

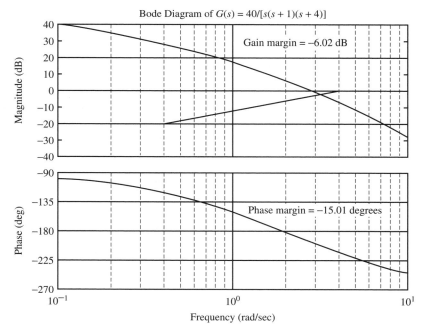

Figure 5–70
Bode diagram of $G(s) = 40/[s(s + 1)(s + 4)]$.

From Figure 5–70, the phase margin of the gain-adjusted, but uncompensated, system is found to be $-15°$, which indicates that this system is unstable. The next step in the design of a lag–lead compensator is to choose a new gain crossover frequency. From the phase-angle curve for $G(j\omega)$, we notice that the phase crossover frequency is $\omega = 2$ rad/sec. We may choose the new gain crossover frequency to be 2 rad/sec so that the phase-lead angle required at $\omega = 2$ rad/sec is about $50°$. A single lag–lead compensator can provide this amount of phase-lead angle quite easily.

Once we choose the gain crossover frequency to be 2 rad/sec, we can determine the corner frequencies of the phase-lag portion of the lag–lead compensator. Let us choose the corner frequency $\omega = 1/T_2$ (which corresponds to the zero of the phase-lag portion of the compensator) to be 1 decade below the new gain crossover frequency, or $\omega = 0.2$ rad/sec. For another corner frequency $\omega = 1/(\beta T_2)$, we need the value of β, which can be determined from a consideration of the lead portion of the compensator, as shown next.

For the lead compensator, the maximum phase-lead angle ϕ_m is given by

$$\sin \phi_m = \frac{\beta - 1}{\beta + 1}$$

Notice that $\beta = 10$ corresponds to $\phi_m = 54.9°$. Since we need a $50°$ phase margin, we may choose $\beta = 10$. (We will be using several degrees less than the maximum angle, $54.9°$.) Thus, we choose

$$\beta = 10$$

Then the corner frequency $\omega = 1/(\beta T_2)$ (which corresponds to the pole of the phase-lag portion of the compensator) becomes

$$\omega = 0.02$$

The transfer function of the phase-lag portion of the lag–lead compensator becomes

$$\frac{s + 0.2}{s + 0.02} = 10\left(\frac{5s + 1}{50s + 1}\right)$$

The phase-lead portion can be determined as follows: Since the new gain crossover frequency is $\omega = 2$ rad/sec, from Figure 5–70, $|G(j2)|$ is found to be 6 dB. Hence, if the lag–lead compensator contributes -6 dB at $\omega = 2$ rad/sec, then the new gain crossover frequency is as desired. From this requirement, it is possible to draw a straight line of slope 20 dB/decade passing through the point (2 rad/sec, -6 dB). (Such a line has been manually drawn in Figure 5–70.) The intersections of this line and the 0-dB line and -20-dB line determine the corner frequencies. From this consideration, the corner frequencies for the lead portion can be determined as $\omega = 0.4$ rad/sec and $\omega = 4$ rad/sec. Thus, the transfer function of the lead portion of the lag–lead compensator becomes

$$\frac{s + 0.4}{s + 4} = \frac{1}{10}\left(\frac{2.5s + 1}{0.25s + 1}\right)$$

Combining the transfer functions of the lag and lead portions of the compensator, we obtain the transfer function $G_c(s)$ of the lag–lead compensator. Since we chose $K_c = 1$, we have

$$G_c(s) = \frac{s + 0.4}{s + 4}\frac{s + 0.2}{s + 0.02} = \frac{(2.5s + 1)(5s + 1)}{(0.25s + 1)(50s + 1)}$$

The Bode diagram of the lag–lead compensator $G_c(s)$ can be obtained by entering MATLAB Program 5–42 into the computer. The resulting plot is shown in Figure 5–71.

MATLAB Program 5–42

```
>> numc = [1   0.6   0.08];
>> denc = [1   4.02   0.08];
>> bode(numc,denc);
>> grid
>> title('Bode Diagram of Lag–Lead Compensator')
```

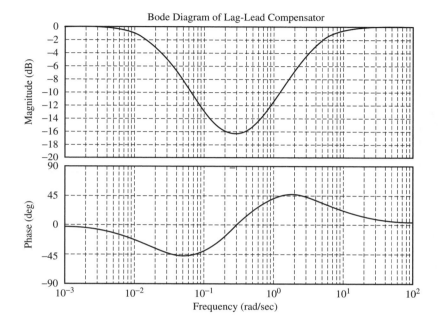

Figure 5–71
Bode diagram of the designed lag–lead compensator.

The open-loop transfer function of the compensated system is

$$G_c(s)G(s) = \frac{(s + 0.4)(s + 0.2)}{(s + 4)(s + 0.02)} \frac{40}{s(s + 1)(s + 4)}$$

$$= \frac{40s^2 + 24s + 3.2}{s^5 + 9.02s^4 + 24.18s^3 + 16.48s^2 + 0.32s}$$

Using MATLAB Program 5–43, we obtain the magnitude and phase-angle curves of the designed open-loop transfer function $G_c(s)G(s)$ as shown in Figure 5–72. Note that the denominator polynomial den 1 was obtained with the conv command, as follows:

```
>> a = [1   4.02   0.08];
>> b = [1   5   4   0];
>> conv(a,b)

ans =

    1.0000    9.0200    24.1800    16.4800    0.320000    0
```

MATLAB Program 5–43

```
>> num1 = [40   24   3.2];
>> den1 = [1   9.02   24.18   16.48   0.32   0];
>> bode(num1,den1)
>> grid
>> title('Bode Diagram of Gc(s)G(s)');
>> [Gm,Pm,wcp,wcg] = margin(num1,den1);
>> GmdB = 20*log10(Gm);
>> [GmdB,Pm,wcp,wcg]

ans =

    11.9189   50.6797   4.3869   1.8618

>> gtext('Gain margin = 11.92 dB')
>> gtext('Phase margin = 50.68 degrees')
```

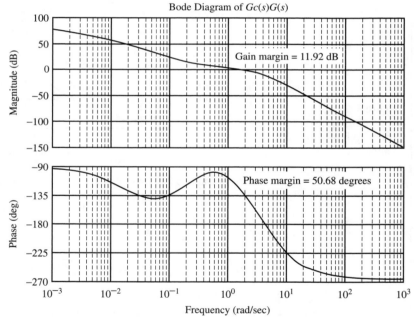

Figure 5–72
Bode diagram of the open-loop transfer function $G_c(s)\,G(s)$ of the compensated system.

Since the phase margin of the compensated system is 50.68°, the gain margin is 11.92 dB, and the static velocity error constant is 10 sec^{-1}, all the requirements are met.

We next investigate the transient-response characteristics of the designed system.

Unit-Step Response: Noting that

$$G_c(s)G(s) = \frac{40(s + 0.4)(s + 0.2)}{(s + 4)(s + 0.02)s(s + 1)(s + 4)}$$

we have

$$\frac{C(s)}{R(s)} = \frac{G_c(s)G(s)}{1 + G_c(s)G(s)}$$

$$= \frac{40(s + 0.4)(s + 0.2)}{(s + 4)(s + 0.02)s(s + 1)(s + 4) + 40(s + 0.4)(s + 0.2)}$$

To determine the denominator polynomial with MATLAB, we may proceed as follows: We define

$$a(s) = (s + 4)(s + 0.02) = s^2 + 4.02s + 0.08$$
$$b(s) = s(s + 1)(s + 4) = s^3 + 5s^2 + 4s$$
$$c(s) = 40(s + 0.4)(s + 0.2) = 40s^2 + 24s + 3.2$$

Then we have

a = [1 4.02 0.08]
b = [1 5 4 0]
c = [40 24 3.2]

Using the following MATLAB program, we obtain the denominator polynomial:

```
>> a = [1   4.02   0.08];
>> b = [1   5   4   0];
>> c = [40   24   3.2];
>> p = [conv(a,b)] + [0   0   0   c]
p =
    1.0000   9.0200   24.1800   56.4800   24.3200   3.2000
```

MATLAB Program 5–44 is used to obtain the unit-step response of the compensated system. The resulting unit-step response curve is shown in Figure 5–73. (Note that the gain-adjusted but uncompensated system is unstable.)

MATLAB Program 5–44

```
>> %*****Unit-step response****
>>
>> num = [40   24   3.2];
>> den = [1   9.02   24.18   56.48   24.32   3.2];
>> t = 0:0.2:40;
>> step(num,den,t)
>> grid
>> title('Unit-Step Response of Compensated System')
```

Unit-Ramp Response: The unit-ramp response of the system may be obtained by entering MATLAB Program 5–45 into the computer. Here, we converted the unit-ramp response of $G_cG/(1 + G_cG)$ into the unit-step response of $G_cG/[s(1 + G_cG)]$. The unit-ramp response curve obtained with this program is shown in Figure 5–74.

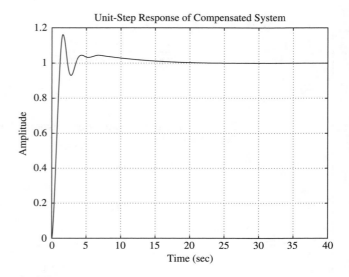

Figure 5–73
Unit-step response curve of the compensated system.

MATLAB Program 5–45

```
>> %*****Unit-ramp response*****
>>
>> num = [40   24   3.2];
>> den = [1   9.02   24.18   56.48   24.32   3.2   0];
>> t = 0:0.05:20;
>> c = step(num,den,t);
>> plot(t,c,'-',t,t,'.')
>> grid
>> title('Unit-Ramp Response of Compensated System')
>> xlabel('Time (sec)')
>> ylabel('Unit-Ramp Input and Output c(t)')
```

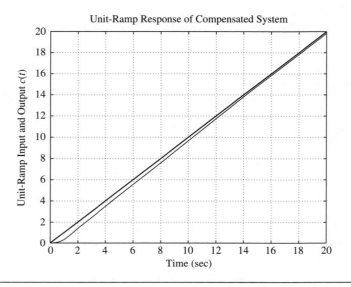

Figure 5–74
Unit-ramp response of the compensated system.

MATLAB Approach to the State-Space Design of Control Systems

6-1 INTRODUCTION

This chapter discusses subjects such as controllability, observability, pole placement, and state observers. MATLAB is useful in checking the controllability and observability conditions of any linear, time-invariant systems. MATLAB is also useful in designing observers and observer controllers. (Quadratic optimal control problems and the design of such systems with MATLAB are presented in Chapter 7.)

6-2 CONTROLLABILITY AND OBSERVABILITY

A system is said to be *controllable at time t_0* if it is possible to transfer the system from any initial state $\mathbf{x}(t_0)$ to any other state in a finite interval of time by means of an unconstrained control vector.

A system is said to be *observable at time t_0* if it is possible to determine the state $\mathbf{x}(t_0)$ of the system from the observation of the output over a finite time interval.

The concepts of controllability and observability were introduced by Kalman. They play an important role in the design of control systems in state space. In fact, the conditions of controllability and observability may govern the existence of a complete solution to the control system design problem. Although most physical systems are controllable and observable, corresponding mathematical models may not possess the properties of controllability and observability. Then it is necessary to know the conditions under which the system is controllable and observable. This section primarily deals with controllability and observability.

In what follows, we shall first derive the condition for complete state controllability followed by discussions of the concept of stabilizability. Then we present the conditions for observability followed by the concept of detectability.

Controllability. Consider the system

$$\dot{\mathbf{x}} = \mathbf{A}\mathbf{x} + \mathbf{B}u \tag{6-1}$$

where $\mathbf{x}$ = state vector (n-vector)
u = control signal (scalar)
$\mathbf{A}$ = $n \times n$ matrix
$\mathbf{B}$ = $n \times 1$ matrix

The system described by Equation (6–1) is said to be *state controllable at* $t = t_0$ if it is possible to construct an unconstrained control signal that will transfer an initial state to any final state in a finite time interval $t_0 \le t \le t_1$. If *every* state is controllable, then the system is said to be *completely state controllable*.

We shall now derive the condition for complete state controllability. Without loss of generality, we can assume that the final state is the origin of the state space and that the initial time is zero, or $t_0 = 0$.

The solution of Equation (6–1) is

$$\mathbf{x}(t) = e^{\mathbf{A}t}\mathbf{x}(0) + \int_0^t e^{\mathbf{A}(t-\tau)}\mathbf{B}u(\tau)\, d\tau$$

Applying the definition of complete state controllability just given, we have

$$\mathbf{x}(t_1) = \mathbf{0} = e^{\mathbf{A}t_1}\mathbf{x}(0) + \int_0^{t_1} e^{\mathbf{A}(t_1-\tau)}\mathbf{B}u(\tau)\, d\tau$$

or

$$\mathbf{x}(0) = -\int_0^{t_1} e^{-\mathbf{A}\tau}\mathbf{B}u(\tau)\, d\tau \tag{6-2}$$

Since $e^{-\mathbf{A}\tau}$ can be written as

$$e^{-\mathbf{A}\tau} = \sum_{k=0}^{n-1} \alpha_k(\tau)\mathbf{A}^k \tag{6-3}*$$

substituting Equation (6–3) into Equation (6–2) gives

$$\mathbf{x}(0) = -\sum_{k=0}^{n-1} \mathbf{A}^k\mathbf{B} \int_0^{t_1} \alpha_k(\tau)u(\tau)\, d\tau \tag{6-4}$$

---

* See Appendix for a derivation of this equation.

Let us put

$$\int_0^{t_1} \alpha_k(\tau) u(\tau) \, d\tau = \beta_k$$

Then Equation (6–4) becomes

$$\mathbf{x}(0) = -\sum_{k=0}^{n-1} \mathbf{A}^k \mathbf{B} \beta_k$$

$$= -[\mathbf{B} \ \vdots \ \mathbf{AB} \ \vdots \ \cdots \ \vdots \ \mathbf{A}^{n-1}\mathbf{B}] \begin{bmatrix} \beta_0 \\ \beta_1 \\ \cdot \\ \cdot \\ \cdot \\ \beta_{n-1} \end{bmatrix} \quad (6\text{–}5)$$

If the system is completely state controllable, then, given any initial state $\mathbf{x}(0)$, Equation (6–5) must be satisfied. This requires that the rank of the $n \times n$ matrix

$$[\mathbf{B} \ \vdots \ \mathbf{AB} \ \vdots \ \cdots \ \vdots \ \mathbf{A}^{n-1}\mathbf{B}]$$

be n.

From the preceding analysis, we can state the condition for complete state controllability as follows: The system given by Equation (6–1) is completely state controllable if and only if the vectors $\mathbf{B}, \mathbf{AB}, \ldots, \mathbf{A}^{n-1}\mathbf{B}$ are linearly independent, or the $n \times n$ matrix

$$[\mathbf{B} \ \vdots \ \mathbf{AB} \ \vdots \ \cdots \ \vdots \ \mathbf{A}^{n-1}\mathbf{B}]$$

is of rank n.

The result just obtained can be extended to the case where the control vector $\mathbf{u}$ is r-dimensional. If the system is described by

$$\dot{\mathbf{x}} = \mathbf{Ax} + \mathbf{Bu}$$

where $\mathbf{u}$ is an r-vector, then it can be proven that the condition for complete state controllability is that the $n \times nr$ matrix

$$[\mathbf{B} \ \vdots \ \mathbf{AB} \ \vdots \ \cdots \ \vdots \ \mathbf{A}^{n-1}\mathbf{B}]$$

be of rank n, or contain n linearly independent column vectors. The matrix

$$[\mathbf{B} \ \vdots \ \mathbf{AB} \ \vdots \ \cdots \ \vdots \ \mathbf{A}^{n-1}\mathbf{B}]$$

is commonly called the *controllability matrix*.

Stabilizability. For a partially controllable system, if the uncontrollable modes are stable and the unstable modes are controllable, then the system is said to be stabilizable. For example, the system defined by

$$\begin{bmatrix} \dot{x}_1 \\ \dot{x}_2 \end{bmatrix} = \begin{bmatrix} 1 & 0 \\ 0 & -1 \end{bmatrix} \begin{bmatrix} x_1 \\ x_2 \end{bmatrix} + \begin{bmatrix} 1 \\ 0 \end{bmatrix} u$$

is not state controllable. The stable mode that corresponds to the eigenvalue -1 is not controllable. The unstable mode that corresponds to the eigenvalue 1 is controllable. Such a system can be made stable by the use of a suitable feedback. Thus, this system is stabilizable.

Observability. Consider the unforced system described by the equations

$$\dot{\mathbf{x}} = \mathbf{A}\mathbf{x} \tag{6-6}$$

$$\mathbf{y} = \mathbf{C}\mathbf{x} \tag{6-7}$$

where $\mathbf{x}$ = state vector (n-vector)
$\mathbf{y}$ = output vector (m-vector)
$\mathbf{A}$ = $n \times n$ matrix
$\mathbf{C}$ = $m \times n$ matrix

The system is said to be *completely observable* if every state $\mathbf{x}(t_0)$ can be determined from the observation of $\mathbf{y}(t)$ over a finite time interval, $t_0 \leq t \leq t_1$. The system is, therefore, completely observable if every transition of the state eventually affects every element of the output vector. The concept of observability is useful in solving the problem of reconstructing unmeasurable state variables from measurable variables in the minimum possible length of time. In this chapter, we treat only linear, time-invariant systems. Therefore, without loss of generality, we can assume that $t_0 = 0$.

The concept of observability is very important because, in practice, the difficulty encountered with state feedback control is that some of the state variables are not accessible for direct measurement, with the result that it becomes necessary to estimate the unmeasurable state variables in order to construct the control signals. It will be shown later that such estimates of state variables are possible if and only if the system is completely observable.

In discussing observability conditions, we consider the unforced system as given by Equations (6–6) and (6–7). The reason for this is as follows: If the system is described by

$$\dot{\mathbf{x}} = \mathbf{A}\mathbf{x} + \mathbf{B}\mathbf{u}$$

$$\mathbf{y} = \mathbf{C}\mathbf{x} + \mathbf{D}\mathbf{u}$$

then

$$\mathbf{x}(t) = e^{\mathbf{A}t}\mathbf{x}(0) + \int_0^t e^{\mathbf{A}(t-\tau)}\mathbf{B}\mathbf{u}(\tau)\,d\tau$$

and

$$\mathbf{y}(t) = \mathbf{C}e^{\mathbf{A}t}\mathbf{x}(0) + \mathbf{C}\int_0^t e^{\mathbf{A}(t-\tau)}\mathbf{B}\mathbf{u}(\tau)\,d\tau + \mathbf{D}\mathbf{u}$$

Since the matrices $\mathbf{A}$, $\mathbf{B}$, $\mathbf{C}$, and $\mathbf{D}$ are known and $\mathbf{u}(t)$ is also known, the last two terms on the right-hand side of this last equation are known quantities. Therefore, they may be subtracted from the observed value of $\mathbf{y}(t)$. Hence, to investigate a necessary and sufficient condition for complete observability, it suffices to consider the system described by Equations (6–6) and (6–7).

Complete observability of continuous-time systems. Consider the system described by Equations (6–6) and (6–7). The output vector $\mathbf{y}(t)$ is

$$\mathbf{y}(t) = \mathbf{C}e^{\mathbf{A}t}\mathbf{x}(0)$$

Since $e^{\mathbf{A}t}$ can be written as

$$e^{\mathbf{A}t} = \sum_{k=0}^{n-1}\alpha_k(t)\mathbf{A}^k$$

we obtain

$$\mathbf{y}(t) = \sum_{k=0}^{n-1}\alpha_k(t)\mathbf{C}\mathbf{A}^k\mathbf{x}(0)$$

or

$$\mathbf{y}(t) = \alpha_0(t)\mathbf{C}\mathbf{x}(0) + \alpha_1(t)\mathbf{C}\mathbf{A}\mathbf{x}(0) + \cdots + \alpha_{n-1}(t)\mathbf{C}\mathbf{A}^{n-1}\mathbf{x}(0) \qquad (6\text{--}8)$$

If the system is completely observable, then, given the output $\mathbf{y}(t)$ over a time interval $0 \le t \le t_1$, $\mathbf{x}(0)$ is uniquely determined from Equation (6–8). It can be shown that this requires the rank of the $nm \times n$ matrix

$$\begin{bmatrix} \mathbf{C} \\ \hline \mathbf{C}\mathbf{A} \\ \hline \cdot \\ \cdot \\ \cdot \\ \hline \mathbf{C}\mathbf{A}^{n-1} \end{bmatrix}$$

to be n.

From the foregoing analysis, we can state the condition for complete observability as follows: The system described by Equations (6–6) and (6–7) is completely observable if and only if the $n \times nm$ matrix

$$[\mathbf{C}^* \vdots \mathbf{A}^*\mathbf{C}^* \vdots \cdots \vdots (\mathbf{A}^*)^{n-1}\mathbf{C}^*]$$

is of rank n or has n linearly independent column vectors. This matrix is called the *observability matrix*.

Detectability. For a partially observable system, if the unobservable modes are stable and the observable modes are unstable, the system is said to be *detectable*. Note that the concept of detectability is dual to the concept of stabilizability.

MATLAB commands for the computation of controllability and observability matrices. MATLAB uses command ctrb for the computation of a controllability matrix and obsv for the computation of an observability matrix, for the system defined by matrices **A**, **B**, **C**, and **D**. Define

$$\text{CONT} = \text{ctrb}(A,B)$$
$$\text{OBSER} = \text{obsv}(A,C)$$

The ranks of CONT and OBSER respectively determine the controllability and observability of the system. If rank(CONT) or rank(OBSER) is less than n, where n is the order of the system, the system is not controllable or not observable, respectively. In terms of transfer functions, if the rank of CONT or OBSER is less than n, there is a cancellation of terms in the numerator and denominator of the transfer function. Whether or not the cancellation occurs between the numerator and denominator of the transfer function, we may use the command minreal(sys) to see if we get a simplified transfer function. (See MATLAB Program 6–1.)

EXAMPLE 6–1 Examine the controllability and observability of the system defined by

$$\dot{\mathbf{x}} = \mathbf{A}\mathbf{x} + \mathbf{B}u$$
$$y = \mathbf{C}\mathbf{x}$$

where

$$\mathbf{A} = \begin{bmatrix} 0 & 1 & 0 \\ 0 & 0 & 1 \\ -6 & -11 & -6 \end{bmatrix} \quad \mathbf{B} = \begin{bmatrix} 0 \\ 0 \\ 1 \end{bmatrix} \quad \mathbf{C} = [5 \quad 6 \quad 1]$$

MATLAB Program 6–1 computes the controllability and observability of this system, which turns out to be controllable, but not observable.

MATLAB Program 6–1

```
>> A = [0   1   0;0   0   1;-6   -11   -6];
>> B = [0;0;1];
>> C = [5   6   1];
>> D = [0];
>> CONT = ctrb(A,B)

CONT =

     0    0    1
     0    1   -6
     1   -6   25

>> rank(CONT)

ans =

     3

>> OBSER = obsv(A,C)

OBSER =

     5    6    1
    -6   -6    0
     0   -6   -6

>> rank(OBSER)

ans =

     2

>> % This system is controllable, but not observable.
>> % This suggests that there is a cancellation of
>> % terms in the numerator and denominator of the
>> % transfer function.
>>
>> [num,den] = ss2tf(A,B,C,D);
>> sys = tf(num,den)

Transfer function:
     s^2 + 6 s + 5
  ---------------------
  s^3 + 6 s^2 + 11 s + 6

>> sys_min = minreal(sys)

Transfer function:
      s + 5
  --------------
   s^2 + 5 s + 6

>> % (s+1) terms in the numerator and denominator of
>> % the transfer function sys are canceled.
```

EXAMPLE 6-2 Consider the system defined by

$$\begin{bmatrix} \dot{x}_1 \\ \dot{x}_2 \end{bmatrix} = \begin{bmatrix} 1 & 0 \\ 0 & -1 \end{bmatrix} \begin{bmatrix} x_1 \\ x_2 \end{bmatrix} + \begin{bmatrix} 1 \\ 0 \end{bmatrix} u$$

$$y = \begin{bmatrix} 1 & 1 \end{bmatrix} \begin{bmatrix} x_1 \\ x_2 \end{bmatrix} + 0u$$

Examine the controllability and observability of the system. MATLAB Program 6–2 demonstrates that the system is not controllable, but may be stabilizable. The system is observable.

MATLAB Program 6–2

```
>> A = [1   0;0  -1];
>> B = [1;0];
>> C = [1  1];
>> D = [0];
>> CONT = ctrb(A,B);
>>      if rank(CONT) == size (A)
            disp('The system is controllable')
         else
            if rank(CONT) == 0
            disp('The system is uncontrollable')
         else
            disp('The system may be stabilizable')
         end
     end

The system may be stabilizable
>>
>> OBSER = obsv(A,C);
>>      if rank(OBSER) == size (A)
            disp('The system is observable.')
         else
            if rank(OBSER) == 0
            disp('The system is unobservable.')
         else
            disp('The system may be detectable.')
         end
     end
The system is observable.
```

6-3 POLE PLACEMENT

In this section, we present a design method commonly called the *pole-placement* or *pole-assignment technique*. We assume that all state variables are measurable and are available for feedback. If the system considered is completely state controllable, then poles of the closed-loop system may be placed at any desired locations by means of state feedback through an appropriate state feedback gain matrix.

The present design technique begins with a determination of the desired closed-loop poles based on the transient-response and/or frequency-response requirements, such as speed, damping ratio, and bandwidth, as well as steady-state requirements.

Let us assume that we decide that the desired closed-loop poles are to be at $s = \mu_1, s = \mu_2, \ldots, s = \mu_n$. By choosing an appropriate gain matrix for state feedback, it is possible to force the system to have closed-loop poles at the desired locations, provided that the original system is completely state controllable.

In this chapter, we limit our discussions to single-input–single-output systems. That is, we assume the control signal $u(t)$ and output signal $y(t)$ to be scalars. In the derivation presented in this section we assume that the reference input $r(t)$ is zero.

Note that when the control signal is a vector quantity, the mathematical aspects of the pole-placement scheme become complicated. We shall not discuss such a case in this book. (When the control signal is a vector quantity, the state feedback gain matrix is not unique: It is possible to choose freely more than n parameters; that is, in addition to being able to place n closed-loop poles properly, we have the freedom to satisfy some or all of the other requirements, if any, of the closed-loop system.)

Design by pole placement. In the conventional approach to the design of a single-input–single-output control system, we design a controller (compensator) such that the dominant closed-loop poles have a desired damping ratio ζ and an undamped natural frequency ω_n. In this approach, the order of the system may be raised by 1 or 2 unless pole–zero cancellation takes place. Also in this approach, we assume that the effects on the responses of nondominant closed-loop poles are negligible.

Different from specifying only dominant closed-loop poles (the conventional design approach), the pole-placement approach specifies all closed-loop poles. (There is a cost associated with placing all closed-loop poles, however, because doing so requires successful measurements of all state variables or else requires the inclusion of a state observer in the system.) There is also the requirement that the system be completely state controllable, in order that the closed-loop poles be placed at arbitrarily chosen locations.

Consider a control system

$$\dot{\mathbf{x}} = \mathbf{A}\mathbf{x} + \mathbf{B}u$$
$$y = \mathbf{C}\mathbf{x} + Du \qquad (6\text{-}9)$$

where **x** = state vector (*n*-vector)
          y = output signal (scalar)
          u = control signal (scalar)
          A = $n \times n$ constant matrix
          B = $n \times 1$ constant matrix
          C = $1 \times n$ constant matrix
          D = constant (scalar)

We shall choose the control signal to be

$$u = -\mathbf{K}\mathbf{x} \qquad (6\text{--}10)$$

This means that the control signal u is determined by an instantaneous state. Such a scheme is called *state feedback*, and the $1 \times n$ matrix **K** is called the *state feedback gain matrix*. We assume that all state variables are available for feedback. In the analysis that follows, we assume that u is unconstrained. A block diagram of this system is shown in Figure 6–1.

This closed-loop system has no input. Its objective is to maintain the zero output. Because of the disturbances that may be present, the output will deviate from zero. The nonzero state will be returned to the zero reference input through the state feedback scheme of the system. Such a system, in which the reference input is always zero, is called a *regulator system*. (Note that if the reference input to the system is always a nonzero constant, the system is also called a regulator system.)

Substituting Equation (6–10) into Equation (6–9) gives

$$\dot{\mathbf{x}}(t) = (\mathbf{A} - \mathbf{B}\mathbf{K})\mathbf{x}(t)$$

The solution of this equation is

$$\mathbf{x}(t) = e^{(\mathbf{A}-\mathbf{B}\mathbf{K})t}\mathbf{x}(0) \qquad (6\text{--}11)$$

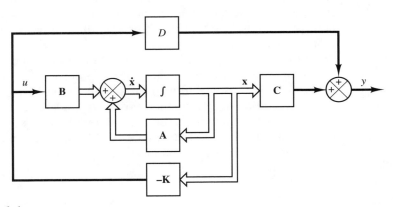

Figure 6–1
Closed-loop control system with $u = -\mathbf{K}\mathbf{x}$.

where $\mathbf{x}(0)$ is the initial state caused by external disturbances. The stability and transient-response characteristics are determined by the eigenvalues of matrix $\mathbf{A} - \mathbf{BK}$. If matrix $\mathbf{K}$ is chosen properly, matrix $\mathbf{A} - \mathbf{BK}$ can be made asymptotically stable, and for all $\mathbf{x}(0) \neq \mathbf{0}$, it is possible to make $\mathbf{x}(t)$ approach $\mathbf{0}$ as t approaches infinity. The eigenvalues of matrix $\mathbf{A} - \mathbf{BK}$ are called the *regulator poles*. If they are placed in the left-half s plane, then $\mathbf{x}(t)$ approaches $\mathbf{0}$ as t approaches infinity. The problem of placing the regulator poles (closed-loop poles) at the desired location is called a *pole-placement problem*.

It can be proven that arbitrary pole placement for a given system is possible if and only if the system is completely state controllable.

Determination of matrix K by direct substitution method. If the system is of low order ($n \leq 3$), the direct substitution of matrix $\mathbf{K}$ into the desired characteristic polynomial may be the simplest approach. For example, if $n = 3$, then we may write the state feedback gain matrix $\mathbf{K}$ as

$$\mathbf{K} = \begin{bmatrix} k_1 & k_2 & k_3 \end{bmatrix}$$

We then substitute this $\mathbf{K}$ matrix into the desired characteristic polynomial $|s\mathbf{I} - \mathbf{A} + \mathbf{BK}|$ and equate it to $(s - \mu_1)(s - \mu_2)(s - \mu_3)$, where $\mu_1, \mu_2,$ and μ_3 are desired closed-loop poles:

$$|s\mathbf{I} - \mathbf{A} + \mathbf{BK}| = (s - \mu_1)(s - \mu_2)(s - \mu_3)$$

Since both sides of this characteristic equation are polynomials in s, it is possible to determine the values of $k_1, k_2,$ and k_3 by equating the coefficients of like powers of s on both sides. This approach is convenient if $n = 2$ or 3. (For $n = 4, 5, 6, \ldots$, the approach may become very tedious.)

Determination of matrix K by Ackermann's formula. There is a well-known formula called Ackermann's formula for the determination of the state feedback gain matrix $\mathbf{K}$. We begin our presentation of this formula with a consideration of the system

$$\dot{\mathbf{x}} = \mathbf{A}\mathbf{x} + \mathbf{B}u$$

where we use the state feedback control $u = -\mathbf{Kx}$. We assume that the system is completely state controllable. We also assume that the desired closed-loop poles are at $s = \mu_1, s = \mu_2, \ldots, s = \mu_n$.

Note that if the system is not completely controllable, matrix $\mathbf{K}$ cannot be determined. (No solution exists.)

Use of the state feedback control

$$u = -\mathbf{Kx}$$

modifies the system equation to

$$\dot{\mathbf{x}} = (\mathbf{A} - \mathbf{BK})\mathbf{x} \quad (6\text{--}12)$$

Let us define

$$\widetilde{\mathbf{A}} = \mathbf{A} - \mathbf{BK}$$

The desired characteristic equation is

$$|s\mathbf{I} - \mathbf{A} + \mathbf{BK}| = |s\mathbf{I} - \widetilde{\mathbf{A}}| = (s - \mu_1)(s - \mu_2)\cdots(s - \mu_n)$$
$$= s^n + \alpha_1 s^{n-1} + \cdots + \alpha_{n-1} s + \alpha_n = 0$$

Since the Cayley–Hamilton theorem states that $\widetilde{\mathbf{A}}$ satisfies its own characteristic equation, we have

$$\phi(\widetilde{\mathbf{A}}) = \widetilde{\mathbf{A}}^n + \alpha_1 \widetilde{\mathbf{A}}^{n-1} + \cdots + \alpha_{n-1}\widetilde{\mathbf{A}} + \alpha_n \mathbf{I} = \mathbf{0} \qquad (6\text{--}13)$$

We shall utilize Equation (6–13) to derive Ackermann's formula. To simplify the derivation, we consider the case where $n = 3$. The derivation can easily be extended to the case of any other positive integer n.

Consider the following identities:

$$\mathbf{I} = \mathbf{I}$$
$$\widetilde{\mathbf{A}} = \mathbf{A} - \mathbf{BK}$$
$$\widetilde{\mathbf{A}}^2 = (\mathbf{A} - \mathbf{BK})^2 = \mathbf{A}^2 - \mathbf{ABK} - \mathbf{BK}\widetilde{\mathbf{A}}$$
$$\widetilde{\mathbf{A}}^3 = (\mathbf{A} - \mathbf{BK})^3 = \mathbf{A}^3 - \mathbf{A}^2\mathbf{BK} - \mathbf{ABK}\widetilde{\mathbf{A}} - \mathbf{BK}\widetilde{\mathbf{A}}^2$$

Multiplying the preceding equations, in order, by $\alpha_3, \alpha_2, \alpha_1,$ and α_0 (where $\alpha_0 = 1$), and adding the results, we obtain

$$\alpha_3 \mathbf{I} + \alpha_2 \widetilde{\mathbf{A}} + \alpha_1 \widetilde{\mathbf{A}}^2 + \widetilde{\mathbf{A}}^3$$
$$= \alpha_3 \mathbf{I} + \alpha_2 (\mathbf{A} - \mathbf{BK}) + \alpha_1 (\mathbf{A}^2 - \mathbf{ABK} - \mathbf{BK}\widetilde{\mathbf{A}}) + \mathbf{A}^3 - \mathbf{A}^2\mathbf{BK}$$
$$-\mathbf{ABK}\widetilde{\mathbf{A}} - \mathbf{BK}\widetilde{\mathbf{A}}^2$$
$$= \alpha_3 \mathbf{I} + \alpha_2 \mathbf{A} + \alpha_1 \mathbf{A}^2 + \mathbf{A}^3 - \alpha_2 \mathbf{BK} - \alpha_1 \mathbf{ABK} - \alpha_1 \mathbf{BK}\widetilde{\mathbf{A}} - \mathbf{A}^2\mathbf{BK}$$
$$-\mathbf{ABK}\widetilde{\mathbf{A}} - \mathbf{BK}\widetilde{\mathbf{A}}^2 \qquad (6\text{--}14)$$

Referring to Equation (6–13), we have

$$\alpha_3 \mathbf{I} + \alpha_2 \widetilde{\mathbf{A}} + \alpha_1 \widetilde{\mathbf{A}}^2 + \widetilde{\mathbf{A}}^3 = \phi(\widetilde{\mathbf{A}}) = \mathbf{0}$$

Also,

$$\alpha_3 \mathbf{I} + \alpha_2 \mathbf{A} + \alpha_1 \mathbf{A}^2 + \mathbf{A}^3 = \phi(\mathbf{A}) \neq \mathbf{0}$$

Substituting the last two equations into Equation (6–14) gives

$$\phi(\widetilde{\mathbf{A}}) = \phi(\mathbf{A}) - \alpha_2 \mathbf{BK} - \alpha_1 \mathbf{BK}\widetilde{\mathbf{A}} - \mathbf{BK}\widetilde{\mathbf{A}}^2 - \alpha_1 \mathbf{ABK} - \mathbf{ABK}\widetilde{\mathbf{A}} - \mathbf{A}^2\mathbf{BK}$$

Since $\phi(\widetilde{\mathbf{A}}) = \mathbf{0}$, we obtain

$$\phi(\mathbf{A}) = \mathbf{B}(\alpha_2 \mathbf{K} + \alpha_1 \mathbf{K}\widetilde{\mathbf{A}} + \mathbf{K}\widetilde{\mathbf{A}}^2) + \mathbf{A}\mathbf{B}(\alpha_1 \mathbf{K} + \mathbf{K}\widetilde{\mathbf{A}}) + \mathbf{A}^2 \mathbf{B}\mathbf{K}$$

$$= [\mathbf{B} \ \vdots \ \mathbf{A}\mathbf{B} \ \vdots \ \mathbf{A}^2 \mathbf{B}] \begin{bmatrix} \alpha_2 \mathbf{K} + \alpha_1 \mathbf{K}\widetilde{\mathbf{A}} + \mathbf{K}\widetilde{\mathbf{A}}^2 \\ \alpha_1 \mathbf{K} + \mathbf{K}\widetilde{\mathbf{A}} \\ \mathbf{K} \end{bmatrix} \quad (6\text{-}15)$$

Because the system is completely state controllable, the inverse of the controllability matrix

$$[\mathbf{B} \ \vdots \ \mathbf{A}\mathbf{B} \ \vdots \ \mathbf{A}^2 \mathbf{B}]$$

exists. Premultiplying both sides of Equation (6–15) by the inverse of the controllability matrix, we obtain

$$[\mathbf{B} \ \vdots \ \mathbf{A}\mathbf{B} \ \vdots \ \mathbf{A}^2 \mathbf{B}]^{-1} \phi(\mathbf{A}) = \begin{bmatrix} \alpha_2 \mathbf{K} + \alpha_1 \mathbf{K}\widetilde{\mathbf{A}} + \mathbf{K}\widetilde{\mathbf{A}}^2 \\ \alpha_1 \mathbf{K} + \mathbf{K}\widetilde{\mathbf{A}} \\ \mathbf{K} \end{bmatrix}$$

Premultiplying both sides of this last equation by $[0 \quad 0 \quad 1]$, we get

$$[0 \ 0 \ 1][\mathbf{B} \ \vdots \ \mathbf{A}\mathbf{B} \ \vdots \ \mathbf{A}^2 \mathbf{B}]^{-1} \phi(\mathbf{A}) = [0 \ 0 \ 1] \begin{bmatrix} \alpha_2 \mathbf{K} + \alpha_1 \mathbf{K}\widetilde{\mathbf{A}} + \mathbf{K}\widetilde{\mathbf{A}}^2 \\ \alpha_1 \mathbf{K} + \mathbf{K}\widetilde{\mathbf{A}} \\ \mathbf{K} \end{bmatrix} = \mathbf{K}$$

which can be rewritten as

$$\mathbf{K} = [0 \ 0 \ 1][\mathbf{B} \ \vdots \ \mathbf{A}\mathbf{B} \ \vdots \ \mathbf{A}^2 \mathbf{B}]^{-1} \phi(\mathbf{A})$$

This last equation gives the required state feedback gain matrix $\mathbf{K}$.

For an arbitrary positive integer n, we have

$$\mathbf{K} = [0 \ 0 \ \cdots \ 0 \ 1][\mathbf{B} \ \vdots \ \mathbf{A}\mathbf{B} \ \vdots \ \cdots \ \vdots \ \mathbf{A}^{n-1}\mathbf{B}]^{-1} \phi(\mathbf{A}) \quad (6\text{-}16)$$

Equation (6–16) is known as *Ackermann's formula for the determination of the state feedback gain matrix* $\mathbf{K}$.

Regulator systems and control systems. Systems that include controllers can be divided into two categories: regulator systems (in which the reference input is constant, including zero) and control systems (in which the reference input is time varying). In what follows, we shall consider regulator systems. Control systems will be treated in Section 6–7.

Choosing the locations of desired closed-loop poles. The first step in the pole-placement design approach is to choose the locations of the desired closed-loop poles. The most frequently used approach is to choose such poles on the basis of experience in root-locus design, placing a dominant pair of closed-loop poles and choosing other poles so that they are far to the left of the dominant closed-loop poles.

Note that if we place the dominant closed-loop poles far from the $j\omega$-axis, so that the system response becomes very fast, the signals in the system become very large, with the result that the system may become nonlinear. This should be avoided.

Another approach is based on the quadratic optimal control approach and determines the desired closed-loop poles such that the system balances between the acceptable response and the amount of control energy required. (See Section 7–2.) Note that requiring a high-speed response implies requiring large amounts of control energy. Also, in general, increasing the speed of the response requires a larger, heavier actuator, which will cost more.

EXAMPLE 6–3 Consider the regulator system shown in Figure 6–2. The plant is given by

$$\dot{\mathbf{x}} = \mathbf{A}\mathbf{x} + \mathbf{B}u$$

where

$$\mathbf{A} = \begin{bmatrix} 0 & 1 & 0 \\ 0 & 0 & 1 \\ -1 & -5 & -6 \end{bmatrix}, \quad \mathbf{B} = \begin{bmatrix} 0 \\ 0 \\ 1 \end{bmatrix}$$

The system uses the state feedback control $u = -\mathbf{K}\mathbf{x}$. Let us choose the desired closed-loop poles at

$$s = -2 + j4, \quad s = -2 - j4, \quad s = -10$$

(We make this choice because we know from experience that such a set of closed-loop poles will result in a reasonable or acceptable transient response.) The desired closed-loop poles determine the desired characteristic equation as follows:

$$(s + 2 - j4)(s + 2 + j4)(s + 10)$$
$$= s^3 + 14s^2 + 60s + 200$$

Determine the state feedback gain matrix $\mathbf{K}$.

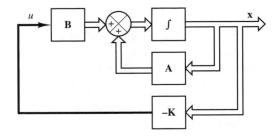

Figure 6–2
Regulator system.

First, we need to check the controllability matrix of the system. Since the controllability matrix **M** is given by

$$\mathbf{M} = [\mathbf{B} \mid \mathbf{AB} \mid \mathbf{A}^2\mathbf{B}] = \begin{bmatrix} 0 & 0 & 1 \\ 0 & 1 & -6 \\ 1 & -6 & 31 \end{bmatrix}$$

we find that $|\mathbf{M}| = -1$, and therefore, rank $(\mathbf{M}) = 3$. Thus, the system is completely state controllable, and arbitrary pole placement is possible.

Next, we solve the problem. We shall demonstrate each of the two methods presented in this chapter.

By defining the desired state feedback gain matrix **K** as

$$\mathbf{K} = \begin{bmatrix} k_1 & k_2 & k_3 \end{bmatrix}$$

and equating $|s\mathbf{I} - \mathbf{A} + \mathbf{BK}|$ with the desired characteristic equation, we obtain

$$|s\mathbf{I} - \mathbf{A} + \mathbf{BK}| = \left| \begin{bmatrix} s & 0 & 0 \\ 0 & s & 0 \\ 0 & 0 & s \end{bmatrix} - \begin{bmatrix} 0 & 1 & 0 \\ 0 & 0 & 1 \\ -1 & -5 & -6 \end{bmatrix} + \begin{bmatrix} 0 \\ 0 \\ 1 \end{bmatrix} \begin{bmatrix} k_1 & k_2 & k_3 \end{bmatrix} \right|$$

$$= \begin{vmatrix} s & -1 & 0 \\ 0 & s & -1 \\ 1 + k_1 & 5 + k_2 & s + 6 + k_3 \end{vmatrix}$$

$$= s^3 + (6 + k_3)s^2 + (5 + k_2)s + 1 + k_1$$

$$= s^3 + 14s^2 + 60s + 200$$

Thus,

$$6 + k_3 = 14, \quad 5 + k_2 = 60, \quad 1 + k_1 = 200$$

from which it follows that

$$k_1 = 199, \quad k_2 = 55, \quad k_3 = 8$$

or

$$\mathbf{K} = \begin{bmatrix} 199 & 55 & 8 \end{bmatrix}$$

The second method is to use Ackermann's formula. From Equation (6–16), we have

$$\mathbf{K} = \begin{bmatrix} 0 & 0 & 1 \end{bmatrix} [\mathbf{B} \mid \mathbf{AB} \mid \mathbf{A}^2\mathbf{B}]^{-1} \phi(\mathbf{A})$$

Since

$$\phi(\mathbf{A}) = \mathbf{A}^3 + 14\mathbf{A}^2 + 60\mathbf{A} + 200\mathbf{I}$$

$$= \begin{bmatrix} 0 & 1 & 0 \\ 0 & 0 & 1 \\ -1 & -5 & -6 \end{bmatrix}^3 + 14 \begin{bmatrix} 0 & 1 & 0 \\ 0 & 0 & 1 \\ -1 & -5 & -6 \end{bmatrix}^2$$

$$+ 60 \begin{bmatrix} 0 & 1 & 0 \\ 0 & 0 & 1 \\ -1 & -5 & -6 \end{bmatrix} + 200 \begin{bmatrix} 1 & 0 & 0 \\ 0 & 1 & 0 \\ 0 & 0 & 1 \end{bmatrix}$$

$$= \begin{bmatrix} 199 & 55 & 8 \\ -8 & 159 & 7 \\ -7 & -43 & 117 \end{bmatrix}$$

and

$$[\mathbf{B} \ \vdots \ \mathbf{AB} \ \vdots \ \mathbf{A}^2\mathbf{B}] = \begin{bmatrix} 0 & 0 & 1 \\ 0 & 1 & -6 \\ 1 & -6 & 31 \end{bmatrix}$$

we obtain

$$\mathbf{K} = \begin{bmatrix} 0 & 0 & 1 \end{bmatrix} \begin{bmatrix} 0 & 0 & 1 \\ 0 & 1 & -6 \\ 1 & -6 & 31 \end{bmatrix}^{-1} \begin{bmatrix} 199 & 55 & 8 \\ -8 & 159 & 7 \\ -7 & -43 & 117 \end{bmatrix}$$

$$= \begin{bmatrix} 0 & 0 & 1 \end{bmatrix} \begin{bmatrix} 5 & 6 & 1 \\ 6 & 1 & 0 \\ 1 & 0 & 0 \end{bmatrix} \begin{bmatrix} 199 & 55 & 8 \\ -8 & 159 & 7 \\ -7 & -43 & 117 \end{bmatrix}$$

$$= \begin{bmatrix} 199 & 55 & 8 \end{bmatrix}$$

As a matter of course, the feedback gain matrix **K** found by the two methods are the same. With this state feedback, the closed-loop poles are placed at $s = -2 \pm j4$ and $s = -10$, as desired.

Comments. It is important to note that the matrix **K** is not unique for a given system, but depends on the locations of the closed-loop poles chosen. (These poles determine the speed and damping of the response.) Note that the selection of the desired closed-loop poles or the desired characteristic equation is a compromise between the rapidity of the response of the error vector and the sensitivity to disturbances and measurement noises. That is, if we increase the speed of error response, then the adverse effects of disturbances and measurement noises generally increase. If the system is of second order, then the system dynamics (response characteristics) can be precisely correlated with the locations of the desired closed-loop poles and the zero(s) of the plant. For higher order systems, the location of the closed-loop poles and the system dynamics (response characteristics) are not easily correlated. Hence, in determining the state feedback gain matrix **K** for a given system, it is desirable to simulate (by computer) the response characteristics of the system for several different matrices **K** (based on several different desired characteristic equations) and to choose the one that gives the best overall system performance.

6–4 SOLVING POLE-PLACEMENT PROBLEMS WITH MATLAB

Pole-placement problems can be solved easily with MATLAB, which has two commands—acker and place—for the computation of the feedback gain matrix **K**. The command acker is based on Ackermann's formula and applies to single-input systems only. The desired closed-loop poles can include multiple poles (poles located at the same place).

If the system involves multiple inputs, then, for a specified set of closed-loop poles, the state-feedback gain matrix **K** is not unique and we have an additional freedom (or freedoms) to choose **K**. There are many approaches to utilize this additional freedom (or freedoms) constructively to determine **K**. One common use is to maximize the stability margin. The pole placement based on this approach is called *robust pole placement,* The MATLAB command for the robust pole placement is place.

Although the command place can be used for both single-input and multiple-input systems, it requires that the multiplicity of poles in the desired closed-loop poles be no greater than the rank of **B**. That is, if matrix **B** is an $n \times 1$ matrix, the command place requires that there be no multiple poles in the set of desired closed-loop poles.

For single-input systems, the commands acker and place yield the same **K**. (For multiple-input systems, one must use the command place instead of acker.)

Note that when the single-input system is barely controllable, some computational problem may occur if the command acker is used. In such a case, the use of the place command is preferred, provided that no multiple poles are involved in the desired set of closed-loop poles.

To use the command acker or place, we first enter the following matrices into the program:

A matrix,   **B** matrix,   **J** matrix

Here, the **J** matrix is the matrix consisting of the desired closed-loop poles such that

$$\mathbf{J} = [\mu_1 \quad \mu_2 \quad \ldots \quad \mu_n]$$

Then we enter

$$K = acker(A,B,J)$$

or

$$K = place(A,B,J)$$

The command eig(A−B*K) may be used to verify that K thus obtained gives the desired eigenvalues.

EXAMPLE 6–4 Consider again the system of Example 6–3. The system equation is

$$\dot{x} = Ax + Bu$$

where

$$A = \begin{bmatrix} 0 & 1 & 0 \\ 0 & 0 & 1 \\ -1 & -5 & -6 \end{bmatrix}, \quad B = \begin{bmatrix} 0 \\ 0 \\ 1 \end{bmatrix}$$

Using state feedback control $u = -Kx$, we desire to have the closed-loop poles at $s = \mu_i \, (i = 1, 2, 3)$, where

$$\mu_1 = -2 + j4, \quad \mu_2 = -2 - j4, \quad \mu_3 = -10$$

Determine the state-feedback gain matrix **K** with MATLAB.

MATLAB programs that generate the matrix **K** are shown in MATLAB Programs 6–3 and 6–4. MATLAB Program 6–3 uses the command acker and MATLAB Program 6–4 uses the command place.

MATLAB Program 6–3

```
>> A = [0  1  0;0  0  1;-1  -5  -6];
>> B = [0;0;1];
>> J = [-2+j*4  -2-j*4  -10];
>> K = acker(A,B,J)

K =

   199    55     8
```

MATLAB Program 6–4

```
>> A = [0  1  0;0  0  1;-1  -5  -6];
>> B = [0;0;1];
>> J = [-2+j*4  -2-j*4  -10];
>> K = place(A,B,J)

K =

   199.0000   55.0000    8.0000
```

EXAMPLE 6–5 Consider the system defined by

$$\begin{bmatrix} \dot{x}_1 \\ \dot{x}_2 \\ \dot{x}_3 \end{bmatrix} = \begin{bmatrix} 0 & 1 & 0 \\ 0 & 0 & 1 \\ -6 & -11 & -6 \end{bmatrix} \begin{bmatrix} x_1 \\ x_2 \\ x_3 \end{bmatrix} + \begin{bmatrix} 0 \\ 0 \\ 10 \end{bmatrix} u$$

$$y = \begin{bmatrix} 1 & 0 & 0 \end{bmatrix} \begin{bmatrix} x_1 \\ x_2 \\ x_3 \end{bmatrix} + 0u$$

Hence,

$$\mathbf{A} = \begin{bmatrix} 0 & 1 & 0 \\ 0 & 0 & 1 \\ -6 & -11 & -6 \end{bmatrix}, \quad \mathbf{B} = \begin{bmatrix} 0 \\ 0 \\ 10 \end{bmatrix}$$

$$\mathbf{C} = \begin{bmatrix} 1 & 0 & 0 \end{bmatrix}, \quad D = [0]$$

Using the state-feedback control $u = -\mathbf{Kx}$, we desire to place the closed-loop poles at

$$s = -2 + j2\sqrt{3}, \quad s = -2 - j2\sqrt{3}, \quad s = -10$$

Obtain the necessary state-feedback gain matrix **K** with MATLAB. (Note that, for the pole placement, matrices **C** and D do not affect the state-feedback gain matrix **K**.)

Two MATLAB programs for obtaining the state-feedback gain matrix **K** are MATLAB Programs 6–5 and 6–6.

MATLAB Program 6–5

```
>> A = [0  1  0;0  0  1;-6  -11  -6];
>> B = [0;0;10];
>> J = [-2+j*2*sqrt(3)   -2-j*2*sqrt(3)   -10];
>> K = acker(A,B,J)

K =

   15.4000    4.5000    0.8000
```

MATLAB Program 6–6

```
>> A = [0  1  0;0  0  1;-6  -11  -6];
>> B = [0;0;10];
>> J = [-2+j*2*sqrt(3)   -2-j*2*sqrt(3)   -10];
>> K = place(A,B,J)

K =

   15.4000    4.5000    0.8000
```

EXAMPLE 6–6 Consider again the system discussed in Example 6–3. We desire that this regulator system have closed-loop poles at

$$s = -2 + j4, \quad s = -2 - j4, \quad s = -10$$

The necessary state feedback gain matrix **K** was obtained in Example 6–3 as follows:

$$\mathbf{K} = \begin{bmatrix} 199 & 55 & 8 \end{bmatrix}$$

Using MATLAB, obtain the response of the system to the following initial condition:

$$\mathbf{x}(0) = \begin{bmatrix} 1 \\ 0 \\ 0 \end{bmatrix}$$

Response to Initial Condition: To obtain the response to the given initial condition $\mathbf{x}(0)$, we substitute $u = -\mathbf{Kx}$ into the plant equation to get

$$\dot{\mathbf{x}} = (\mathbf{A} - \mathbf{BK})\mathbf{x}, \qquad \mathbf{x}(0) = \begin{bmatrix} 1 \\ 0 \\ 0 \end{bmatrix}$$

To plot the response curves (x_1 versus t, x_2 versus t, and x_3 versus t), we may use the command initial. We first define the state-space equations for the system as

$$\dot{\mathbf{x}} = (\mathbf{A} - \mathbf{BK})\mathbf{x} + \mathbf{Iu}$$
$$\mathbf{y} = \mathbf{Ix} + \mathbf{Iu}$$

where we included $\mathbf{u}$ (a three-dimensional input vector). The vector $\mathbf{u}$ is taken to be $\mathbf{0}$ in the computation of the response to the initial condition. Then we define

$$\text{sys} = \text{ss}(\mathbf{A} - \mathbf{BK}, \text{eye}(3), \text{eye}(3), \text{eye}(3))$$

and use the initial command as follows:

$$x = \text{initial}(\text{sys}, [1;0;0], t)$$

In this command, t is the duration we want to use, such as

$$t = 0:0.01:4;$$

Then we obtain x1, x2, and x3 as

$$x1 = [1 \quad 0 \quad 0]*x';$$
$$x2 = [0 \quad 1 \quad 0]*x';$$
$$x3 = [0 \quad 0 \quad 1]*x';$$

and we use the plot command. These commands are included in MATLAB Program 6–7, which plots the response curves shown in Figure 6–3.

MATLAB Program 6–7

```
>> % Response to initial condition:
>>
>> A = [0  1  0;0  0  1;−1  −5  −6];
>> B = [0;0;1];
>> K = [199   55   8];
>> sys = ss(A−B*K,eye(3),eye(3),eye(3));
>> t = 0:0:01:4;
>> x = initial(sys,[1;0;0],t);
>> x1 = [1  0  0]*x';
>> x2 = [0  1  0]*x';
>> x3 = [0  0  1]*x';
>> subplot(3,1,1);plot(t,x1),grid
>> title('Response to Initial Condition')
>> ylabel('state variable x_1')
>> subplot(3,1,2);plot(t,x2),grid
>> ylabel('state variable x_2')
>> subplot(3,1,3);plot(t,x3),grid
>> xlabel('t(sec)')
>> ylabel('state variable x_3')
```

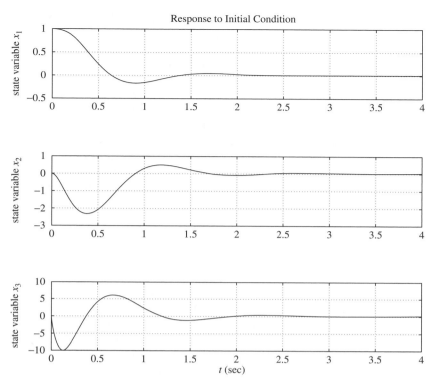

Figure 6–3
Response to initial condition.

6–5 DESIGN OF STATE OBSERVERS WITH MATLAB

In the pole-placement approach to the design of control systems, we assumed that all state variables are available for feedback. In practice, however, not all state variables are so available. Thus, we need to estimate the values of unavailable state variables. This kind of estimation is commonly called *observation*. A device (or a computer program) that estimates or observes the state variables is called a *state observer,* or simply an *observer*. If the state observer observes all state variables of the system, regardless of whether some are available for direct measurement, the state observer is called a *full-order state observer*. There are times when this will not be necessary—when we will need observation of only the unmeasurable state variables, but not of those that are directly measurable as well. For example, since the output variables are observable and are linearly related to the state variables, we need not observe all state variables, but observe only $n - m$ state variables, where n is the dimension of the state vector and m is the dimension of the output vector.

An observer that estimates fewer than n state variables, where n is the dimension of the state vector, is called a *reduced-order state observer* or, simply, a *reduced-order observer*. If the order of the reduced-order state observer is the minimum possible, the observer is called a *minimum-order state observer* or *minimum-order observer*. In this chapter, we shall discuss both the full-order state observer and the minimum-order state observer.

State observer. A state observer estimates the state variables on the basis of measurements of the output and control variables. Here, the concept of observability discussed in Section 6–2 plays an important role. As we shall see later, state observers can be designed if and only if the observability condition is satisfied.

In the subsequent discussions of state observers, we shall use the notation $\tilde{\mathbf{x}}$ to designate the observed state vector. In many practical cases, the observed state vector $\tilde{\mathbf{x}}$ is used in the state feedback to generate the desired control vector.

Consider the plant defined by

$$\dot{\mathbf{x}} = \mathbf{A}\mathbf{x} + \mathbf{B}u \tag{6-17}$$

$$y = \mathbf{C}\mathbf{x} \tag{6-18}$$

The observer is a subsystem used to reconstruct the state vector of the plant. The mathematical model of the observer is basically the same as that of the plant, except that we include an additional term which includes the estimation error to compensate for inaccuracies in matrices **A** and **B** and the absence of the initial error. The estimation error or observation error is the difference between the measured output and the estimated output. The initial error is the difference between the initial state and the estimated initial state. Thus, we define the mathematical model of the observer to be

$$\dot{\tilde{\mathbf{x}}} = \mathbf{A}\tilde{\mathbf{x}} + \mathbf{B}u + \mathbf{K}_e(y - \mathbf{C}\tilde{\mathbf{x}})$$
$$= (\mathbf{A} - \mathbf{K}_e\mathbf{C})\tilde{\mathbf{x}} + \mathbf{B}u + \mathbf{K}_e y \tag{6-19}$$

where $\tilde{\mathbf{x}}$ is the estimated state and $\mathbf{C}\tilde{\mathbf{x}}$ is the estimated output. The inputs to the observer are the output y and the control input u. Matrix $\mathbf{K}_e$, which is called the *observer gain matrix*, is a weighting matrix to the correction term involving the difference between the measured output y and the estimated output $\mathbf{C}\tilde{\mathbf{x}}$. This term continuously corrects the output from the model and improves the performance of the observer. Figure 6–4 shows the block diagram of the system and the full-order state observer.

Full-order state observer. The order of the state observer that will be discussed here is the same as that of the plant. Suppose that the plant is defined by Equations (6–17) and (6–18) and the observer model is defined by Equation (6–19).

To obtain the observer error equation, let us subtract Equation (6–19) from Equation (6–17):

$$\dot{\mathbf{x}} - \dot{\tilde{\mathbf{x}}} = \mathbf{A}\mathbf{x} - \mathbf{A}\tilde{\mathbf{x}} - \mathbf{K}_e(\mathbf{C}\mathbf{x} - \mathbf{C}\tilde{\mathbf{x}})$$
$$= (\mathbf{A} - \mathbf{K}_e\mathbf{C})(\mathbf{x} - \tilde{\mathbf{x}}) \tag{6-20}$$

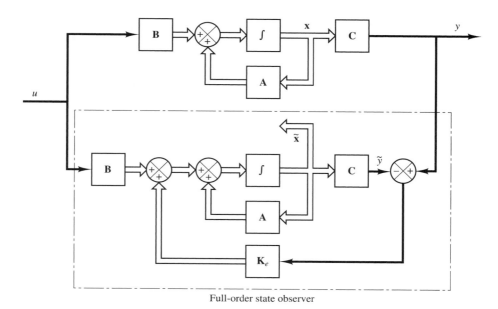

Figure 6–4
Block diagram of system and full-order state observer when input u and output y are scalars.

We define the difference between $\mathbf{x}$ and $\widetilde{\mathbf{x}}$ as the error vector $\mathbf{e}$, or

$$\mathbf{e} = \mathbf{x} - \widetilde{\mathbf{x}}$$

Then Equation (6–20) becomes

$$\dot{\mathbf{e}} = (\mathbf{A} - \mathbf{K}_e \mathbf{C})\mathbf{e} \qquad (6\text{–}21)$$

From Equation (6–21), we see that the dynamic behavior of the error vector is determined by the eigenvalues of matrix $\mathbf{A} - \mathbf{K}_e \mathbf{C}$. If that matrix is a stable matrix, the error vector will converge to zero for any initial error vector $\mathbf{e}(0)$. That is, $\widetilde{\mathbf{x}}(t)$ will converge to $\mathbf{x}(t)$ regardless of the values of $\mathbf{x}(0)$ and $\widetilde{\mathbf{x}}(0)$. If the eigenvalues of matrix $\mathbf{A} - \mathbf{K}_e \mathbf{C}$ are chosen in such a way that the dynamic behavior of the error vector is asymptotically stable and is adequately fast, then any error vector will tend to zero (the origin) with adequate speed.

It can be proved that if the plant is completely observable, then it is possible to choose matrix $\mathbf{K}_e$ such that $\mathbf{A} - \mathbf{K}_e \mathbf{C}$ has arbitrarily desired eigenvalues. That is, the observer gain matrix $\mathbf{K}_e$ can be determined to yield the desired matrix $\mathbf{A} - \mathbf{K}_e \mathbf{C}$. We shall discuss this matter in what follows.

Dual problem. The problem of designing a full-order observer becomes that of determining the observer gain matrix $\mathbf{K}_e$ such that the error dynamics defined by Equation (6–21) are asymptotically stable with sufficient speed of response. (The asymptotic stability and the speed of response of the error dynamics are determined by the eigenvalues of matrix $\mathbf{A} - \mathbf{K}_e\mathbf{C}$.) Hence, the design of the full-order observer becomes that of determining an appropriate $\mathbf{K}_e$ such that $\mathbf{A} - \mathbf{K}_e\mathbf{C}$ has desired eigenvalues. Thus, the problem here becomes the same as the pole-placement problem we discussed in Section 6–3. In fact, the two problems are mathematically the same, a property called *duality*.

Consider the system defined by

$$\dot{\mathbf{x}} = \mathbf{A}\mathbf{x} + \mathbf{B}u$$
$$y = \mathbf{C}\mathbf{x}$$

In designing the full-order state observer, we may solve the dual problem; that is, we may solve the pole-placement problem for the dual system

$$\dot{\mathbf{z}} = \mathbf{A}^*\mathbf{z} + \mathbf{C}^*v$$
$$n = \mathbf{B}^*\mathbf{z}$$

assuming the control signal v to be

$$v = -\mathbf{K}\mathbf{z}$$

If the dual system is completely state controllable, then the state feedback gain matrix $\mathbf{K}$ can be determined such that matrix $\mathbf{A}^* - \mathbf{C}^*\mathbf{K}$ will yield a set of the desired eigenvalues.

If $\mu_1, \mu_2, \ldots, \mu_n$ are the desired eigenvalues of the state observer matrix, then, by taking the same μ_i's as the desired eigenvalues of the state-feedback gain matrix of the dual system, we obtain

$$|s\mathbf{I} - (\mathbf{A}^* - \mathbf{C}^*\mathbf{K})| = (s - \mu_1)(s - \mu_2)\cdots(s - \mu_n)$$

Noting that the eigenvalues of $\mathbf{A}^* - \mathbf{C}^*\mathbf{K}$ and those of $\mathbf{A} - \mathbf{K}^*\mathbf{C}$ are the same, we have

$$|s\mathbf{I} - (\mathbf{A}^* - \mathbf{C}^*\mathbf{K})| = |s\mathbf{I} - (\mathbf{A} - \mathbf{K}^*\mathbf{C})|$$

Comparing the characteristic polynomial $|s\mathbf{I} - (\mathbf{A} - \mathbf{K}^*\mathbf{C})|$ and the characteristic polynomial $|s\mathbf{I} - (\mathbf{A} - \mathbf{K}_e\mathbf{C})|$ for the observer system [see Equation (6–19)], we find that $\mathbf{K}_e$ and $\mathbf{K}^*$ are related by

$$\mathbf{K}_e = \mathbf{K}^*$$

Thus, using the matrix $\mathbf{K}$ determined by the pole-placement approach in the dual system, we can find the observer gain matrix $\mathbf{K}_e$ for the original system from the relationship $\mathbf{K}_e = \mathbf{K}^*$.

Necessary and sufficient condition for state observation. As discussed, a necessary and sufficient condition for the determination of the observer gain matrix $\mathbf{K}_e$ for the desired eigenvalues of $\mathbf{A} - \mathbf{K}_e\mathbf{C}$ is that the dual of the original system

$$\dot{\mathbf{z}} = \mathbf{A}^*\mathbf{z} + \mathbf{C}^*v$$

be completely state controllable. The complete state controllability condition for this dual system is that the rank of

$$[\mathbf{C}^* \mid \mathbf{A}^*\mathbf{C}^* \mid \cdots \mid (\mathbf{A}^*)^{n-1}\mathbf{C}^*]$$

be n. This is the condition for complete observability of the original system defined by Equations (6–17) and (6–18). Accordingly, a necessary and sufficient condition for the observation of the state of the system defined by Equations (6–17) and (6–18) is that the system be completely observable.

Once we select the desired eigenvalues (or desired characteristic equation), the full-order state observer can be designed, provided that the plant is completely observable. The desired eigenvalues of the characteristic equation should be chosen so that the state observer responds at least two to five times faster than the closed-loop system considered. As stated earlier, the equation for the full-order state observer is

$$\dot{\tilde{\mathbf{x}}} = (\mathbf{A} - \mathbf{K}_e\mathbf{C})\tilde{\mathbf{x}} + \mathbf{B}u + \mathbf{K}_e y \qquad (6\text{–}22)$$

Note that thus far we have assumed the matrices $\mathbf{A}$, $\mathbf{B}$, and $\mathbf{C}$ in the observer to be exactly the same as those of the physical plant. If there are discrepancies in $\mathbf{A}$, $\mathbf{B}$, and $\mathbf{C}$ in the observer and in the physical plant, the dynamics of the observer error are no longer governed by Equation (6–21). This means that the error may not approach zero as expected. Therefore, we need to choose $\mathbf{K}_e$ so that the observer is stable and the error remains acceptably small in the presence of small modeling errors.

Direct substitution approach to obtain state observer gain matrix $\mathbf{K}_e$.
As in the case of pole placement, if the system is of low order, then direct substitution of matrix $\mathbf{K}_e$ into the desired characteristic polynomial may be simpler. For example, if $\mathbf{x}$ is a 3-vector, then we write the observer gain matrix $\mathbf{K}_e$ as

$$\mathbf{K}_e = \begin{bmatrix} k_{e1} \\ k_{e2} \\ k_{e3} \end{bmatrix}$$

We substitute this $\mathbf{K}_e$ matrix into the desired characteristic polynomial:

$$|s\mathbf{I} - (\mathbf{A} - \mathbf{K}_e\mathbf{C})| = (s - \mu_1)(s - \mu_2)(s - \mu_3)$$

By equating the coefficients of like powers of s on both sides of this last equation, we can determine the values of k_{e1}, k_{e2}, and k_{e3}. This approach is convenient if $n = 1, 2$, or 3, where n is the dimension of the state vector $\mathbf{x}$. (Although the approach can be used when $n = 4, 5, 6, \ldots$, the computations involved may become tedious.)

Another approach to the determination of the state observer gain matrix $\mathbf{K}_e$ is to use Ackermann's formula. This approach is presented next.

Ackermann's formula. Consider the system defined by

$$\dot{\mathbf{x}} = \mathbf{A}\mathbf{x} + \mathbf{B}u \qquad (6\text{–}23)$$

$$y = \mathbf{C}\mathbf{x} \qquad (6\text{–}24)$$

In Section 6–3, we derived Ackermann's formula for pole placement for the system defined by Equation (6–23). The result was given by Equation (6–16), rewritten thus:

$$\mathbf{K} = \begin{bmatrix} 0 & 0 & \cdots & 0 & 1 \end{bmatrix}[\mathbf{B} \mid \mathbf{AB} \mid \cdots \mid \mathbf{A}^{n-1}\mathbf{B}]^{-1}\phi(\mathbf{A})$$

For the dual of the system defined by Equations (6–23) and (6–24), namely,

$$\dot{\mathbf{z}} = \mathbf{A}^*\mathbf{z} + \mathbf{C}^*v$$
$$n = \mathbf{B}^*\mathbf{z}$$

the preceding Ackermann's formula for pole placement is modified to

$$\mathbf{K} = \begin{bmatrix} 0 & 0 & \cdots & 0 & 1 \end{bmatrix}[\mathbf{C}^* \mid \mathbf{A}^*\mathbf{C}^* \mid \cdots \mid (\mathbf{A}^*)^{n-1}\mathbf{C}^*]^{-1}\phi(\mathbf{A}^*) \quad (6\text{--}25)$$

As stated earlier, the state observer gain matrix $\mathbf{K}_e$ is given by $\mathbf{K}^*$, where $\mathbf{K}$ is found from Equation (6–25). Thus,

$$\mathbf{K}_e = \mathbf{K}^* = \phi(\mathbf{A}^*)^* \begin{bmatrix} \mathbf{C} \\ \mathbf{CA} \\ \cdot \\ \cdot \\ \cdot \\ \mathbf{CA}^{n-2} \\ \mathbf{CA}^{n-1} \end{bmatrix}^{-1} \begin{bmatrix} 0 \\ 0 \\ \cdot \\ \cdot \\ \cdot \\ 0 \\ 1 \end{bmatrix} = \phi(\mathbf{A}) \begin{bmatrix} \mathbf{C} \\ \mathbf{CA} \\ \cdot \\ \cdot \\ \cdot \\ \mathbf{CA}^{n-2} \\ \mathbf{CA}^{n-1} \end{bmatrix}^{-1} \begin{bmatrix} 0 \\ 0 \\ \cdot \\ \cdot \\ \cdot \\ 0 \\ 1 \end{bmatrix} \quad (6\text{--}26)$$

where $\phi(s)$ is the desired characteristic polynomial for the state observer, or

$$\phi(s) = (s - \mu_1)(s - \mu_2)\cdots(s - \mu_n)$$

where $\mu_1, \mu_2, \ldots, \mu_n$ are the desired eigenvalues. Equation (6–26) is called *Ackermann's formula for the determination of the observer gain matrix* $\mathbf{K}_e$.

Comments on selecting the best $\mathbf{K}_e$. Notice in Figure 6–4 that the feedback signal through the observer gain matrix $\mathbf{K}_e$ serves as a correction signal to the plant model to account for the unknowns in the plant. If significant unknowns are involved, the feedback signal through the matrix $\mathbf{K}_e$ should be relatively large. However, if the output signal is contaminated significantly by disturbances and measurement noises, then the output y is not reliable and the feedback signal through the matrix $\mathbf{K}_e$ should be relatively small. In determining the matrix $\mathbf{K}_e$, we should carefully examine the effects of disturbances and noises involved in the output y.

Remember that the observer gain matrix $\mathbf{K}_e$ depends on the desired characteristic equation:

$$(s - \mu_1)(s - \mu_2)\cdots(s - \mu_n) = 0$$

The choice of a set of $\mu_1, \mu_2, \ldots, \mu_n$ is, in many instances, not unique. As a general rule, however, the observer poles must be two to five times faster than the controller poles to make sure that the observation error (the estimation error) converges to zero quickly. This means that the observer estimation error decays two to five times faster than does the state vector **x**. Such faster decay of the observer error compared with the desired dynamics makes the controller poles dominate the system response.

It is important to note that if sensor noise is considerable, we may choose the observer poles to be slower than two times the controller poles, so that the bandwidth of the system will become lower and smooth the noise. In this case, the system response will be strongly influenced by the observer poles. If the observer poles are located to the right of the controller poles in the left-half s plane, the system response will be dominated by the observer poles rather than by the control poles.

In the design of the state observer, it is desirable to determine several observer gain matrices $\mathbf{K}_e$ based on several different desired characteristic equations. For each of the several different matrices $\mathbf{K}_e$, simulation tests must be run to evaluate the resulting system performance. Then we select the best $\mathbf{K}_e$ from the viewpoint of overall system performance. In many practical cases, the selection of the best matrix $\mathbf{K}_e$ boils down to a compromise between speedy response and sensitivity to disturbances and noises.

EXAMPLE 6-7 Consider the system

$$\dot{\mathbf{x}} = \mathbf{A}\mathbf{x} + \mathbf{B}u$$
$$y = \mathbf{C}\mathbf{x}$$

where

$$\mathbf{A} = \begin{bmatrix} 0 & 20.6 \\ 1 & 0 \end{bmatrix}, \quad \mathbf{B} = \begin{bmatrix} 0 \\ 1 \end{bmatrix}, \quad \mathbf{C} = \begin{bmatrix} 0 & 1 \end{bmatrix}$$

We use the observed state feedback such that

$$u = -\mathbf{K}\widetilde{\mathbf{x}}$$

Design a full-order state observer, assuming that the system configuration is identical to that shown in Figure 6–4. Assume that the desired eigenvalues of the observer matrix are

$$\mu_1 = -10, \quad \mu_2 = -10$$

The design of the state observer reduces to the determination of an appropriate observer gain matrix $\mathbf{K}_e$.

Let us examine the observability matrix. The rank of

$$[\mathbf{C}^* \vdots \mathbf{A}^*\mathbf{C}^*] = \begin{bmatrix} 0 & 1 \\ 1 & 0 \end{bmatrix}$$

is 2. Hence, the system is completely observable and the determination of the desired observer gain matrix is possible. We shall solve this problem by two methods.

Method 1: From Equation (6–21), namely,

$$\dot{\mathbf{e}} = (\mathbf{A} - \mathbf{K}_e\mathbf{C})\mathbf{e}$$

the characteristic equation for the observer becomes

$$|s\mathbf{I} - \mathbf{A} + \mathbf{K}_e\mathbf{C}| = 0$$

Define

$$\mathbf{K}_e = \begin{bmatrix} k_{e1} \\ k_{e2} \end{bmatrix}$$

Then the characteristic equation becomes

$$\left| \begin{bmatrix} s & 0 \\ 0 & s \end{bmatrix} - \begin{bmatrix} 0 & 20.6 \\ 1 & 0 \end{bmatrix} + \begin{bmatrix} k_{e1} \\ k_{e2} \end{bmatrix} [0 \quad 1] \right| = \left| \begin{matrix} s & -20.6 + k_{e1} \\ -1 & s + k_{e2} \end{matrix} \right|$$

$$= s^2 + k_{e2}s - 20.6 + k_{e1} = 0 \tag{6–27}$$

Since the desired characteristic equation is

$$s^2 + 20s + 100 = 0$$

by comparing Equation (6–27) with this last equation, we obtain

$$k_{e1} = 120.6, \qquad k_{e2} = 20$$

or

$$\mathbf{K}_e = \begin{bmatrix} 120.6 \\ 20 \end{bmatrix}$$

Method 2: We shall use Ackermann's formula given by Equation (6–26):

$$\mathbf{K}_e = \phi(\mathbf{A}) \begin{bmatrix} \mathbf{C} \\ \mathbf{CA} \end{bmatrix}^{-1} \begin{bmatrix} 0 \\ 1 \end{bmatrix}$$

where

$$\phi(s) = (s - \mu_1)(s - \mu_2) = s^2 + 20s + 100$$

Thus,

$$\phi(\mathbf{A}) = \mathbf{A}^2 + 20\mathbf{A} + 100\mathbf{I}$$

and

$$\mathbf{K}_e = (\mathbf{A}^2 + 20\mathbf{A} + 100\mathbf{I}) \begin{bmatrix} 0 & 1 \\ 1 & 0 \end{bmatrix}^{-1} \begin{bmatrix} 0 \\ 1 \end{bmatrix}$$

$$= \begin{bmatrix} 120.6 & 412 \\ 20 & 120.6 \end{bmatrix} \begin{bmatrix} 0 & 1 \\ 1 & 0 \end{bmatrix} \begin{bmatrix} 0 \\ 1 \end{bmatrix} = \begin{bmatrix} 120.6 \\ 20 \end{bmatrix}$$

As a matter of course, we get the same $\mathbf{K}_e$ regardless of the method employed.

The equation for the full-order state observer is given by Equation (6–19),

$$\dot{\tilde{\mathbf{x}}} = (\mathbf{A} - \mathbf{K}_e\mathbf{C})\tilde{\mathbf{x}} + \mathbf{B}u + \mathbf{K}_e y$$

or

$$\begin{bmatrix} \dot{\tilde{x}}_1 \\ \dot{\tilde{x}}_2 \end{bmatrix} = \begin{bmatrix} 0 & -100 \\ 1 & -20 \end{bmatrix}\begin{bmatrix} \tilde{x}_1 \\ \tilde{x}_2 \end{bmatrix} + \begin{bmatrix} 0 \\ 1 \end{bmatrix}u + \begin{bmatrix} 120.6 \\ 20 \end{bmatrix}y$$

Effects of the addition of the observer on a closed-loop system. In the pole-placement design process, we assumed that the actual state $\mathbf{x}(t)$ was available for feedback. In practice, however, the actual state $\mathbf{x}(t)$ may not be measurable, so we will need to design an observer and use the observed state $\tilde{\mathbf{x}}(t)$ for feedback as shown in Figure 6–5. The design process, therefore, becomes a two-stage process, the first stage being the determination of the feedback gain matrix $\mathbf{K}$ to yield the desired characteristic equation and the second stage being the determination of the observer gain matrix $\mathbf{K}_e$ to yield the desired observer characteristic equation.

Let us now investigate the effects of the use of the observed state $\tilde{\mathbf{x}}(t)$, rather than the actual state $\mathbf{x}(t)$, on the characteristic equation of a closed-loop control system.

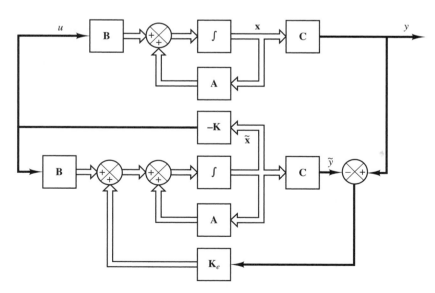

Figure 6–5
Observed-state feedback control system.

Consider the completely state controllable and completely observable system defined by the equations

$$\dot{\mathbf{x}} = \mathbf{Ax} + \mathbf{B}u$$
$$y = \mathbf{Cx}$$

For the state-feedback control based on the observed state $\tilde{\mathbf{x}}$,

$$u = -\mathbf{K}\tilde{\mathbf{x}}$$

With this control, the state equation becomes

$$\dot{\mathbf{x}} = \mathbf{Ax} - \mathbf{BK}\tilde{\mathbf{x}} = (\mathbf{A} - \mathbf{BK})\mathbf{x} + \mathbf{BK}(\mathbf{x} - \tilde{\mathbf{x}}) \tag{6-28}$$

The difference between the actual state $\mathbf{x}(t)$ and the observed state $\tilde{\mathbf{x}}(t)$ has been defined as the error $\mathbf{e}(t)$:

$$\mathbf{e}(t) = \mathbf{x}(t) - \tilde{\mathbf{x}}(t)$$

Substitution of the error vector $\mathbf{e}(t)$ into Equation (6–28) gives

$$\dot{\mathbf{x}} = (\mathbf{A} - \mathbf{BK})\mathbf{x} + \mathbf{BK}\mathbf{e} \tag{6-29}$$

Note that the observer error equation was given by Equation (6–21), repeated here:

$$\dot{\mathbf{e}} = (\mathbf{A} - \mathbf{K}_e\mathbf{C})\mathbf{e} \tag{6-30}$$

Combining Equations (6–29) and (6–30), we obtain

$$\begin{bmatrix} \dot{\mathbf{x}} \\ \dot{\mathbf{e}} \end{bmatrix} = \begin{bmatrix} \mathbf{A} - \mathbf{BK} & \mathbf{BK} \\ \mathbf{0} & \mathbf{A} - \mathbf{K}_e\mathbf{C} \end{bmatrix} \begin{bmatrix} \mathbf{x} \\ \mathbf{e} \end{bmatrix} \tag{6-31}$$

Equation (6–31) describes the dynamics of the observed-state feedback control system. The characteristic equation for the system is

$$\begin{vmatrix} s\mathbf{I} - \mathbf{A} + \mathbf{BK} & -\mathbf{BK} \\ \mathbf{0} & s\mathbf{I} - \mathbf{A} + \mathbf{K}_e\mathbf{C} \end{vmatrix} = 0$$

or

$$|s\mathbf{I} - \mathbf{A} + \mathbf{BK}||s\mathbf{I} - \mathbf{A} + \mathbf{K}_e\mathbf{C}| = 0$$

Notice that the closed-loop poles of the observed-state feedback control system consist of the poles due to the pole-placement design alone and the poles due to the observer design alone. This means that the pole-placement design and the observer design are independent of each other. They can be designed separately and combined to form the observed-state feedback control system. Note that, if the order of the plant is n, then the observer is also of nth order (if the full-order state observer is used), and the resulting characteristic equation for the entire closed-loop system becomes of order $2n$.

6-6 MINIMUM-ORDER OBSERVERS

The observers discussed thus far are designed to reconstruct all the state variables. In practice, some of the state variables may be measured accurately. Such accurately measurable state variables need not be estimated.

Suppose that the state vector $\mathbf{x}$ is an n-vector and the output vector $\mathbf{y}$ is an m-vector that can be measured. Since m output variables are linear combinations of the state variables, m state variables need not be estimated. Instead, we need to estimate only $n - m$ state variables. Then the reduced-order observer becomes an $(n - m)$th-order observer. Such an observer is the minimum-order observer. Figure 6-6 shows the block diagram of a system with a minimum-order observer.

It is important to note, however, that if the measurement of output variables involves significant noise and is relatively inaccurate, then the use of the full-order observer may result in better system performance.

To present the basic idea of the minimum-order observer without undue mathematical complications, we shall present the case where the output is a scalar (i.e., $m = 1$) and derive the state equation for the minimum-order observer [which in this case is an $(n - 1)$th-order observer]. Consider the system

$$\dot{\mathbf{x}} = \mathbf{A}\mathbf{x} + \mathbf{B}u \quad (6\text{-}32)$$

$$y = \mathbf{C}\mathbf{x} \quad (6\text{-}33)$$

where the state vector $\mathbf{x}$ can be partitioned into two parts: x_a (a scalar) and $\mathbf{x}_b$ [an $(n - 1)$-vector]. Here the state variable x_a is equal to the output y—and thus can

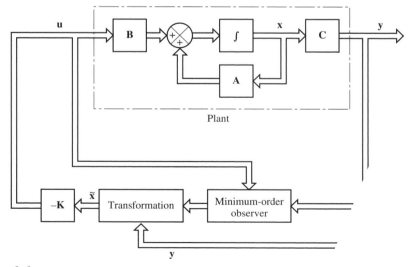

Figure 6-6
Observed-state feedback control system with a minimum-order ob

be directly measured—and $\mathbf{x}_b$ is the unmeasurable portion of the state vector. Then the partitioned state and output equations become

$$\left[\begin{array}{c}\dot{x}_a \\ \hline \dot{\mathbf{x}}_b\end{array}\right] = \left[\begin{array}{c|c}A_{aa} & \mathbf{A}_{ab} \\ \hline \mathbf{A}_{ba} & \mathbf{A}_{bb}\end{array}\right]\left[\begin{array}{c}x_a \\ \hline \mathbf{x}_b\end{array}\right] + \left[\begin{array}{c}B_a \\ \hline \mathbf{B}_b\end{array}\right]u \tag{6-34}$$

$$y = [1 \mid 0]\left[\begin{array}{c}x_a \\ \hline \mathbf{x}_b\end{array}\right] \tag{6-35}$$

where A_{aa} = scalar
$\mathbf{A}_{ab} = 1 \times (n - 1)$ matrix
$\mathbf{A}_{ba} = (n - 1) \times 1$ matrix
$\mathbf{A}_{bb} = (n - 1) \times (n - 1)$ matrix
B_a = scalar
$\mathbf{B}_b = (n - 1) \times 1$ matrix

From Equation (6–34), the equation for the measured portion of the state becomes

$$\dot{x}_a = A_{aa}x_a + \mathbf{A}_{ab}\mathbf{x}_b + B_a u$$

or

$$\dot{x}_a - A_{aa}x_a - B_a u = \mathbf{A}_{ab}\mathbf{x}_b \tag{6-36}$$

The quantities on the left-hand side of Equation (6–36) can be measured. That equation acts as the output equation. In designing the minimum-order observer, which is an $(n - 1)$th-order observer, we take the left-hand side of Equation (6–36) to consist of known quantities. Thus, this equation relates the measurable quantities and unmeasurable quantities of the state.

From Equation (6–34), the equation for the unmeasured portion of the state becomes

$$\dot{\mathbf{x}}_b = \mathbf{A}_{ba}x_a + \mathbf{A}_{bb}\mathbf{x}_b + \mathbf{B}_b u \tag{6-37}$$

Noting that the terms $\mathbf{A}_{ba}x_a$ and $\mathbf{B}_b u$ represent known quantities, we see that Equation (6–37) describes the dynamics of the unmeasured portion of the state.

In what follows, we shall present a method for designing a minimum-order observer. The design procedure can be simplified if we utilize the design technique developed for the full-order state observer.

Let us compare the state equation for the full-order observer with that for the minimum-order observer. The state equation for the full-order observer is

$$\dot{\mathbf{x}} = \mathbf{A}\mathbf{x} + \mathbf{B}u$$

and the "state equation" for the minimum-order observer is

$$\dot{\mathbf{x}}_b = \mathbf{A}_{bb}\mathbf{x}_b + \mathbf{A}_{ba}x_a + \mathbf{B}_b u$$

The output equation for the full-order observer is

$$y = \mathbf{C}\mathbf{x}$$

and the "output equation" for the minimum-order observer is

$$\dot{x}_a - A_{aa}x_a - B_a u = \mathbf{A}_{ab}\mathbf{x}_b$$

The design of the minimum-order observer can be carried out as follows: First, note that the observer equation for the full-order observer was given by Equation (6–19), which we repeat here:

$$\dot{\tilde{\mathbf{x}}} = (\mathbf{A} - \mathbf{K}_e\mathbf{C})\tilde{\mathbf{x}} + \mathbf{B}u + \mathbf{K}_e y \qquad (6\text{–}38)$$

Then, substituting the quantities listed in Table 6–1 into Equation (6–38), we obtain

$$\dot{\tilde{\mathbf{x}}}_b = (\mathbf{A}_{bb} - \mathbf{K}_e\mathbf{A}_{ab})\tilde{\mathbf{x}}_b + \mathbf{A}_{ba}x_a + \mathbf{B}_b u + \mathbf{K}_e(\dot{x}_a - A_{aa}x_a - B_a u) \qquad (6\text{–}39)$$

where the state observer gain matrix $\mathbf{K}_e$ is an $(n - 1) \times 1$ matrix. Notice in Equation (6–39) that, in order to estimate $\tilde{\mathbf{x}}_b$, we need the derivative of x_a. This presents a difficulty, because differentiation amplifies noise. If $x_a\,(= y)$ is noisy, the use of $\dot{x}_a$ is unacceptable. To avoid this difficulty, we eliminate $\dot{x}_a$ in the following way: First we rewrite Equation (6–39) as

$$\begin{aligned}\dot{\tilde{\mathbf{x}}}_b - \mathbf{K}_e \dot{x}_a &= (\mathbf{A}_{bb} - \mathbf{K}_e\mathbf{A}_{ab})\tilde{\mathbf{x}}_b + (\mathbf{A}_{ba} - \mathbf{K}_e A_{aa})y + (\mathbf{B}_b - \mathbf{K}_e B_a)u \\ &= (\mathbf{A}_{bb} - \mathbf{K}_e\mathbf{A}_{ab})(\tilde{\mathbf{x}}_b - \mathbf{K}_e y) \\ &\quad + [(\mathbf{A}_{bb} - \mathbf{K}_e\mathbf{A}_{ab})\mathbf{K}_e + \mathbf{A}_{ba} - \mathbf{K}_e A_{aa}]y \\ &\quad + (\mathbf{B}_b - \mathbf{K}_e B_a)u \qquad (6\text{–}40)\end{aligned}$$

Table 6–1 List of Necessary Substitutions for Writing the Observer Equation the Minimum-Order State Observer, which is an $(n - 1)$th-Order Observer

Full-Order State Observer	Minimum-Order State Observer
$\tilde{\mathbf{x}}$	$\tilde{\mathbf{x}}_b$
$\mathbf{A}$	$\mathbf{A}_{bb}$
$\mathbf{B}u$	$\mathbf{A}_{ba}x_a + \mathbf{B}_b u$
y	$\dot{x}_a - A_{aa}x_a -$
$\mathbf{C}$	$\mathbf{A}_{ab}$
$\mathbf{K}_e$ ($n \times 1$ matrix)	$\mathbf{K}_e$ [$(n - 1) \times$

Next, we define

$$\mathbf{x}_b - \mathbf{K}_e y = \mathbf{x}_b - \mathbf{K}_e x_a = \boldsymbol{\eta}$$

and

$$\tilde{\mathbf{x}}_b - \mathbf{K}_e y = \tilde{\mathbf{x}}_b - \mathbf{K}_e x_a = \tilde{\boldsymbol{\eta}} \qquad (6\text{--}41)$$

Then Equation (6–40) becomes

$$\dot{\tilde{\boldsymbol{\eta}}} = (\mathbf{A}_{bb} - \mathbf{K}_e \mathbf{A}_{ab})\tilde{\boldsymbol{\eta}} + [(\mathbf{A}_{bb} - \mathbf{K}_e \mathbf{A}_{ab})\mathbf{K}_e \\ + \mathbf{A}_{ba} - \mathbf{K}_e \mathbf{A}_{aa}]y + (\mathbf{B}_b - \mathbf{K}_e \mathbf{B}_a)u \qquad (6\text{--}42)$$

Now we define

$$\hat{\mathbf{A}} = \mathbf{A}_{bb} - \mathbf{K}_e \mathbf{A}_{ab}$$
$$\hat{\mathbf{B}} = \hat{\mathbf{A}} \mathbf{K}_e + \mathbf{A}_{ba} - \mathbf{K}_e \mathbf{A}_{aa}$$
$$\hat{\mathbf{F}} = \mathbf{B}_b - \mathbf{K}_e \mathbf{B}_a$$

Then Equation (6–42) becomes

$$\dot{\tilde{\boldsymbol{\eta}}} = \hat{\mathbf{A}} \tilde{\boldsymbol{\eta}} + \hat{\mathbf{B}} y + \hat{\mathbf{F}} u \qquad (6\text{--}43)$$

Equation (6–43) and Equation (6–41) together define the minimum-order observer.

Since

$$y = \begin{bmatrix} 1 & \vdots & \mathbf{0} \end{bmatrix} \begin{bmatrix} x_a \\ \mathbf{x}_b \end{bmatrix}$$

$$\tilde{\mathbf{x}} = \begin{bmatrix} x_a \\ \tilde{\mathbf{x}}_b \end{bmatrix} = \begin{bmatrix} y \\ \tilde{\mathbf{x}}_b \end{bmatrix} = \begin{bmatrix} \mathbf{0} \\ \mathbf{I}_{n-1} \end{bmatrix} [\tilde{\mathbf{x}}_b - \mathbf{K}_e y] + \begin{bmatrix} 1 \\ \mathbf{K}_e \end{bmatrix} y$$

where $\mathbf{0}$ is a row vector consisting of $(n - 1)$ zeros, if we define

$$\hat{\mathbf{C}} = \begin{bmatrix} \mathbf{0} \\ \mathbf{I}_{n-1} \end{bmatrix}, \qquad \hat{\mathbf{D}} = \begin{bmatrix} 1 \\ \mathbf{K}_e \end{bmatrix}$$

then we can write $\tilde{\mathbf{x}}$ in terms of $\tilde{\boldsymbol{\eta}}$ and y as follows:

$$\tilde{\mathbf{x}} = \hat{\mathbf{C}} \tilde{\boldsymbol{\eta}} + \hat{\mathbf{D}} y \qquad (6\text{--}44)$$

This equation gives the transformation from $\tilde{\boldsymbol{\eta}}$ to $\tilde{\mathbf{x}}$.

Figure 6–7 shows the block diagram of the observed-state feedback control system with the minimum-order observer [which, in this case, is an $(n - 1)$th-order observer], based on Equations (6–32), (6–33), (6–43), and (6–44) and on $u = -\mathbf{K}\tilde{\mathbf{x}}$.

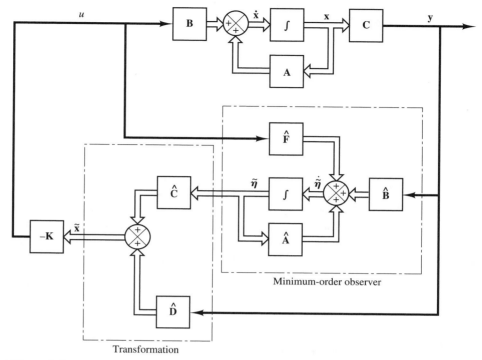

Figure 6–7
System with observed-state feedback, where the observer is the minimum-order observer.

Next, we shall derive the observer error equation. Using Equation (6–36), we can modify Equation (6–39) to yield

$$\dot{\widetilde{\mathbf{x}}}_b = (\mathbf{A}_{bb} - \mathbf{K}_e \mathbf{A}_{ab})\widetilde{\mathbf{x}}_b + \mathbf{A}_{ba} x_a + \mathbf{B}_b u + \mathbf{K}_e \mathbf{A}_{ab} \mathbf{x}_b \tag{6-45}$$

Subtracting Equation (6–45) from Equation (6–37), we obtain

$$\dot{\mathbf{x}}_b - \dot{\widetilde{\mathbf{x}}}_b = (\mathbf{A}_{bb} - \mathbf{K}_e \mathbf{A}_{ab})(\mathbf{x}_b - \widetilde{\mathbf{x}}_b) \tag{6-46}$$

We define

$$\mathbf{e} = \mathbf{x}_b - \widetilde{\mathbf{x}}_b = \boldsymbol{\eta} - \widetilde{\boldsymbol{\eta}}$$

Then Equation (6–46) becomes

$$\dot{\mathbf{e}} = (\mathbf{A}_{bb} - \mathbf{K}_e \mathbf{A}_{ab})\mathbf{e} \tag{6-47}$$

This is the error equation for the minimum-order observer, which is an $(n-1)$th-order observer in this case. Note that $\mathbf{e}$ is an $(n-1)$-vector.

The error dynamics can be chosen as desired by following the technique developed for the full-order observer, provided that the rank of matrix

$$\begin{bmatrix} \mathbf{A}_{ab} \\ \mathbf{A}_{ab}\mathbf{A}_{bb} \\ \vdots \\ \mathbf{A}_{ab}\mathbf{A}_{bb}^{n-2} \end{bmatrix}$$

is $n - 1$. (This is the complete observability condition applicable to the minimum-order observer, which is an $(n - 1)$th-order observer in the present case.)

The characteristic equation for the minimum-order observer, which is an $(n - 1)$th-order observer, is obtained from Equation (6–47) as

$$|s\mathbf{I} - \mathbf{A}_{bb} + \mathbf{K}_e\mathbf{A}_{ab}| = (s - \mu_1)(s - \mu_2)\cdots(s - \mu_{n-1})$$
$$= s^{n-1} + \hat{\alpha}_1 s^{n-2} + \cdots + \hat{\alpha}_{n-2} s + \hat{\alpha}_{n-1} = 0 \quad (6\text{–}48)$$

where $\mu_1, \mu_2, \ldots, \mu_{n-1}$ are desired eigenvalues for the minimum-order observer. The observer gain matrix $\mathbf{K}_e$ can be determined by first choosing the desired eigenvalues for the minimum-order observer [that is, by placing the roots of the characteristic equation, Equation (6–48), at the desired locations] and then using the procedure developed for the full-order observer with appropriate modifications.

If Ackermann's formula given by Equation (6–26) is to be used, then it should be modified to

$$\mathbf{K}_e = \phi(\mathbf{A}_{bb}) \begin{bmatrix} \mathbf{A}_{ab} \\ \mathbf{A}_{ab}\mathbf{A}_{bb} \\ \vdots \\ \mathbf{A}_{ab}\mathbf{A}_{bb}^{n-3} \\ \mathbf{A}_{ab}\mathbf{A}_{bb}^{n-2} \end{bmatrix}^{-1} \begin{bmatrix} 0 \\ 0 \\ \vdots \\ 0 \\ 1 \end{bmatrix} \quad (6\text{–}49)$$

where

$$\phi(\mathbf{A}_{bb}) = \mathbf{A}_{bb}^{n-1} + \hat{\alpha}_1 \mathbf{A}_{bb}^{n-2} + \cdots + \hat{\alpha}_{n-2}\mathbf{A}_{bb} + \hat{\alpha}_{n-1}\mathbf{I}$$

Observed-state feedback control system with minimum-order observer. We have shown that the closed-loop poles of the observed-state feedback control system with full-order state observer consist of the poles due to the pole-placement design alone, plus the poles due to the observer design alone. Hence, the pole-placement design and the full-order observer design are independent of each other.

For the observed-state feedback control system with minimum-order observer, the same conclusion applies. The system characteristic equation can be derived as

$$|s\mathbf{I} - \mathbf{A} + \mathbf{BK}||s\mathbf{I} - \mathbf{A}_{bb} + \mathbf{K}_e\mathbf{A}_{ab}| = 0 \qquad (6\text{–}50)$$

The closed-loop poles of the observed-state feedback control system with a minimum-order observer comprise the closed-loop poles due to pole placement [the eigenvalues of matrix $(\mathbf{A} - \mathbf{BK})$] and the closed-loop poles due to the minimum-order observer [the eigenvalues of matrix $(\mathbf{A}_{bb} - \mathbf{K}_e\mathbf{A}_{ab})$]. Therefore, as with the full-order state observer, the pole-placement design and the design of the minimum-order observer are independent of each other.

Determining the observer gain matrix $\mathbf{K}_e$ with MATLAB. Because of the duality of pole placement and observer design, the same algorithm can be applied to both the pole-placement problem and the observer-design problem. Thus, the commands acker and place can be used to determine the observer gain matrix $\mathbf{K}_e$.

The closed-loop poles of the observer are the eigenvalues of matrix $\mathbf{A} - \mathbf{K}_e\mathbf{C}$. The closed-loop poles of the pole placement are the eigenvalues of matrix $\mathbf{A} - \mathbf{BK}$.

Referring to the duality relationship between the pole-placement problem and the observer-design problem, we can determine $\mathbf{K}_e$ by considering the pole-placement problem for the dual system. That is, we determine $\mathbf{K}_e$ by placing the eigenvalues of $\mathbf{A}^* - \mathbf{C}^*\mathbf{K}_e$ at the desired place. Since $\mathbf{K}_e = \mathbf{K}^*$, for the full-order observer we use the command

$$K_e = \text{acker}(A',C',L)'$$

where L is the vector of the desired eigenvalues for the observer. Similarly, for the full-order observer, we may use

$$K_e = \text{place}(A',C',L)'$$

provided that L does not include multiple poles. [In the preceding commands, prime (') indicates the conjugate transpose.] For the minimum-order (or reduced-order) observers, we use the command

$$K_e = \text{acker}(Abb',Aab',L)'$$

or the command

$$K_e = \text{place}(Abb',Aab',L)'$$

EXAMPLE 6-8 Consider the system

$$\dot{\mathbf{x}} = \mathbf{A}\mathbf{x} + \mathbf{B}u$$
$$y = \mathbf{C}\mathbf{x}$$

where

$$\mathbf{A} = \begin{bmatrix} 0 & 1 & 0 \\ 0 & 0 & 1 \\ -6 & -11 & -6 \end{bmatrix}, \quad \mathbf{B} = \begin{bmatrix} 0 \\ 0 \\ 1 \end{bmatrix}, \quad \mathbf{C} = \begin{bmatrix} 1 & 0 & 0 \end{bmatrix}$$

Let us assume that we want to place the closed-loop poles at

$$s_1 = -2 + j2\sqrt{3}, \quad s_2 = -2 - j2\sqrt{3}, \quad s_3 = -6$$

Then the necessary state-feedback gain matrix $\mathbf{K}$ can be obtained as follows:

$$\mathbf{K} = \begin{bmatrix} 90 & 29 & 4 \end{bmatrix}$$

(See MATLAB Program 6–8 for a MATLAB computation of this matrix $\mathbf{K}$.)

Next, let us assume that the output y can be measured accurately, so that state variable x_1 (which is equal to y) need not be estimated. Let us design a minimum-order observer of second order. We choose the desired observer poles to be at

$$s = -10, \quad s = -10$$

From Equation (6–48), the characteristic equation for the minimum-order observer is

$$|s\mathbf{I} - \mathbf{A}_{bb} + \mathbf{K}_e \mathbf{A}_{ab}| = (s - \mu_1)(s - \mu_2)$$
$$= (s + 10)(s + 10)$$
$$= s^2 + 20s + 100 = 0$$

In what follows, we shall use Ackermann's formula given by Equation (6–49), or

$$\mathbf{K}_e = \phi(\mathbf{A}_{bb}) \begin{bmatrix} \mathbf{A}_{ab} \\ \hdashline \mathbf{A}_{ab}\mathbf{A}_{bb} \end{bmatrix}^{-1} \begin{bmatrix} 0 \\ 1 \end{bmatrix} \quad (6\text{–}51)$$

where

$$\phi(\mathbf{A}_{bb}) = \mathbf{A}_{bb}^2 + \hat{\alpha}_1 \mathbf{A}_{bb} + \hat{\alpha}_2 \mathbf{I} = \mathbf{A}_{bb}^2 + 20\mathbf{A}_{bb} + 100\mathbf{I}$$

Since

$$\widetilde{\mathbf{x}} = \begin{bmatrix} x_a \\ \hline \widetilde{\mathbf{x}}_b \end{bmatrix} = \begin{bmatrix} x_1 \\ \hline \widetilde{x}_2 \\ \widetilde{x}_3 \end{bmatrix}, \quad \mathbf{A} = \begin{bmatrix} 0 & | & 1 & 0 \\ \hline 0 & | & 0 & 1 \\ -6 & | & -11 & -6 \end{bmatrix}, \quad \mathbf{B} = \begin{bmatrix} 0 \\ \hline 0 \\ 1 \end{bmatrix}$$

we have

$$A_{aa} = 0, \quad \mathbf{A}_{ab} = \begin{bmatrix} 1 & 0 \end{bmatrix}, \quad \mathbf{A}_{ba} = \begin{bmatrix} 0 \\ -6 \end{bmatrix}$$

$$\mathbf{A}_{bb} = \begin{bmatrix} 0 & 1 \\ -11 & -6 \end{bmatrix}, \quad B_a = 0, \quad \mathbf{B}_b = \begin{bmatrix} 0 \\ 1 \end{bmatrix}$$

Equation (6–51) now becomes

$$\mathbf{K}_e = \left\{ \begin{bmatrix} 0 & 1 \\ -11 & -6 \end{bmatrix}^2 + 20 \begin{bmatrix} 0 & 1 \\ -11 & -6 \end{bmatrix} + 100 \begin{bmatrix} 1 & 0 \\ 0 & 1 \end{bmatrix} \right\} \begin{bmatrix} 1 & 0 \\ 0 & 1 \end{bmatrix}^{-1} \begin{bmatrix} 0 \\ 1 \end{bmatrix}$$

$$= \begin{bmatrix} 89 & 14 \\ -154 & 5 \end{bmatrix} \begin{bmatrix} 0 \\ 1 \end{bmatrix} = \begin{bmatrix} 14 \\ 5 \end{bmatrix}$$

(A MATLAB computation of this $\mathbf{K}_e$ is given in MATLAB Program 6–8.)

MATLAB Program 6–8

```
>> A = [0  1  0;0  0  1;-6  -11  -6];
>> B = [0;0;1];
>> J = [-2+j*2*sqrt(3)  -2-j*2*sqrt(3)  -6];
>> K = acker(A,B,J)

K =

    90.0000   29.0000    4.0000

>> Abb = [0  1;-11  -6];
>> Aab = [1  0];
>> L = [-10  -10];
>> Ke = acker(Abb',Aab',L)'

Ke =

    14
     5
```

From Equations (6–41) and (6–42), the equation for the minimum-order observer (which is a second-order observer) can be given by

$$\dot{\tilde{\boldsymbol{\eta}}} = (\mathbf{A}_{bb} - \mathbf{K}_e\mathbf{A}_{ab})\tilde{\boldsymbol{\eta}} + [(\mathbf{A}_{bb} - \mathbf{K}_e\mathbf{A}_{ab})\mathbf{K}_e + \mathbf{A}_{ba} - \mathbf{K}_e\mathbf{A}_{aa}]y + (\mathbf{B}_b - \mathbf{K}_e\mathbf{B}_a)u \quad (6\text{–}52)$$

where

$$\tilde{\boldsymbol{\eta}} = \tilde{\mathbf{x}}_b - \mathbf{K}_e y = \tilde{\mathbf{x}}_b - \mathbf{K}_e x_1$$

Noting that

$$\mathbf{A}_{bb} - \mathbf{K}_e\mathbf{A}_{ab} = \begin{bmatrix} 0 & 1 \\ -11 & -6 \end{bmatrix} - \begin{bmatrix} 14 \\ 5 \end{bmatrix}\begin{bmatrix} 1 & 0 \end{bmatrix} = \begin{bmatrix} -14 & 1 \\ -16 & -6 \end{bmatrix}$$

we see that Equation (6–52) for the minimum-order observer becomes

$$\begin{bmatrix} \dot{\tilde{\eta}}_2 \\ \dot{\tilde{\eta}}_3 \end{bmatrix} = \begin{bmatrix} -14 & 1 \\ -16 & -6 \end{bmatrix}\begin{bmatrix} \tilde{\eta}_2 \\ \tilde{\eta}_3 \end{bmatrix} + \left\{ \begin{bmatrix} -14 & 1 \\ -16 & -6 \end{bmatrix}\begin{bmatrix} 14 \\ 5 \end{bmatrix} \right.$$
$$\left. + \begin{bmatrix} 0 \\ -6 \end{bmatrix} - \begin{bmatrix} 14 \\ 5 \end{bmatrix}0 \right\}y + \left\{ \begin{bmatrix} 0 \\ 1 \end{bmatrix} - \begin{bmatrix} 14 \\ 5 \end{bmatrix}0 \right\}u$$

or

$$\begin{bmatrix} \dot{\tilde{\eta}}_2 \\ \dot{\tilde{\eta}}_3 \end{bmatrix} = \begin{bmatrix} -14 & 1 \\ -16 & -6 \end{bmatrix}\begin{bmatrix} \tilde{\eta}_2 \\ \tilde{\eta}_3 \end{bmatrix} + \begin{bmatrix} -191 \\ -260 \end{bmatrix}y + \begin{bmatrix} 0 \\ 1 \end{bmatrix}u$$

where

$$\begin{bmatrix} \tilde{\eta}_2 \\ \tilde{\eta}_3 \end{bmatrix} = \begin{bmatrix} \tilde{x}_2 \\ \tilde{x}_3 \end{bmatrix} - \mathbf{K}_e y$$

or

$$\begin{bmatrix} \tilde{x}_2 \\ \tilde{x}_3 \end{bmatrix} = \begin{bmatrix} \tilde{\eta}_2 \\ \tilde{\eta}_3 \end{bmatrix} + \mathbf{K}_e x_1$$

If the observed-state feedback is used, then the control signal u becomes

$$u = -\mathbf{K}\tilde{\mathbf{x}} = -\mathbf{K}\begin{bmatrix} x_1 \\ \tilde{x}_2 \\ \tilde{x}_3 \end{bmatrix}$$

where $\mathbf{K}$ is the state feedback gain matrix. Figure 6–8 is a block diagram showing the configuration of the system with observed-state feedback, where the observer is the minimum-order observer.

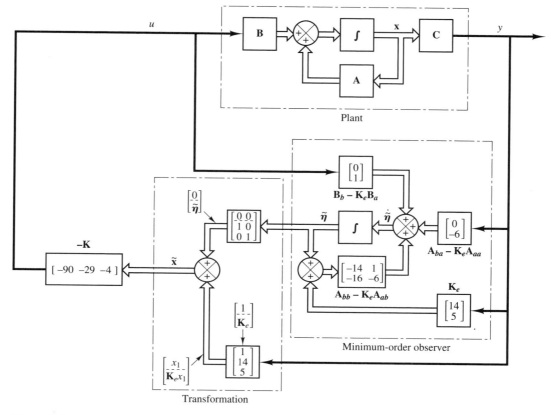

Figure 6–8
System with observed state feedback, where the observer is the minimum-order observer. This system is designed in Example 6–8.

EXAMPLE 6–9 Consider the system shown in Figure 6–9. Using the pole-placement-with-observer approach, design a regulator system that will maintain the zero position ($y_1 = 0$ and $y_2 = 0$) in the presence of disturbances. Choose the desired closed-loop poles for the pole-placement part to be

$$s = -2 + j2\sqrt{3}, \qquad s = -2 - j2\sqrt{3}, \qquad s = -10, \qquad s = -10$$

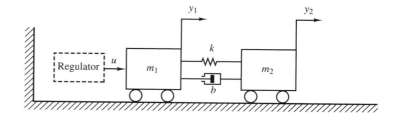

Figure 6–9
Mechanical system.

Section 6–6 / Minimum-Order Observers

and the desired poles for the minimum-order observer to be

$$s = -15, \quad s = -16$$

(This is a fourth-order system with two outputs y_1 and y_2, so the minimum-order observer is a second-order observer.)

First, determine the state feedback gain matrix **K** and observer gain matrix $\mathbf{K}_e$. Then obtain the response of the system to an arbitrary initial condition—for example,

$$y_1(0) = 0.1, \quad y_2(0) = 0, \quad \dot{y}_1(0) = 0, \quad \dot{y}_2(0) = 0$$
$$e_1(0) = 0.1, \quad e_2(0) = 0.05$$

where e_1 and e_2 are defined by

$$e_1 = y_1 - \tilde{y}_1$$
$$e_2 = y_2 - \tilde{y}_2$$

Assume that $m_1 = 1$ kg, $m_2 = 2$ kg, $k = 36$ N/m, and $b = 0.6$ N-s/m.
The equations for the system are

$$m_1 \ddot{y}_1 = k(y_2 - y_1) + b(\dot{y}_2 - \dot{y}_1) + u$$
$$m_2 \ddot{y}_2 = k(y_1 - y_2) + b(\dot{y}_1 - \dot{y}_2)$$

Substituting the given numerical values for m_1, m_2, k, and b and simplifying, we obtain

$$\ddot{y}_1 = -36y_1 + 36y_2 - 0.6\dot{y}_1 + 0.6\dot{y}_2 + u$$
$$\ddot{y}_2 = 18y_1 - 18y_2 + 0.3\dot{y}_1 - 0.3\dot{y}_2$$

Let us choose the state variables as follows:

$$x_1 = y_1$$
$$x_2 = y_2$$
$$x_3 = \dot{y}_1$$
$$x_4 = \dot{y}_2$$

Then the state-space equations become

$$\begin{bmatrix} \dot{x}_1 \\ \dot{x}_2 \\ \dot{x}_3 \\ \dot{x}_4 \end{bmatrix} = \begin{bmatrix} 0 & 0 & 1 & 0 \\ 0 & 0 & 0 & 1 \\ -36 & 36 & -0.6 & 0.6 \\ 18 & -18 & 0.3 & -0.3 \end{bmatrix} \begin{bmatrix} x_1 \\ x_2 \\ x_3 \\ x_4 \end{bmatrix} + \begin{bmatrix} 0 \\ 0 \\ 1 \\ 0 \end{bmatrix} u$$

$$\begin{bmatrix} y_1 \\ y_2 \end{bmatrix} = \begin{bmatrix} 1 & 0 & 0 & 0 \\ 0 & 1 & 0 & 0 \end{bmatrix} \begin{bmatrix} x_1 \\ x_2 \\ x_3 \\ x_4 \end{bmatrix}$$

We define

$$\mathbf{A} = \left[\begin{array}{cc|cc} 0 & 0 & 1 & 0 \\ 0 & 0 & 0 & 1 \\ \hline -36 & 36 & -0.6 & 0.6 \\ 18 & -18 & 0.3 & -0.3 \end{array}\right] = \left[\begin{array}{c|c} \mathbf{A}_{aa} & \mathbf{A}_{ab} \\ \hline \mathbf{A}_{ba} & \mathbf{A}_{bb} \end{array}\right], \quad \mathbf{B} = \left[\begin{array}{c} 0 \\ 0 \\ \hline 1 \\ 0 \end{array}\right] = \left[\begin{array}{c} \mathbf{B}_a \\ \hline \mathbf{B}_b \end{array}\right]$$

The state feedback gain matrix **K** and observer gain matrix $\mathbf{K}_e$ can be obtained easily with MATLAB as follows:

$$\mathbf{K} = \begin{bmatrix} 130.4444 & -41.5556 & 23.1000 & 15.4185 \end{bmatrix}$$

$$\mathbf{K}_e = \begin{bmatrix} 14.4 & 0.6 \\ 0.3 & 15.7 \end{bmatrix}$$

(See MATLAB Program 6–9.)

MATLAB Program 6–9

```
>> A = [0  0  1  0;0  0  0  1;−36  36  −0.6  0.6;18  −18  0.3  −0.3];
>> B = [0;0;1;0];
>> J = [−2+j*2*sqrt(3)   −2−j*2*sqrt(3)   −10   −10];
>> K = acker(A,B,J)

K =

   130.4444   −41.5556   23.1000   15.4185

>> Aab = [1  0;0  1];
>> Abb = [−0.6  0.6;0.3  −0.3];
>> L = [−15  −16];
>> Ke = place(Abb',Aab',L)'
Ke =
    14.4000    0.6000
     0.3000   15.7000
```

Response to Initial Condition: Next, we obtain the response of the designed system to the given initial condition. Since

$$\dot{\mathbf{x}} = \mathbf{A}\mathbf{x} + \mathbf{B}u$$
$$u = -\mathbf{K}\tilde{\mathbf{x}}$$
$$\tilde{\mathbf{x}} = \begin{bmatrix} \mathbf{x}_a \\ \tilde{\mathbf{x}}_b \end{bmatrix} = \begin{bmatrix} \mathbf{y} \\ \tilde{\mathbf{x}}_b \end{bmatrix}$$

we have

$$\dot{\mathbf{x}} = \mathbf{A}\mathbf{x} - \mathbf{B}\mathbf{K}\tilde{\mathbf{x}} = (\mathbf{A} - \mathbf{B}\mathbf{K})\mathbf{x} + \mathbf{B}\mathbf{K}(\mathbf{x} - \tilde{\mathbf{x}}) \qquad (6\text{--}53)$$

Note that

$$\mathbf{x} - \tilde{\mathbf{x}} = \begin{bmatrix} \mathbf{x}_a \\ \hline \mathbf{x}_b \end{bmatrix} - \begin{bmatrix} \mathbf{x}_a \\ \hline \tilde{\mathbf{x}}_b \end{bmatrix} = \begin{bmatrix} \mathbf{0} \\ \hline \mathbf{x}_b - \tilde{\mathbf{x}}_b \end{bmatrix} = \begin{bmatrix} \mathbf{0} \\ \hline \mathbf{e} \end{bmatrix} = \begin{bmatrix} \mathbf{0} \\ \hline \mathbf{I} \end{bmatrix}\mathbf{e} = \mathbf{F}\mathbf{e}$$

where

$$F = \begin{bmatrix} 0 \\ \hline I \end{bmatrix}$$

Then Equation (6–53) can be written as

$$\dot{x} = (A - BK)x + BKFe \qquad (6\text{–}54)$$

Since, from Equation (6–47), we have

$$\dot{e} = (A_{bb} - K_e A_{ab})e \qquad (6\text{–}55)$$

by combining Equations (6–54) and (6–55) into one equation, we get

$$\begin{bmatrix} \dot{x} \\ \hline \dot{e} \end{bmatrix} = \begin{bmatrix} A - BK & BKF \\ \hline 0 & A_{bb} - K_e A_{ab} \end{bmatrix} \begin{bmatrix} x \\ e \end{bmatrix}$$

The state matrix here is a 6 × 6 matrix. The response of the system to the given initial condition can be obtained easily with MATLAB. (See MATLAB Program 6–10.) The resulting response curves are shown in Figure 6–10. The response curves seem to be acceptable.

MATLAB Program 6–10

```
>> % Response to initial condition
>>
>> A = [0  0  1  0;0  0  0  1;−36  36  −0.6  0.6;18  −18  0.3  −0.3];
>> B = [0;0;1;0];
>> K = [130.4444  −41.5556  23.1000  15.4185];
>> Ke = [14.4  0.6;0.3  15.7];
>> F = [0  0;0  0;1  0;0  1];
>> Aab = [1  0;0  1];
>> Abb = [−0.6  0.6;0.3  −0.3];
>> AA = [A − B*K   B*K*F;zeros(2,4)   Abb − Ke*Aab];
>> sys = ss(AA,eye(6),eye(6),eye(6));
>> t = 0:0.01:4;
>> y = initial(sys,[0.1;0;0;0;0.1;0.05],t);
>> x1 = [1  0  0  0  0  0]*y';
>> x2 = [0  1  0  0  0  0]*y';
>> x3 = [0  0  1  0  0  0]*y';
>> x4 = [0  0  0  1  0  0]*y';
>> e1 = [0  0  0  0  1  0]*y';
>> e2 = [0  0  0  0  0  1]*y';
>>
>> subplot(3,2,1); plot(t,x1); grid; title('Response to initial condition'),
>> xlabel('t (sec)'); ylabel('x_1')
>> subplot(3,2,2); plot(t,x2); grid; title('Response to initial condition'),
>> xlabel('t (sec)'); ylabel('x_2')
>> subplot(3,2,3); plot(t,x3); grid; xlabel('t (sec)'); ylabel('x_3')
>> subplot(3,2,4); plot(t,x4); grid; xlabel('t (sec)'); ylabel('x_4')
>> subplot(3,2,5); plot(t,e1); grid; xlabel('t (sec)');ylabel('e_1')
>> subplot(3,2,6); plot(t,e2); grid; xlabel('t (sec)'); ylabel('e_2')
```

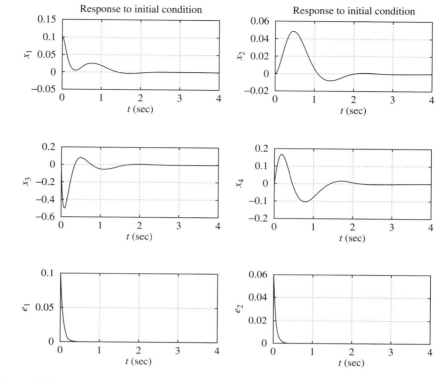

Figure 6–10
Response curves to initial condition.

6–7 OBSERVER CONTROLLERS

Transfer function of the observer-based controller. Consider the plant defined by

$$\dot{\mathbf{x}} = \mathbf{A}\mathbf{x} + \mathbf{B}u$$
$$y = \mathbf{C}\mathbf{x}$$

Suppose that the plant is completely observable. Suppose also that we use observed-state feedback control $u = -\mathbf{K}\tilde{\mathbf{x}}$. Then the equations for the observer are given by

$$\dot{\tilde{\mathbf{x}}} = (\mathbf{A} - \mathbf{K}_e\mathbf{C} - \mathbf{B}\mathbf{K})\tilde{\mathbf{x}} + \mathbf{K}_e y \tag{6–56}$$

$$u = -\mathbf{K}\tilde{\mathbf{x}} \tag{6–57}$$

where Equation (6–56) is obtained by substituting $u = -\mathbf{K}\tilde{\mathbf{x}}$ into Equation (6–19).

Taking the Laplace transform of Equation (6–56), assuming a zero initial condition, and solving for $\tilde{\mathbf{X}}(s)$, we obtain

$$\tilde{\mathbf{X}}(s) = (s\mathbf{I} - \mathbf{A} + \mathbf{K}_e\mathbf{C} + \mathbf{BK})^{-1}\mathbf{K}_e Y(s)$$

Substituting this $\tilde{\mathbf{X}}(s)$ into the Laplace transform of Equation (6–57) yields

$$U(s) = -\mathbf{K}(s\mathbf{I} - \mathbf{A} + \mathbf{K}_e\mathbf{C} + \mathbf{BK})^{-1}\mathbf{K}_e Y(s) \qquad (6\text{–}58)$$

Then the transfer function $U(s)/Y(s)$ can be obtained as

$$\frac{U(s)}{Y(s)} = -\mathbf{K}(s\mathbf{I} - \mathbf{A} + \mathbf{K}_e\mathbf{C} + \mathbf{BK})^{-1}\mathbf{K}_e$$

Figure 6–11 shows the block diagram representation of the system. Notice that the transfer function

$$\mathbf{K}(s\mathbf{I} - \mathbf{A} + \mathbf{K}_e\mathbf{C} + \mathbf{BK})^{-1}\mathbf{K}_e$$

acts as a controller for the system. Hence, we call the transfer function

$$\frac{U(s)}{-Y(s)} = \frac{\text{num}}{\text{den}} = \mathbf{K}(s\mathbf{I} - \mathbf{A} + \mathbf{K}_e\mathbf{C} + \mathbf{BK})^{-1}\mathbf{K}_e \qquad (6\text{–}59)$$

the *observer-based controller transfer function* or, simply, the *observer-controller transfer function*.

Note that the observer-controller matrix

$$\mathbf{A} - \mathbf{K}_e\mathbf{C} - \mathbf{BK}$$

may or may not be stable, although $\mathbf{A} - \mathbf{BK}$ and $\mathbf{A} - \mathbf{K}_e\mathbf{C}$ are chosen to be stable. In fact, in some cases the matrix $\mathbf{A} - \mathbf{K}_e\mathbf{C} - \mathbf{BK}$ may be poorly stable or even unstable.

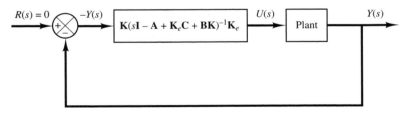

Figure 6–11
Block diagram representation of system with a observer-controller.

EXAMPLE 6–10 Consider the design of a regulator system for the plant

$$\dot{\mathbf{x}} = \mathbf{A}\mathbf{x} + \mathbf{B}u \tag{6-60}$$

$$y = \mathbf{C}\mathbf{x} \tag{6-61}$$

where

$$\mathbf{A} = \begin{bmatrix} 0 & 1 \\ 20.6 & 0 \end{bmatrix}, \quad \mathbf{B} = \begin{bmatrix} 0 \\ 1 \end{bmatrix}, \quad \mathbf{C} = \begin{bmatrix} 1 & 0 \end{bmatrix}$$

Suppose that we use the pole-placement approach to the design of the system and that the desired closed-loop poles for this system are at $s = \mu_i \, (i = 1, 2)$, where $\mu_1 = -1.8 + j2.4$ and $\mu_2 = -1.8 - j2.4$. The state-feedback gain matrix $\mathbf{K}$ for this case can be obtained as follows:

$$\mathbf{K} = \begin{bmatrix} 29.6 & 3.6 \end{bmatrix}$$

Using this state-feedback gain matrix $\mathbf{K}$, we find that the control signal u is given by

$$u = -\mathbf{K}\mathbf{x} = -\begin{bmatrix} 29.6 & 3.6 \end{bmatrix} \begin{bmatrix} x_1 \\ x_2 \end{bmatrix}$$

Suppose that we use the observed-state feedback control instead of the actual-state feedback control, or

$$u = -\mathbf{K}\tilde{\mathbf{x}} = -\begin{bmatrix} 29.6 & 3.6 \end{bmatrix} \begin{bmatrix} \tilde{x}_1 \\ \tilde{x}_2 \end{bmatrix}$$

where we choose the observer poles to be at

$$s = -8, \quad s = -8$$

Obtain the observer gain matrix $\mathbf{K}_e$ and draw a block diagram of the observed-state feedback control system. Then obtain the transfer function $U(s)/[-Y(s)]$ for the observer controller, and draw another block diagram with the observer controller as a series controller in the feedforward path. Finally, obtain the response of the system to the following initial condition:

$$\mathbf{x}(0) = \begin{bmatrix} 1 \\ 0 \end{bmatrix}, \quad \mathbf{e}(0) = \mathbf{x}(0) - \tilde{\mathbf{x}}(0) = \begin{bmatrix} 0.5 \\ 0 \end{bmatrix}$$

To begin, we define

$$\mathbf{K}_e = \begin{bmatrix} k_{e1} \\ k_{e2} \end{bmatrix}$$

Substituting this $\mathbf{K}_e$ into the desired characteristic equation gives

$$|s\mathbf{I} - (\mathbf{A} - \mathbf{K}_e\mathbf{C})| = (s - \mu_1)(s - \mu_2) = (s + 8)(s + 8)$$
$$= s^2 + 16s + 64$$

Hence,

$$\left| \begin{bmatrix} s & 0 \\ 0 & s \end{bmatrix} - \begin{bmatrix} 0 & 1 \\ 20.6 & 0 \end{bmatrix} + \begin{bmatrix} k_{e1} \\ k_{e2} \end{bmatrix} \begin{bmatrix} 1 & 0 \end{bmatrix} \right|$$

$$= \left| \begin{bmatrix} s + k_{e1} & -1 \\ -20.6 + k_{e2} & s \end{bmatrix} \right| = (s + k_{e1})s - (20.6 - k_{e2})$$

$$= s^2 + 16s + 64$$

or

$$k_{e1} = 16, \quad k_{e2} = 84.6$$

That is,

$$\mathbf{K}_e = \begin{bmatrix} k_{e1} \\ k_{e2} \end{bmatrix} = \begin{bmatrix} 16 \\ 84.6 \end{bmatrix} \quad (6\text{--}62)$$

Equation (6–62) gives the observer gain matrix $\mathbf{K}_e$. The observer equation is given by Equation (6–22):

$$\dot{\tilde{\mathbf{x}}} = (\mathbf{A} - \mathbf{K}_e \mathbf{C})\tilde{\mathbf{x}} + \mathbf{B}u + \mathbf{K}_e y \quad (6\text{--}63)$$

Since

$$u = -\mathbf{K}\tilde{\mathbf{x}}$$

Equation (6–63) becomes

$$\dot{\tilde{\mathbf{x}}} = (\mathbf{A} - \mathbf{K}_e \mathbf{C} - \mathbf{B}\mathbf{K})\tilde{\mathbf{x}} + \mathbf{K}_e y$$

or

$$\begin{bmatrix} \dot{\tilde{x}}_1 \\ \dot{\tilde{x}}_2 \end{bmatrix} = \left\{ \begin{bmatrix} 0 & 1 \\ 20.6 & 0 \end{bmatrix} - \begin{bmatrix} 16 \\ 84.6 \end{bmatrix} \begin{bmatrix} 1 & 0 \end{bmatrix} - \begin{bmatrix} 0 \\ 1 \end{bmatrix} \begin{bmatrix} 29.6 & 3.6 \end{bmatrix} \right\} \begin{bmatrix} \tilde{x}_1 \\ \tilde{x}_2 \end{bmatrix} + \begin{bmatrix} 16 \\ 84.6 \end{bmatrix} y$$

$$= \begin{bmatrix} -16 & 1 \\ -93.6 & -3.6 \end{bmatrix} \begin{bmatrix} \tilde{x}_1 \\ \tilde{x}_2 \end{bmatrix} + \begin{bmatrix} 16 \\ 84.6 \end{bmatrix} y$$

The block diagram of the system with observed-state feedback is shown in Figure 6–12(a). From Equation (6–59), the transfer function of the observer controller is

$$\frac{U(s)}{-Y(s)} = \mathbf{K}(s\mathbf{I} - \mathbf{A} + \mathbf{K}_e\mathbf{C} + \mathbf{B}\mathbf{K})^{-1}\mathbf{K}_e$$

$$= \begin{bmatrix} 29.6 & 3.6 \end{bmatrix} \begin{bmatrix} s + 16 & -1 \\ 93.6 & s + 3.6 \end{bmatrix}^{-1} \begin{bmatrix} 16 \\ 84.6 \end{bmatrix}$$

$$= \frac{778.2s + 3690.7}{s^2 + 19.6s + 151.2}$$

The same transfer function can be obtained with MATLAB. For example, MATLAB Program 6–11 produces the transfer function of the observer controller when the observer is of full order. [Refer to Equations (6–56) and (6–57).] Figure 6–12(b) shows a block diagram of the system.

The dynamics of the observed-state feedback control system just designed can be described by the following equations: For the plant,

$$\begin{bmatrix} \dot{x}_1 \\ \dot{x}_2 \end{bmatrix} = \begin{bmatrix} 0 & 1 \\ 20.6 & 0 \end{bmatrix} \begin{bmatrix} x_1 \\ x_2 \end{bmatrix} + \begin{bmatrix} 0 \\ 1 \end{bmatrix} u$$

$$y = \begin{bmatrix} 1 & 0 \end{bmatrix} \begin{bmatrix} x_1 \\ x_2 \end{bmatrix}$$

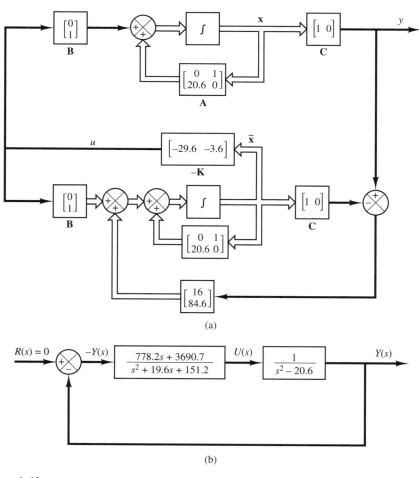

Figure 6–12
(a) Block diagram of system with observed-state feedback; (b) block diagram of transfer function system.

Section 6–7 / Observer Controllers

> **MATLAB Program 6–11**
>
> ```
> >> % Obtaining transfer function of observer controller—full-order observer
> >>
> >> A = [0   1;20.6   0];
> >> B = [0;1];
> >> C = [1   0];
> >> K = [29.6   3.6];
> >> Ke = [16;84.6];
> >> AA = A - Ke*C - B*K;
> >> BB = Ke;
> >> CC = K;
> >> DD = 0;
> >> [num,den] = ss2tf(AA,BB,CC,DD)
> ```
>
> num =
>
>   1.0e+003*
>
>   0   0.7782   3.6907
>
> den =
>
>   1.0000   19.6000   151.2000

For the observer,

$$\begin{bmatrix} \dot{\tilde{x}}_1 \\ \dot{\tilde{x}}_2 \end{bmatrix} = \begin{bmatrix} -16 & 1 \\ -93.6 & -3.6 \end{bmatrix} \begin{bmatrix} \tilde{x}_1 \\ \tilde{x}_2 \end{bmatrix} + \begin{bmatrix} 16 \\ 84.6 \end{bmatrix} y$$

$$u = -\begin{bmatrix} 29.6 & 3.6 \end{bmatrix} \begin{bmatrix} \tilde{x}_1 \\ \tilde{x}_2 \end{bmatrix}$$

The system, as a whole, is of fourth order. The characteristic equation for the system is

$$|s\mathbf{I} - \mathbf{A} + \mathbf{BK}||s\mathbf{I} - \mathbf{A} + \mathbf{K}_e\mathbf{C}| = (s^2 + 3.6s + 9)(s^2 + 16s + 64)$$
$$= s^4 + 19.6s^3 + 130.6s^2 + 374.4s + 576 = 0$$

The characteristic equation can also be obtained from the block diagram of the system shown in Figure 6–12(b). Since the closed-loop transfer function is

$$\frac{Y(s)}{U(s)} = \frac{778.2s + 3690.7}{(s^2 + 19.6s + 151.2)(s^2 - 20.6) + 778.2s + 3690.7}$$

the characteristic equation is

$$(s^2 + 19.6s + 151.2)(s^2 - 20.6) + 778.2s + 3690.7$$
$$= s^4 + 19.6s^3 + 130.6s^2 + 374.4s + 576 = 0$$

As a matter of course, the characteristic equation is the same for the system in state-space representation and in transfer-function representation.

Finally, we shall obtain the response of the system to the following initial condition:

$$x(0) = \begin{bmatrix} 1 \\ 0 \end{bmatrix}, \quad e(0) = \begin{bmatrix} 0.5 \\ 0 \end{bmatrix}$$

Referring to Equation (6–31), we can determine the response to the initial condition from

$$\begin{bmatrix} \dot{x} \\ \dot{e} \end{bmatrix} = \begin{bmatrix} A - BK & BK \\ 0 & A - K_eC \end{bmatrix} \begin{bmatrix} x \\ e \end{bmatrix}, \quad \begin{bmatrix} x(0) \\ e(0) \end{bmatrix} = \begin{bmatrix} 1 \\ 0 \\ 0.5 \\ 0 \end{bmatrix}$$

A MATLAB Program to obtain the response is shown in MATLAB Program 6–12. The resulting response curves are shown in Figure 6–13.

MATLAB Program 6–12

```
>> A = [0   1;20.6   0];
>> B = [0;1];
>> C = [1   0];
>> K = [29.6   3.6];
>> Ke = [16;84.6];
>> sys = ss([A − B*K   B*K;zeros(2,2)   A − Ke*C],eye(4),eye(4),eye(4));
>> t = 0:0.01:4;
>> z = initial(sys,[1;0;0.5;0],t);
>> x1 = [1   0   0   0]*z';
>> x2 = [0   1   0   0]*z';
>> e1 = [0   0   1   0]*z';
>> e2 = [0   0   0   1]*z';
>>
>> subplot(2,2,1);plot(t,x1),grid
>> title('Response to Initial Condition')
>> ylabel('state variable x_1')
>>
>> subplot(2,2,2);plot(t,x2),grid
>> title('Response to Initial Condition')
>> ylabel('state variable x_2')
>>
>> subplot(2,2,3);plot(t,e1),grid
>> xlabel('t (sec)'),ylabel('error state variable e_1')
>>
>> subplot(2,2,4);plot(t,e2),grid
>> xlabel('t (sec)'),ylabel('error state variable e_2')
```

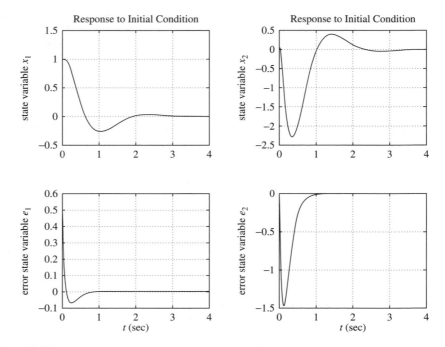

Figure 6–13
Response curves to initial condition.

Transfer function of minimum-order observer-based controller. In the minimum-order observer equation given by Equation (6–42), namely,

$$\dot{\tilde{\eta}} = (A_{bb} - K_e A_{ab})\tilde{\eta} + [(A_{bb} - K_e A_{ab})K_e + A_{ba} - K_e A_{aa}]y + (B_b - K_e B_a)u$$

we define,

$$\hat{A} = A_{bb} - K_e A_{ab}$$
$$\hat{B} = \hat{A}K_e + A_{ba} - K_e A_{aa}$$
$$\hat{F} = B_b - K_e B_a$$

as we did in the case of the derivation of Equation (6–43). Then the following three equations define the minimum-order oberver:

$$\dot{\tilde{\eta}} = \hat{A}\tilde{\eta} + \hat{B}y + \hat{F}u \quad (6\text{--}64)$$

$$\tilde{\eta} = \tilde{x}_b - K_e y \quad (6\text{--}65)$$

$$u = -K\tilde{x} \quad (6\text{--}66)$$

Since Equation (6–66) can be rewritten as

$$u = -\mathbf{K}\tilde{\mathbf{x}} = -[K_a \quad \mathbf{K}_b]\begin{bmatrix} y \\ \tilde{\mathbf{x}}_b \end{bmatrix} = -K_a y - \mathbf{K}_b \tilde{\mathbf{x}}_b$$
$$= -\mathbf{K}_b \tilde{\boldsymbol{\eta}} - (K_a + \mathbf{K}_b \mathbf{K}_e)y \qquad (6\text{–}67)$$

by substituting Equation (6–67) into Equation (6–64), we obtain

$$\dot{\tilde{\boldsymbol{\eta}}} = \hat{\mathbf{A}}\tilde{\boldsymbol{\eta}} + \hat{\mathbf{B}}y + \hat{\mathbf{F}}[-\mathbf{K}_b \tilde{\boldsymbol{\eta}} - (K_a + \mathbf{K}_b \mathbf{K}_e)y]$$
$$= (\hat{\mathbf{A}} - \hat{\mathbf{F}}\mathbf{K}_b)\tilde{\boldsymbol{\eta}} + [\hat{\mathbf{B}} - \hat{\mathbf{F}}(K_a + \mathbf{K}_b \mathbf{K}_e)]y \qquad (6\text{–}68)$$

We define

$$\tilde{\mathbf{A}} = \hat{\mathbf{A}} - \hat{\mathbf{F}}\mathbf{K}_b$$
$$\tilde{\mathbf{B}} = \hat{\mathbf{B}} - \hat{\mathbf{F}}(K_a + \mathbf{K}_b \mathbf{K}_e)$$
$$\tilde{\mathbf{C}} = -\mathbf{K}_b$$
$$\tilde{D} = -(K_a + \mathbf{K}_b \mathbf{K}_e)$$

Then Equations (6–68) and (6–67) can be written, respectively, as

$$\dot{\tilde{\boldsymbol{\eta}}} = \tilde{\mathbf{A}}\tilde{\boldsymbol{\eta}} + \tilde{\mathbf{B}}y \qquad (6\text{–}69)$$

$$u = \tilde{\mathbf{C}}\tilde{\boldsymbol{\eta}} + \tilde{D}y \qquad (6\text{–}70)$$

Equations (6–69) and (6–70) define the minimum-order observer-based controller. By considering u as the output and $-y$ as the input, we can write

$$U(s) = [\tilde{\mathbf{C}}(s\mathbf{I} - \tilde{\mathbf{A}})^{-1}\tilde{\mathbf{B}} + \tilde{D}]Y(s)$$
$$= -[\tilde{\mathbf{C}}(s\mathbf{I} - \tilde{\mathbf{A}})^{-1}\tilde{\mathbf{B}} + \tilde{D}][-Y(s)]$$

Since the input to the observer controller is $-Y(s)$, rather than $Y(s)$, the transfer function of the observer controller is

$$\frac{U(s)}{-Y(s)} = \frac{\text{num}}{\text{den}} = -[\tilde{\mathbf{C}}(s\mathbf{I} - \tilde{\mathbf{A}})^{-1}\tilde{\mathbf{B}} + \tilde{D}] \qquad (6\text{–}71)$$

This transfer function can be easily obtained with the following MATLAB command:

$$[\text{num,den}] = \text{ss2tf(Atilde, Btilde, }-\text{Ctilde, }-\text{Dtilde)} \qquad (6\text{–}72)$$

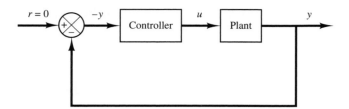

Figure 6–14
Regulator system.

Design of regulator system with observer. Consider the regulator system shown in Figure 6–14. The reference input is zero. The system is defined by

$$\begin{bmatrix} \dot{x}_1 \\ \dot{x}_2 \\ \dot{x}_3 \end{bmatrix} = \begin{bmatrix} 0 & 1 & 0 \\ 0 & 0 & 1 \\ 0 & -24 & -10 \end{bmatrix} \begin{bmatrix} x_1 \\ x_2 \\ x_3 \end{bmatrix} + \begin{bmatrix} 0 \\ 10 \\ -80 \end{bmatrix} u$$

$$y = \begin{bmatrix} 1 & 0 & 0 \end{bmatrix} \begin{bmatrix} x_1 \\ x_2 \\ x_3 \end{bmatrix} + [0]u$$

Using the pole-placement approach, design a controller such that when the system is subjected to the initial condition

$$\mathbf{x}(0) = \begin{bmatrix} 1 \\ 0 \\ 0 \end{bmatrix}, \quad \mathbf{e}(0) = \begin{bmatrix} 1 \\ 0 \end{bmatrix}$$

where **x** is the state vector for the plant and **e** is the observer error vector, the maximum undershoot of $y(t)$ is 25 to 35% and the settling time is about 4 sec. Assume that we use the minimum-order observer. (We assume that only the output y is measurable.)

We shall use the following design procedure:

1. Derive a transfer function of the plant.
2. Choose the desired closed-loop poles for pole placement. Choose the desired observer poles.
3. Determine the state feedback gain matrix **K** and the observer gain matrix $\mathbf{K}_e$.
4. Using the gain matrices **K** and $\mathbf{K}_e$ obtained in step 3, derive the transfer function of the observer controller. If it is a stable controller, check the response to the given initial condition. If the response is not acceptable, adjust the closed-loop pole location and/or observer pole location until an acceptable response is obtained.

Design step 1: We shall derive the transfer function of the plant, which may be obtained with the use of the command

[num,den] = ss2tf(A,B,C,D)

MATLAB Program 6–13 produces the transfer function representation of the plant as follows:

$$\frac{Y(s)}{U(s)} = \frac{10s + 20}{s^3 + 10s^2 + 24s}$$

$$= \frac{10(s+2)}{s(s+4)(s+6)}$$

MATLAB Program 6–13

```
>> A = [0   1   0;0   0   1;0   -24   -10];
>> B = [0;10;-80];
>> C = [1   0   0];
>> D = [0];
>> [num,den] = ss2tf(A,B,C,D)

num =

    0   -0.0000   10.0000   20.0000

den =

    1.0000   10.0000   24.0000   0
```

Design step 2: As the first trial, let us choose the desired closed-loop poles at

$$s = -1 + j2, \qquad s = -1 - j2, \qquad s = -5$$

and choose the desired observer poles at

$$s = -10, \qquad s = -10$$

Design step 3: We shall use MATLAB to compute the state feedback gain matrix **K** and the observer gain matrix **K**$_e$. MATLAB Program 6–14 produces these two matrices. In the program, matrices **J** and **L** represent the desired closed-loop poles for pole placement and the desired poles for the observer, respectively. The matrices **K** and **K**$_e$ are obtained as

$$\mathbf{K} = \begin{bmatrix} 1.25 & 1.25 & 0.19375 \end{bmatrix}$$

$$\mathbf{K}_e = \begin{bmatrix} 10 \\ -24 \end{bmatrix}$$

MATLAB Program 6–14

```
>> % Obtaining the state feedback gain matrix K
>>
>> A = [0   1   0;0   0   1;0   -24   -10];
>> B = [0;10;-80];
>> C = [1   0   0];
>> J = [-1+j*2   -1-j*2   -5];
>> K = acker(A,B,J)

K =

    1.2500    1.2500    0.19375

>> % Obtaining the observer gain matrix Ke
>>
>> Aaa = 0;Aab = [1 0];Aba = [0;0];Abb = [0   1;-24   -10];Ba = 0;Bb = [10;-80];
>> L = [-10   -10];
>> Ke = acker(Abb',Aab',L)'

Ke =

    10
   -24
```

Design step 4: We shall determine the transfer function of the observer controller. From Equation (6–71), the transfer function of the observer controller can be given by

$$G_c(s) = \frac{U(s)}{-Y(s)} = \frac{\text{num}}{\text{den}} = -[\widetilde{\mathbf{C}}(s\mathbf{I} - \widetilde{\mathbf{A}})^{-1}\widetilde{\mathbf{B}} + \widetilde{D}]$$

MATLAB Program 6–15 calculates the transfer function of the observer controller:

$$G_c(s) = \frac{9.1s^2 + 73.5s + 125}{s^2 + 17s - 30}$$

$$= \frac{9.1(s + 5.6425)(s + 2.4344)}{(s + 18.6119)(s - 1.6119)}$$

We define the system with this observer controller as System 1. Figure 6–15 shows the block diagram of System 1.

MATLAB Program 6–15

```
>> % Determination of transfer function of observer controller
>>
>> A = [0   1   0;0   0   1;0   -24   -10];
>> B = [0;10;-80];
>> Aaa = 0;Aab = [1   0]; Aba = [0;0]; Abb = [0   1;-24   -10];
>> Ba = 0;Bb = [10;-80];
>> Ka = 1.25;Kb = [1.25   0.19375];
>> Ke = [10;-24];
>> Ahat = Abb - Ke*Aab;
>> Bhat = Ahat*Ke + Aba - Ke*Aaa;
>> Fhat = Bb - Ke*Ba;
>> Atilde = Ahat - Fhat*Kb;
>> Btilde = Bhat - Fhat*(Ka + Kb*Ke);
>> Ctilde = -Kb;
>> Dtilde = -(Ka + Kb*Ke);
>> [num,den] = ss2tf(Atilde,Btilde,-Ctilde,-Dtilde)

num =
   9.1000   73.5000   125.0000

den =
   1.0000   17.0000   -30.0000
```

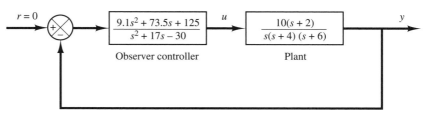

Figure 6–15
Block diagram of System 1.

The observer controller has a pole in the right-half s plane ($s = 1.6119$). The existence of an open-loop right-half s plane pole in the observer controller means that the system is open-loop unstable, although the closed-loop system is stable. The latter property can be seen from the characteristic equation for the system:

$$|s\mathbf{I} - \mathbf{A} + \mathbf{BK}| \cdot |s\mathbf{I} - \mathbf{A}_{bb} + \mathbf{K}_e\mathbf{A}_{ab}|$$
$$= s^5 + 27s^4 + 255s^3 + 1025s^2 + 2000s + 2500$$
$$= (s + 1 + j2)(s + 1 - j2)(s + 5)(s + 10)(s + 10) = 0$$

(See MATLAB Program 6–16 for the calculation of the characteristic equation.)

MATLAB Program 6–16

```
>> % Obtaining the characteristic equation
>>
>> [num1,den1] = ss2tf(A−B*K,eye(3),eye(3),eye(3),1);
>> [num2,den2] = ss2tf(Abb−Ke*Aab,eye(2),eye(2),eye(2),1);
>> charact_eq = conv(den1,den2)
>>

charact_eq =

   1.0e+003*

   0.0010   0.0270   0.2550   1.0250   2.0000   2.5000
```

A disadvantage of using an unstable controller is that the system becomes unstable if its dc gain becomes small. Such a control system is neither desirable nor acceptable. Hence, to get a satisfactory system, we need to modify the closed-loop pole location or the observer pole location (or both).

Second trial: Let us keep the desired closed-loop poles for pole placement as before, but modify the observer pole locations as follows:

$$s = -4.5, \quad s = -4.5$$

Thus,

$$\mathbf{L} = \begin{bmatrix} -4.5 & -4.5 \end{bmatrix}$$

Using MATLAB, we find the new $\mathbf{K}_e$ to be

$$\mathbf{K}_e = \begin{bmatrix} -1 \\ 6.25 \end{bmatrix}$$

Next, we shall obtain the transfer function of the observer controller. MATLAB Program 6–17 produces this transfer function as follows:

$$G_c(s) = \frac{1.2109s^2 + 11.2125s + 25.3125}{s^2 + 6s + 2.1406}$$

$$= \frac{1.2109(s + 5.3582)(s + 3.9012)}{(s + 5.619)(s + 0.381)}$$

> **MATLAB Program 6–17**
>
> ```
> >> % Determination of transfer function of observer controller.
> >>
> >> A = [0   1   0;0   0   1;0  -24  -10];
> >> B = [0;10;-80];
> >> Aaa = 0;Aab = [1   0];Aba = [0;0];Abb = [0   1;-24  -10];
> >> Ba = 0; Bb = [10;-80];
> >> Ka = 1.25; Kb = [1.25   0.19375];
> >> Ke = [-1;6.25];
> >> Ahat = Abb - Ke*Aab;
> >> Bhat = Ahat*Ke + Aba - Ke*Aaa;
> >> Fhat = Bb - Ke*Ba;
> >> Atilde = Ahat - Fhat*Kb;
> >> Btilde = Bhat - Fhat*(Ka + Kb*Ke);
> >> Ctilde = -Kb;
> >> Dtilde = -(Ka + Kb*Ke);
> >> [num,den] = ss2tf(Atilde,Btilde,-Ctilde,-Dtilde)
>
> num =
>     1.2109   11.2125   25.3125
> den =
>     1.0000    6.0000    2.1406
> ```

Notice that this is a stable controller. We define the system with this observer controller as System 2. We shall proceed to obtain the response of System 2 to the given initial condition

$$\mathbf{x}(0) = \begin{bmatrix} 1 \\ 0 \\ 0 \end{bmatrix}, \qquad \mathbf{e}(0) = \begin{bmatrix} 1 \\ 0 \end{bmatrix}$$

Substituting $u = -\mathbf{K}\tilde{\mathbf{x}}$ into the state-space equation for the plant, we obtain

$$\dot{\mathbf{x}} = \mathbf{A}\mathbf{x} - \mathbf{B}\mathbf{K}\tilde{\mathbf{x}} = \mathbf{A}\mathbf{x} - \mathbf{B}\mathbf{K}\begin{bmatrix} x_a \\ \tilde{\mathbf{x}}_b \end{bmatrix} = \mathbf{A}\mathbf{x} - \mathbf{B}\mathbf{K}\begin{bmatrix} x_a \\ \mathbf{x}_b - \mathbf{e} \end{bmatrix}$$

$$= \mathbf{A}\mathbf{x} - \mathbf{B}\mathbf{K}\left\{\mathbf{x} - \begin{bmatrix} 0 \\ \mathbf{e} \end{bmatrix}\right\} = \mathbf{A}\mathbf{x} - \mathbf{B}\mathbf{K}\mathbf{x} + \mathbf{B}[K_a \quad \mathbf{K}_b]\begin{bmatrix} 0 \\ \mathbf{e} \end{bmatrix} \qquad (6\text{--}73)$$

The error equation for the minimum-order observer is

$$\dot{\mathbf{e}} = (\mathbf{A}_{bb} - \mathbf{K}_e\mathbf{A}_{ab})\mathbf{e} \qquad (6\text{--}74)$$

Combining Equations (6–73) and (6–74), we get

$$\begin{bmatrix} \dot{\mathbf{x}} \\ \dot{\mathbf{e}} \end{bmatrix} = \begin{bmatrix} \mathbf{A} - \mathbf{B}\mathbf{K} & \mathbf{B}\mathbf{K}_b \\ \mathbf{0} & \mathbf{A}_{bb} - \mathbf{K}_e\mathbf{A}_{ab} \end{bmatrix} \begin{bmatrix} \mathbf{x} \\ \mathbf{e} \end{bmatrix}$$

with the initial condition

$$\begin{bmatrix} \mathbf{x}(0) \\ \mathbf{e}(0) \end{bmatrix} = \begin{bmatrix} 1 \\ 0 \\ 0 \\ 1 \\ 0 \end{bmatrix}$$

MATLAB Program 6–18 produces the response to the given initial condition. The response curves are shown in Figure 6–16. They seem to be acceptable.

MATLAB Program 6–18

```
>> % Response to initial condition.
>>
>> A = [0   1   0;0   0   1;0   −24   −10];
>> B = [0;10;−80];
>> K = [1.25   1.25   0.19375];
>> Kb = [1.25   0.19375];
>> Ke = [−1;6.25];
>> Aab = [1   0]; Abb = [0   1;−24   −10];
>> AA = [A − B*K   B*Kb;zeros(2,3)   Abb − Ke*Aab];
>> sys = ss(AA,eye(5),eye(5),eye(5));
>> t = 0:0.01:8;
>> x = initial(sys,[1;0;0;1;0],t);
>> x1 = [1   0   0   0   0]*x';
>> x2 = [0   1   0   0   0]*x';
>> x3 = [0   0   1   0   0]*x';
>> e1 = [0   0   0   1   0]*x';
>> e2 = [0   0   0   0   1]*x';
>>
>> subplot(3,2,1);plot(t,x1);grid
>> xlabel('t (sec)');ylabel('x_1')
>>
>> subplot(3,2,2);plot(t,x2);grid
>> xlabel('t (sec)');ylabel('x_2')
>>
>> subplot(3,2,3);plot(t,x3);grid
>> xlabel('t (sec)');ylabel('x_3')
>>
>> subplot(3,2,4);plot(t,e1);grid
>> xlabel('t (sec)');ylabel('e_1')
>>
>> subplot(3,2,5);plot(t,e2);grid
>> xlabel('t (sec)');ylabel('e_2')
```

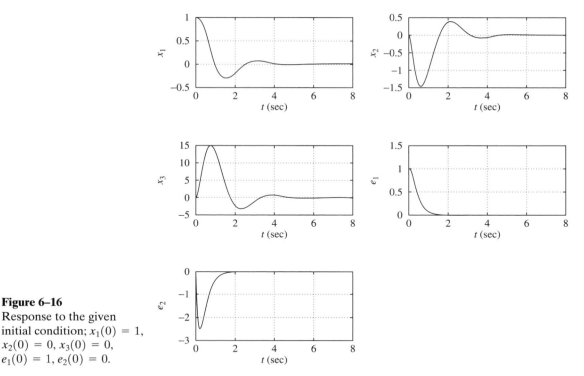

Figure 6–16
Response to the given initial condition; $x_1(0) = 1$, $x_2(0) = 0$, $x_3(0) = 0$, $e_1(0) = 1$, $e_2(0) = 0$.

Next, we shall check the frequency-response characteristics. The Bode diagram of the open-loop system just designed is shown in Figure 6–17. The phase margin is about 40° and the gain margin is $+\infty$ dB. The Bode diagram of the closed-loop system is shown in Figure 6–18. The bandwidth of the system is approximately 3.8 rad/sec.

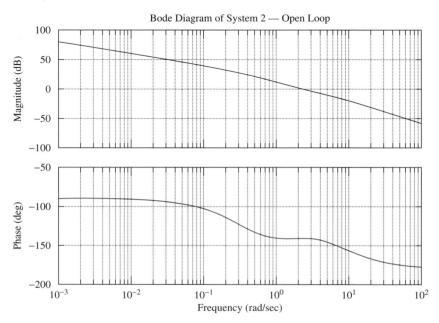

Figure 6–17
Bode diagram for the open-loop transfer function of System 2.

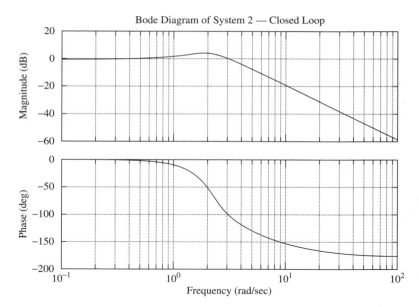

Figure 6–18
Bode diagram for the closed-loop transfer function of System 2.

Finally, we shall compare the root-locus plots of the first system with $L = \begin{bmatrix} -10 & -10 \end{bmatrix}$ and the second system with $L = \begin{bmatrix} -4.5 & -4.5 \end{bmatrix}$. The plot for the first system, given in Figure 6–19(a), shows that the system is unstable for a small dc gain and becomes stable for a large dc gain. The plot for the second system, shown in Figure 6–19(b), demonstrates that the system is stable for any positive dc gain.

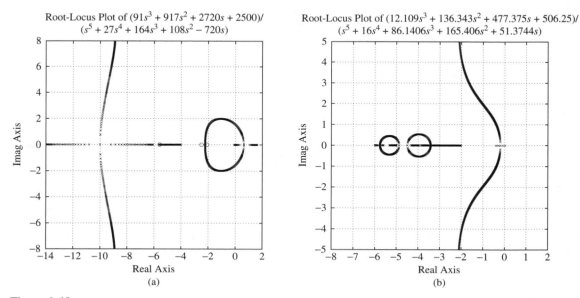

Figure 6–19
(a) Root-locus plot of the system with observer poles at $s = -10$ and $s = -10$; (b) root-locus plot of the system with observer poles at $s = -4.5$ and $s = -4.5$.

EXAMPLE 6-11 Consider the system shown in Figure 6–20. Design both the full-order and minimum-order observers for the plant. Assume that the desired closed-loop poles for the pole-placement part are located at

$$s = -2 + j2\sqrt{3}, \quad s = -2 - j2\sqrt{3}$$

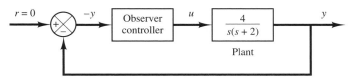

Figure 6–20
Regulator system.

Assume also that the desired observer poles are located at

(a) $s = -8$, $s = -8$ for the full-order observer

(b) $s = -8$ for the minimum-order observer

Compare the responses to the following initial conditions:

(a) For the full-order observer,

$$x_1(0) = 1, \quad x_2(0) = 0, \quad e_1(0) = 1, \quad e_2(0) = 0$$

(b) For the minimum-order observer,

$$x_1(0) = 1, \quad x_2(0) = 0, \quad e_1(0) = 1$$

Also, compare the bandwidths of both systems.

We first determine the state-space representation of the plant. By defining state variables x_1 and x_2 as

$$x_1 = y$$
$$x_2 = \dot{y}$$

we obtain

$$\begin{bmatrix} \dot{x}_1 \\ \dot{x}_2 \end{bmatrix} = \begin{bmatrix} 0 & 1 \\ 0 & -2 \end{bmatrix} \begin{bmatrix} x_1 \\ x_2 \end{bmatrix} + \begin{bmatrix} 0 \\ 4 \end{bmatrix} u$$

$$y = \begin{bmatrix} 1 & 0 \end{bmatrix} \begin{bmatrix} x_1 \\ x_2 \end{bmatrix}$$

For the pole-placement part, we determine the state feedback gain matrix $\mathbf{K}$. Using MATLAB, we find $\mathbf{K}$ to be

$$\mathbf{K} = \begin{bmatrix} 4 & 0.5 \end{bmatrix}$$

(See MATLAB Program 6–19.)

Next, we determine the observer gain matrix $\mathbf{K}_e$ for the full-order observer. Using MATLAB, we find that

$$\mathbf{K}_e = \begin{bmatrix} 14 \\ 36 \end{bmatrix}$$

(See MATLAB Program 6–19.)

MATLAB Program 6–19

```
>> % Obtaining matrices K and Ke.
>>
>> A = [0   1;0  −2];
>> B = [0;4];
>> C = [1   0];
>> J = [−2+j*2*sqrt(3)   −2−j*2*sqrt(3)];
>> L = [−8   −8];
>> K = acker(A,B,J)

K =
    4.0000    0.5000

>> Ke = acker(A',C',L)'

Ke =
    14
    36
```

Now we find the response of the system to the given initial condition. Referring to Equation (6–31), we have

$$\begin{bmatrix} \dot{\mathbf{x}} \\ \dot{\mathbf{e}} \end{bmatrix} = \begin{bmatrix} \mathbf{A} - \mathbf{BK} & \mathbf{BK} \\ \mathbf{0} & \mathbf{A} - \mathbf{K}_e\mathbf{C} \end{bmatrix} \begin{bmatrix} \mathbf{x} \\ \mathbf{e} \end{bmatrix}$$

This equation defines the dynamics of the designed system using the full-order observer. MATLAB Program 6–20 produces the response to the given initial condition. The resulting response curves are shown in Figure 6–21.

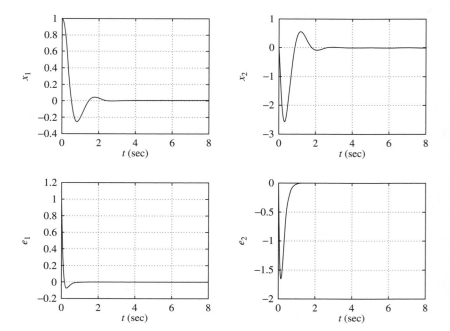

Figure 6–21
Response curves to initial condition.

MATLAB Program 6–20

```
>> % Response to initial condition—full-order observer
>>
>> A = [0   1;0   −2];
>> B = [0;4];
>> C = [1   0];
>> K = [4   0.5];
>> Ke = [14;36];
>> AA = [A − B*K   B*K;zeros(2,2)   A − Ke*C];
>> sys = ss(AA,eye(4),eye(4),eye(4));
>> t = 0:0.01:8;
>> x = initial(sys,[1;0;1;0],t);
>> x1 = [1   0   0   0]*x';
>> x2 = [0   1   0   0]*x';
>> e1 = [0   0   1   0]*x';
>> e2 = [0   0   0   1]*x';
>>
>> subplot(2,2,1);plot(t,x1);grid
>> xlabel('t(sec)');ylabel('x_1')
>>
>> subplot(2,2,2);plot(t,x2);grid
>> xlabel('t(sec)');ylabel('x_2')
>>
>> subplot(2,2,3);plot(t,e1);grid
>> xlabel('t(sec)');ylabel('e_1')
>>
>> subplot(2,2,4);plot(t,e2);grid
>> xlabel('t (sec)');ylabel('e_2')
```

MATLAB Program 6–21 produces the transfer function of the observer controller:

$$\frac{\text{num}}{\text{den}} = \frac{74s + 256}{s^2 + 18s + 108} = \frac{74(s + 3.4595)}{(s + 9 + j5.1962)(s + 9 - j5.1962)}$$

MATLAB Program 6–22 produces the observer gain matrix K_e for the minimum-order observer:

$$K_e = 6$$

MATLAB Program 6–21

```
>> % Determination of transfer function of observer controller—full-order observer
>>
>> A = [0   1;0   −2];
>> B = [0;4];
>> C = [1   0];
>> K = [4   0.5];
>> Ke = [14;36];
>> [num,den] = ss2tf(A − Ke*C − B*K,Ke,K,0)

num =

    0   74.0000   256.0000

den =

    1   18   108
```

MATLAB Program 6–22

```
>> % Obtaining Ke—minimum-order observer
>>
>> Aab = [1];
>> Abb = [−2];
>> LL = [−8];
>> Ke = acker(Abb',Aab',LL)'

Ke =
      6
```

The response of the system with minimum-order observer to the initial condition can be obtained as follows: Substituting $u = -\mathbf{K}\widetilde{\mathbf{x}}$ into the plant equation

$$\dot{\mathbf{x}} = \mathbf{A}\mathbf{x} + \mathbf{B}u$$

we find that

$$\dot{\mathbf{x}} = \mathbf{A}\mathbf{x} - \mathbf{B}\mathbf{K}\widetilde{\mathbf{x}} = \mathbf{A}\mathbf{x} - \mathbf{B}\mathbf{K}\mathbf{x} + \mathbf{B}\mathbf{K}(\mathbf{x} - \widetilde{\mathbf{x}})$$

$$= (\mathbf{A} - \mathbf{B}\mathbf{K})\mathbf{x} + \mathbf{B}[K_a \quad K_b]\begin{bmatrix} 0 \\ e \end{bmatrix}$$

or

$$\dot{\mathbf{x}} = (\mathbf{A} - \mathbf{B}\mathbf{K})\mathbf{x} + \mathbf{B}K_b e$$

The error equation is

$$\dot{e} = (A_{bb} - K_e A_{ab})e$$

Hence, the system dynamics are defined by

$$\begin{bmatrix} \dot{\mathbf{x}} \\ \dot{e} \end{bmatrix} = \begin{bmatrix} \mathbf{A} - \mathbf{BK} & \mathbf{B}K_b \\ 0 & A_{bb} - K_e A_{ab} \end{bmatrix} \begin{bmatrix} \mathbf{x} \\ e \end{bmatrix}$$

On the basis of this last equation, MATLAB Program 6–23 produces the response to the given initial condition. The resulting response curves are shown in Figure 6–22.

MATLAB Program 6–23

```
>> % Response to intial condition—minimum-order observer
>>
>> A = [0   1;0   −2];
>> B = [0;4];
>> K = [4   0.5];
>> Kb = 0.5;
>> Ke = 6;
>> Aab = 1;Abb = −2;
>> AA = [A − B*K   B*Kb;zeros(1,2)   Abb − Ke*Aab];
>> sys = ss(AA,eye(3),eye(3),eye(3));
>> t = 0:0.01:8;
>> x = initial(sys,[1;0;1],t);
>> x1 = [1   0   0]*x';
>> x2 = [0   1   0]*x';
>> e = [0   0   1]*x';
>>
>> subplot(2,2,1);plot(t,x1);grid
>> xlabel('t(sec)');ylabel('x_1')
>>
>> subplot(2,2,2);plot(t,x2);grid
>> xlabel('t(sec)');ylabel('x_2')
>>
>> subplot(2,2,3);plot(t,e);grid
>> xlabel('t(sec)');ylabel('e')
```

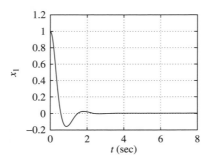
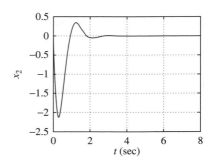
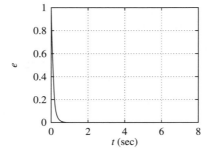

Figure 6–22
Response curves to initial condition.

The transfer function of the observer controller when the system uses the minimum-order observer can be obtained with MATLAB Program 6–24. The result is

$$\frac{\text{num}}{\text{den}} = \frac{7s + 32}{s + 10} = \frac{7(s + 4.5714)}{s + 10}$$

MATLAB Program 6–24

```
>> % Determination of transfer function of observer controller—minimum-order observer
>>
>> A = [0  1;0  −2];
>> B = [0;4];
>> Aaa = 0;Aab = 1;Aba = 0;Abb = −2;
>> Ba = 0;Bb = 4;
>> Ka = 4;Kb = 0.5;
>> Ke = 6;
>> Ahat = Abb − Ke*Aab;
>> Bhat = Ahat*Ke + Aba − Ke*Aaa;
>> Fhat = Bb − Ke*Ba;
>> Atilde = Ahat − Fhat*Kb;
>> Btilde = Bhat − Fhat*(Ka + Kb*Ke);
>> Ctilde = −Kb;
>> Dtilde = −(Ka + Kb*Ke);
>> [num,den] = ss2tf(Atilde,Btilde,−Ctilde,−Dtilde)

num =
    7   32

den =
    1   10
```

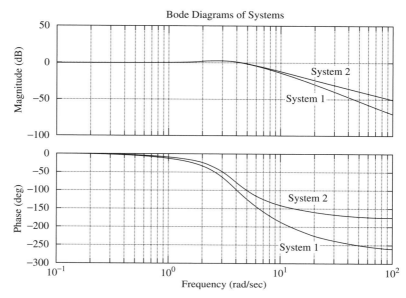

Figure 6–23
Bode diagrams of System 1 (system with full-order observer) and System 2 (system with minimum-order observer). System 1 = $(296s + 1024)/(s^4 + 20s^3 + 144s^2 + 512s + 1024)$, System 2 = $(28s + 128)/(s^3 + 12s^2 + 48s + 128)$

The observer controller is clearly a lead compensator.

The Bode diagrams of System 1 (the closed-loop system with full-order observer) and of System 2 (the closed-loop system with minimum-order observer) are shown in Figure 6–23. Clearly, the bandwidth of System 2 is wider than that of System 1. System 1 has a better high-frequency noise-rejection characteristic than System 2.

Comments

1. In designing regulator systems, if the dominant controller poles are placed far to the left of the $j\omega$-axis, the elements of the state feedback gain matrix **K** will become large. Large gain values will make the actuator output become large, so saturation may take place. Then the designed system will not behave as intended.

   In designing a system by the pole-placement approach, several different sets of desired closed-loop poles need be considered, the response characteristics compared, and the best one chosen.

2. If not all state variables can be measured, an observer must be incorporated to estimate the unmeasurable state variables.

3. Adding an observer to the system generally reduces the stability margin. In some cases, an observer controller may have zero(s) in the right-half s plane, which means that the controller may be stable, but of nonminimum

phase. In other cases, the controller may have pole(s) in the right-half s plane—that is, the controller is unstable. Then the designed system may become conditionally stable. Note that an unstable observer controller is not acceptable.

4. If the observer controller becomes unstable, move the observer poles to the right in the left-half s plane until the observer controller becomes stable. Also, the desired closed-loop pole locations may need to be modified.

5. If the observer poles are placed far to the left of the $j\omega$-axis, the bandwidth of the observer will increase and will cause noise problems. If there is a serious noise problem, the observer poles should not be placed too far to the left of the $j\omega$-axis. The general requirement is that the bandwidth should be sufficiently low that the sensor noise will not become a problem.

    The bandwidth of the system with the minimum-order observer is higher than that of the system with the full-order observer, provided that the multiple observer poles are placed at the same location for both observers. If sensor noise is a serious problem, a full-order observer is recommended.

6. When the system is designed by the pole-placement-with-observer approach, it is advisable to check the stability margins (phase margin and gain margin) by a frequency-response method. If the system designed has poor stability margins, it is possible that the designed system may become unstable if the mathematical model involves uncertainties.

7. For nth-order systems, classical design methods (root-locus and frequency-response methods) yield low-order compensators (first or second order). Since the observer-based controllers are nth order (or mth order if the minimum-order observer is used) for an nth-order system, the designed system will become $2n$th order [or $(n + m)$th order]. Since lower order compensators are cheaper than higher order ones, the designer should first apply classical methods and, if no suitable compensators can be determined, then try the pole-placement-with-observer design approach presented in this chapter.

Some Optimization Problems Solved with MATLAB

7-1 COMPUTATIONAL APPROACH TO OBTAINING OPTIMAL SETS OF PARAMETER VALUES

In this section, we shall explore how to use MATLAB to obtain optimal sets of parameter values of control systems to satisfy transient-response specifications. We shall present several examples illustrating the MATLAB approach to obtaining optimal solutions.

EXAMPLE 7-1 Consider the control system shown in Figure 7–1(a). The controller is a PID controller. The transfer function of the PID controller is given by

$$G_c(s) = K_P\left(1 + \frac{1}{T_i s} + T_d s\right)$$

$$= \frac{K(s + \alpha)(s + \beta)}{s}$$

Frequently, we choose $\alpha = \beta$ so that $G_c(s)$ may be written as

$$G_c(s) = \frac{K(s + a)^2}{s} \qquad (7\text{–}1)$$

with this choice, the number of parameters of the PID controller becomes 2 (K and a) instead of 3 (K, α, and β). This simplification makes the optimization process simpler.

375

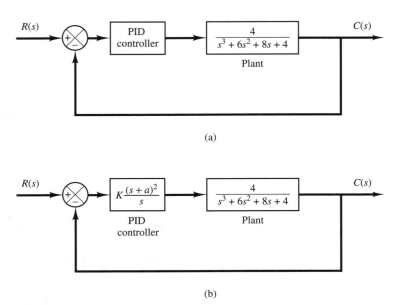

Figure 7–1
(a) PID-controlled system; (b) PID-controlled system with a simplified PID controller.

We assume here that the PID controller has the transfer function given by Equation (7–1), as shown in Figure 7–1(b).

We seek a combination of K and a such that the closed-loop system is underdamped and the maximum overshoot in the unit-step response is less than 15% but more than 10%. (Other conditions can be included—for example, that the settling time be less than a specified value; see Example 7–5.)

To solve this problem with MATLAB, we first specify the region to search for appropriate K and a. We then write a MATLAB program that, in the unit-step response, will find a combination of K and a which will satisfy the criterion that the maximum overshoot is less than 15% but more than 10%.

Note that the gain K should not be too large, so as to avoid the possibility that the system require an unnecessarily large power unit.

Assume that the region to search for K and a is

$$3 \le K \le 5 \quad \text{and} \quad 0.1 \le a \le 3$$

If a solution does not exist in this region, then we need to expand it. In some problems, there is no solution, no matter what the search region might be.

In the computational approach, we need to determine the step size for each of K and a. In the actual design process, we need to choose step sizes small enough. However, in this example, to avoid an overly large number of computations, we choose the step sizes to be reasonable—say, 0.2 for K and 0.1 for a.

There may be more than one possible combination of K and a that will satisfy the given condition. Here, however, we shall obtain the first solution. (For approaches to finding all possible sets of solutions in the given search region, see Examples 7–3 through 7–5.)

We proceed by writing a MATLAB program with nested loops that begin with the lowest values of K and a and step toward the highest, because we would like to make the value of the gain K smaller rather than larger. (In some problems, we may write a MATLAB program to begin with the highest values of given parameters and step toward the lowest.) We shall write a MATLAB program to plot the unit-step response curve of the system, using the first-found solution of K and a. Such a first-found solution gives the smallest value of the gain K for the power unit for the system configuration considered.

To solve the present problem, we may write many different MATLAB programs. Here, we present two of them: MATLAB Program 7–1 and MATLAB Program 7–2.

MATLAB Program 7–1: In MATLAB Program 7–1, we write the numerator and denominator of the closed-loop system in terms of K and a. The closed-loop transfer function of the system is

$$\frac{C(s)}{R(s)} = \frac{4K(s+a)^2}{s(s^3 + 6s^2 + 8s + 4) + 4K(s+a)^2}$$

$$= \frac{4Ks^2 + 8Kas + 4Ka^2}{s^4 + 6s^3 + (8 + 4K)s^2 + (4 + 8Ka)s + 4Ka^2}$$

For each set of K and a values chosen, we examine the unit-step response curve if the condition $m < 1.15$ and $m > 1.10$, where m is the maximum value of the output, is satisfied.

MATLAB Program 7–1

```
>> t = 0:0.01:8;
>> for K = 3:0.2:5;
     for a = 0.1:0.1:3;
       num = [4*K   8*K*a   4*K*a^2];
       den = [1   6   8+4*K   4+8*K*a   4*K*a^2];
       y = step(num,den,t);
       m = max(y);
       if m<1.15 & m>1.10;
       break; % Breaks the inner loop.
       end
     end
       if m<1.15 & m>1.10;
       break; % Breaks the outer loop.
       end
     end
>> solution = [K   a   m]

solution =

   3.0000   1.0000   1.1469

>> plot(t,y) % Unit-step response curve is shown in Figure 7–2.
>> grid
>> title('Unit-Step Response')
>> xlabel('t sec')
>> ylabel('Output y(t)')
```

In solving this problem, we use two for loops. We start the program with the outer loop to vary the K values. Next, we vary the a values in the inner loop. If the condition is satisfied, we break the inner loop and terminate the process of varying the a values. Then we break the outer loop to terminate the process of varying the K values and finish up the searching process.

When MATLAB finishes its computations for the first-found set of K and a that satisfies the condition, we plot the response curve, shown in Figure 7–2.

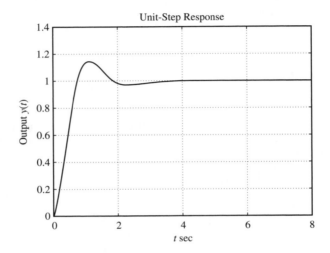

Figure 7–2
Unit-step response curve for the case where $K = 3$ and $a = 1$.

Note that, depending on the system, the region over which the search for K and a is conducted, the step sizes used, it may take from a few seconds to a minute or more for MATLAB to compute the desired set of values. In the program, the statement

$$\text{solution} = [K \quad a \quad m]$$

displays the chosen set of values of K, a, and m.

MATLAB Program 7–2: MATLAB Program 7–2 is basically the same as MATLAB Program 7–1. We use two loops (the outer loop for varying K values and the inner loop for varying a values). We define the PID controller by num1 and den1 and obtain the transfer-function expression tf1 for the controller. Then we define the plant by num2 and den2 and obtain the transfer-function expression tf2 for the plant. The open-loop transfer function for the system is obtained by multiplying tf1 and tf2.

MATLAB Program 7-2

```
>> t = 0:0.01:8;
>> for K = 3:0.2:5;  %  Starts the outer loop to vary the K values.
       for a = 0.1:0.1:3;  % Starts the inner loop to vary the a values.
           num1 = K*[1  2*a  a^2];
           den1 = [1   0];
            tf1 = tf(num1,den1);
           num2 = [4];
           den2 = [1   6   8   4];
           tf2 = tf(num2,den2);
           tf3 = tf1*tf2;
           sys = feedback(tf3,1);
            y = step(sys,t);
            m = max(y);
           if m<1.15  &  m>1.10;
           plot(t,y); % See Figure 7-3 for the unit-step response curve obtained.
           grid;
           title('Unit-Step Response')
           xlabel('t sec')
           ylabel('Output y(t)')
           solution = [K   a   m]
            break; % Breaks the inner loop
          end
          end
       if m<1.15  &  m>1.10;
       break;  %  Breaks the outer loop.
      end
      end

solution =

   3.0000    1.0000    1.1469

>> KK = num2str(K); % String value of K to be printed on plot.
>> aa = num2str(a); % String value of a to be printed on plot.
>> text(2.20,0.73,'K = '), text(2.75,0.73,KK)
>> text(2.20,0.53,'a = '), text(2.75,0.53,aa)
```

Using the feedback command, we obtain the closed-loop transfer function sys. Then we get the unit-step response of the closed-loop system. If the maximum output m is less than 1.15, but more than 1.10, this MATLAB program automatically displays the first-found solution of K, a, and m. Figure 7-3 is the unit-step response curve obtained with the program.

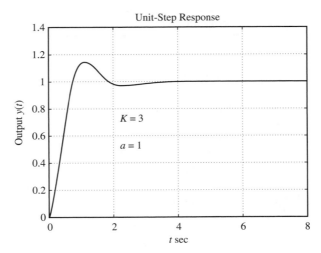

Figure 7–3
Unit-step response curve obtained with MATLAB Program 7–2.

EXAMPLE 7-2 In this example, we consider the case where there is no optimal solution. Consider again the system shown in Figure 7–1(b). Let us assume that we wish to find a combination of K and a such that the maximum output m satisfies the condition

$$m < 2.60 \quad \text{and} \quad m > 2.55$$

We assume that the search region for K and a is

$$2 \leq K \leq 5 \quad \text{and} \quad 0.1 \leq a \leq 3$$

The computation step sizes are chosen to be 0.2 for K and 0.1 for a.

A MATLAB program to solve this problem is MATLAB Program 7–3, which is similar to MATLAB Program 7–1, except that the condition for m is different. According to MATLAB program 7–3, the solution is

$$K = 5.0000, \quad a = 3.0000, \quad m = 2.9784$$

Obviously, the condition for m is not satisfied and the solution given is not the real solution, but is the last search point. The way the program is written, it assumes that the search point which corresponds to the break point (the point at which breaking the inner loop and outer loop occur) is the solution. So, if a real solution does not exist and the last point is tested, this last search point is regarded as the solution because the search process ends (that is, both inner and outer loops break) at that point.

```
MATLAB Program 7-3
>> t= 0:0.01:8;
>> for K = 3:0.2:5;
     for a = 0.1:0.1:3;
       num = [4*K   8*K*a   4*K*a^2];
       den = [1   6   8+4*K   4+8*K*a   4*K*a^2];
         y = step(num,den,t);
         m = max(y);
         if m<2.60 & m>2.55;
         break
         end
       end
         if m<2.60 & m>2.55;
         break
         end
     end
>> solution = [K   a   m]

solution =

    5.0000    3.0000    2.9784

>>
>> % Note that the point K = 5, a = 3 corresponds to the last
>> % search point. It is not the solution to this problem.
```

Of course, it is obvious that $m = 2.9784$ does not satisfy the required condition ($m < 2.60$ and $m > 2.55$), so the point $K = 5, a = 3$ (the last search point) is not the solution point.

In the MATLAB program, the search point at which breaking the inner and outer loops occurs is treated as the solution. The last search point corresponds to the point where the inner and outer loops break. So, unless a solution was found in an earlier search process, the last search point will appear as if it is a solution.

To avoid this situation, we may add NaN (the code for "not a number") at the end of the search regions for K and a. For example, we may define

$$K = [3:0.2:5 \quad \text{NaN}]$$
$$a = [0.1:0.1:3 \quad \text{NaN}]$$

This means that

$$K = 3, 3.2, 3.4, \ldots, 4.8, 5, \text{NaN}$$
$$a = 0.1, 0.2, 0.3, \ldots, 2.9, 3, \text{NaN}$$

See the MATLAB outputs shown below.

```
>> K = [3:0.5:5]
K =
    3.0000    3.5000    4.0000    4.5000    5.0000
>> K = [3:0.5:5   NaN]
K =
    3.0000    3.5000    4.0000    4.5000    5.0000   NaN
```

If NaN is added in the search for K and a, then the last search point in the search procedure becomes

$$[K \quad a] = [NaN \quad NaN]$$

So, if the real solution does not exist, we get

$$solution = [NaN \quad NaN \quad NaN]$$

(See MATLAB Program 7–4.)

MATLAB Program 7–4

```
>> t = 0:0.01:8;
>> for K = [3:0.2:5   NaN];
     for a = [0.1:0.1:3   NaN];
       num = [4*K   8*K*a   4*K*a^2];
       den = [1   6   8+4*K   4+8*K*a   4*K*a^2];
         y = step(num,den,t);
         m = max(y);
         if m<2.60 & m>2.55;
         break
       end
     end
         if m<2.60 & m>2.55;
         break
       end
     end
>> solution = [K   a   m]

solution =
   NaN   NaN   NaN
```

Since the real solution does not exist in this case, the last search point K = NaN, a = NaN appears as if it is the solution. The solution [K a m] is expressed as

$$[K \quad a \quad m] = [NaN \quad NaN \quad NaN]$$

Clearly, it indicates that the solution that satisfies the given condition does not exist.

For the case where a real solution exists, the inclusion or noninclusion of NaN in the search for K and a does not make any difference as far as getting the real solution is concerned. For example, in MATLAB Program 7–5, the condition on m is

$$m < 1.20 \quad \text{and} \quad m > 1.15$$

and the solution exists.

In MATLAB Program 7–6, the condition on m is

$$m < 1.20 \quad \text{and} \quad m > 1.18$$

and the solution also exists.

Notice that, in comparing these two MATLAB programs, if the condition on m is made more strict, the optimal K value (where K is the gain constant of the PID controller) becomes larger. In many practical cases, a controller with a smaller K is cheaper to build (or buy) and is desirable. So the condition on m should not be more restrictive than is necessary.

MATLAB Program 7–5

```
>> t = 0:0.01:8;
>> for K = [3:0.2:5   NaN];
     for a = [0.1:0.1:3   NaN];
       num = [4*K   8*K*a   4*K*a^2];
       den = [1   6   8+4*K   4+8*K*a   4*K*a^2];
         y = step(num,den,t);
         m = max(y);
         if m<1.20 & m>1.15;
         break
       end
     end
         if m<1.20 & m>1.15;
         break
     end
   end
>> solution = [K   a   m]

solution =

    3.0000   1.1000   1.1979
```

Section 7–1 / Computational Approach to Obtaining Optimal Sets

MATLAB Program 7-6

```
>> t = 0:0.01:8;
>> for K = [3:0.2:5   NaN];
     for a = [0.2:0.2:2.2   NaN];
        num = [4*K   8*K*a   4*K*a^2];
        den = [1   6   8+4*K   4+8*K*a   4*K*a^2];
        y = step(num,den,t);
        m = max(y);
        if m<1.20 & m >1.18;
        break
        end
     end
        if m<1.20 & m >1.18;
        break
        end
     end
>> solution = [K   a   m]

solution =

   3.8000   1.0000   1.1888
```

In Examples 7–1 and 7–2, we wrote MATLAB programs to find the first set of parameters that satisfy the given specifications. For any particular problem, there may be more than one set of parameters that satisfy the specifications. In what follows, we shall present methods for obtaining all sets of parameters that satisfy a given set of specifications.

EXAMPLE 7-3 Consider the same system as in Example 7–1, except that the problem here is to find all sets of K and a that will satisfy the given specification that the maximum overshoot in the unit-step response be between 10% and 20%.

The system is shown in Figure 7–1(b). Assume that the search region here is defined by

$$3 \leq K \leq 5, \quad 0.1 \leq a \leq 3$$

We choose the step sizes for computation to be 0.2 for K and 0.1 for a. We shall obtain all sets of K and a combination that will satisfy the following condition:

$$m < 1.20 \quad \text{and} \quad m > 1.10$$

We can write many different MATLAB programs to solve this problem. Here, we present one such program: MATLAB Program 7–7. Note that this program is different from MATLAB Program 7–3 in that the last search point is not a solution unless it satisfies the condition on m.

In MATLAB Program 7–7, the statement

$$\text{solution}(k,:) = [K(i) \quad a(j) \quad m]$$

will produce a table of [K   a   m] values. (In the current system, there are 24 sets of K and a combination that exhibit m < 1.20 and m > 1.10—that is, the maximum overshoot is between 10% and 20%.) To sort out the solution sets in the order of the magnitude of the maximum overshoot (starting from the smallest value of m in the table and ending with the largest value), we use the command

$$\text{sortsolution} = \text{sortrows}(\text{solution}, 3)$$

(This means that the solutions are sorted out according to column 3.)

If we want to plot the response curve with the overshoot closest to 15%, we use the K and a values shown in the 13th row in the sortsolution. That is, we enter the commands

$$K = \text{sortsolution}(13,1)$$
$$a = \text{sortsolution}(13,2)$$

and use the step command. If we want to plot the unit-step response curve of the system with any set of K and a values in the sorted table, we specify the K and a values by entering the sortsolution commands

$$K = \text{sortsolution}(n,1)$$
$$a = \text{sortsolution}(n,2)$$

where n is the row number in the sorted table.

MATLAB Program 7–7

```
>> t = 0:0.01:8;
>> K = 3:0.2:5;
>> a = 0.1:0.1:3;
>> k = 0;
>> for i = 1:11;
     for j = 1:30;
       num = [4*K(i)   8*K(i)*a(j)   4*K(i)*a(j)^2];
       den = [1   6   8+4*K(i)   4+8*K(i)*a(j)   4*K(i)*a(j)^2];
       y = step(num,den,t);
       m = max(y);
       if m<1.20 & m>1.10;
       k = k+1;
       solution(k,:) = [K(i)   a(j)   m];
     end
   end
 end
```

```
>> solution

solution =

    3.0000    1.0000    1.1469
    3.0000    1.1000    1.1979
    3.2000    0.9000    1.1065
    3.2000    1.0000    1.1581
    3.4000    0.9000    1.1181
    3.4000    1.0000    1.1688
    3.6000    0.9000    1.1291
    3.6000    1.0000    1.1791
    3.8000    0.9000    1.1396
    3.8000    1.0000    1.1888
    4.0000    0.8000    1.1003
    4.0000    0.9000    1.1497
    4.0000    1.0000    1.1982
    4.2000    0.8000    1.1107
    4.2000    0.9000    1.1594
    4.4000    0.8000    1.1208
    4.4000    0.9000    1.1687
    4.6000    0.8000    1.1304
    4.6000    0.9000    1.1776
    4.8000    0.8000    1.1396
    4.8000    0.9000    1.1862
    5.0000    0.7000    1.1019
    5.0000    0.8000    1.1485
    5.0000    0.9000    1.1945

>> sortsolution = sortrows(solution,3)

sortsolution =

    4.0000    0.8000    1.1003
    5.0000    0.7000    1.1019
    3.2000    0.9000    1.1065
    4.2000    0.8000    1.1107
    3.4000    0.9000    1.1181
    4.4000    0.8000    1.1208
    3.6000    0.9000    1.1291
    4.6000    0.8000    1.1304
    4.8000    0.8000    1.1396
    3.8000    0.9000    1.1396
    3.0000    1.0000    1.1469
    5.0000    0.8000    1.1485
    4.0000    0.9000    1.1497
    3.2000    1.0000    1.1581
```

```
        4.2000    0.9000    1.1594
        4.4000    0.9000    1.1687
        3.4000    1.0000    1.1688
        4.6000    0.9000    1.1776
        3.6000    1.0000    1.1791
        4.8000    0.9000    1.1862
        3.8000    1.0000    1.1888
        5.0000    0.9000    1.1945
        3.0000    1.1000    1.1979
        4.0000    1.0000    1.1982

>> % Plot the response with the overshoot closest to 15%
>>
>> K = sortsolution(13,1)

K =

    4

>> a = sortsolution(13,2)

a =

    0.9000

>> num = [4*K    8*K*a    4*K*a^2];
>> den = [1   6   8+4*K    4+8*K*a    4*K*a^2];
>>
>> num

num =

   16.0000   28.8000   12.9600

>> den

den =

    1.0000    6.0000    24.0000    32.8000    12.9600

>> y = step(num,den,t);
>> plot(t,y)
>> grid
>> title('Unit-Step Response')
>> xlabel('t sec')
>> ylabel('Output y')
```

The unit-step response curve with $K = 4$ and $a = 0.9$ is shown in Figure 7–4.

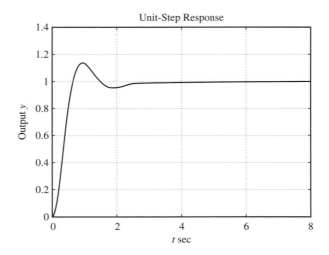

Figure 7–4
Unit-step response curve of the system with $K = 4$ and $a = 0.9$. (The maximum overshoot is 14.97%.)

EXAMPLE 7–4 Consider the system shown in Figure 7–5. We shall obtain all sets of K and a values which satisfy the specification that the maximum overshoot in the unit-step response be less than 10%. (This means that overdamped systems are included.) Assume the search region to be

$$2 \le K \le 3, \qquad 0.5 \le a \le 1.5$$

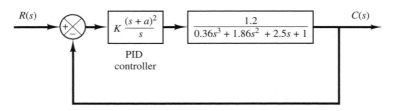

Figure 7–5
PID-controlled system.

In the actual design process, the step size should be sufficiently small. In this example, however, we choose a fairly large step size to make the total number of search points reasonable. Thus, we choose the step size of both K and a to be 0.2.

We can write many different MATLAB programs to solve this problem. Here, we present one possible such program: MATLAB Program 7–8.

MATLAB Program 7-8

```
>> %'K' and 'a' values to test
>>
>> K = [2.0   2.2   2.4   2.6   2.8   3.0];
>> a = [0.5   0.7   0.9   1.1   1.3   1.5];
>>
>> % Evaluate closed-loop unit-step response at each 'K' and 'a' combination
>> % that will yield a maximum overshoot less than 10%
>>
>> t = 0:0.01:5;
>> g = tf([1.2],[0.36   1.86   2.5   1]);
>> k = 0;
>> for i = 1:6;
      for j = 1:6;
        gc = tf(K(i)*[1   2*a(j)   a(j)^2],[1   0]); % controller
        G = gc*g/(1 + gc*g); % closed-loop transfer function
        y = step(G,t);
        m = max(y);
        if m < 1.10
        k = k+1;
        solution(k,:) = [K(i)   a(j)   m];
        end
      end
    end
>> solution % Print solution table

solution =

    2.0000    0.5000    0.9002
    2.0000    0.7000    0.9807
    2.0000    0.9000    1.0614
    2.2000    0.5000    0.9114
    2.2000    0.7000    0.9837
    2.2000    0.9000    1.0772
    2.4000    0.5000    0.9207
    2.4000    0.7000    0.9859
    2.4000    0.9000    1.0923
    2.6000    0.5000    0.9283
    2.6000    0.7000    0.9877
    2.8000    0.5000    0.9348
    2.8000    0.7000    1.0024
    3.0000    0.5000    0.9402
    3.0000    0.7000    1.0177

>> sortsolution = sortrows(solution,3) % Print solution table sorted by
>>                                     % column 3
```

```
sortsolution =
    2.0000    0.5000    0.9002
    2.2000    0.5000    0.9114
    2.4000    0.5000    0.9207
    2.6000    0.5000    0.9283
    2.8000    0.5000    0.9348
    3.0000    0.5000    0.9402
    2.0000    0.7000    0.9807
    2.2000    0.7000    0.9837
    2.4000    0.7000    0.9859
    2.6000    0.7000    0.9877
    2.8000    0.7000    1.0024
    3.0000    0.7000    1.0177
    2.0000    0.9000    1.0614
    2.2000    0.9000    1.0772
    2.4000    0.9000    1.0923
>> % Plot the response with the largest overshoot that is less than 10%
>>
>> K = sortsolution(15,1)
K =
    2.4000
>> a = sortsolution(15,2)
a =
    0.9000
>> gc = tf(K*[1    2*a    a^2],[1    0]);
>> G = gc*g/(1 + gc*g);
>> step(G,t)
>> grid % See Figure 7-6
>>
>> % If you wish to plot the response with the smallest overshoot that is
>> % greater than 0%, then enter the following values of 'K' and 'a'
>>
>> K = sortsolution(11,1)
K =
    2.8000
>> a = sortsolution(11,2)
a =
    0.7000
>> gc = tf(K*[1    2*a    a^2],[1    0]);
>> G = gc*g/(1 + gc*g);
>> step(G,t)
>> grid % See Figure 7-7
```

If we want to plot the unit-step response curve of the last set of the K and a values in the sorted table, we enter the commands

$$K = \text{sortsolution}(15,1)$$
$$a = \text{sortsolution}(15,2)$$

Then we get the transfer-function expression and use the step command. The resulting unit-step response curve is shown in Figure 7–6. Similarly, to plot the unit-step response curve with the smallest overshoot that is greater than 0% found in the sorted table, we enter the commands

$$K = \text{sortsolution}(11,1)$$
$$a = \text{sortsolution}(11,2)$$

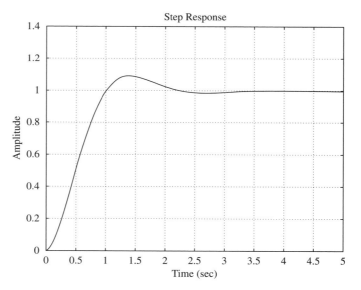

Figure 7–6
Unit-step response curve of the system with $K = 2.4$ and $a = 0.9$. (The maximum overshoot is 9.23%.)

Again, we get the transfer-function expression and use the step command. (The resulting unit-step response curve is shown in Figure 7–7.)

Note that if the specification were that the maximum overshoot be between 10% and 5%, there would be three sets of solutions:

$$K = 2.0000, \quad a = 0.9000, \quad m = 1.0614$$
$$K = 2.2000, \quad a = 0.9000, \quad m = 1.0772$$
$$K = 2.4000, \quad a = 0.9000, \quad m = 1.0923$$

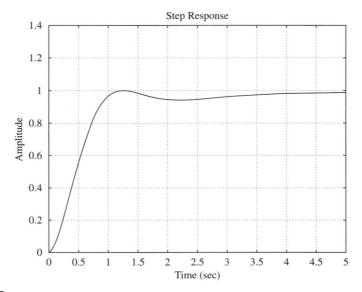

Figure 7–7
Unit-step response curve of the system with $K = 2.8$ and $a = 0.7$. (The maximum overshoot is 0.24%.)

Unit-step response curves for these three cases are shown in Figure 7–8. Notice that the system with a larger gain K has a smaller rise time and a larger maximum overshoot. Which one of these three systems is best depends on the designer's objective.

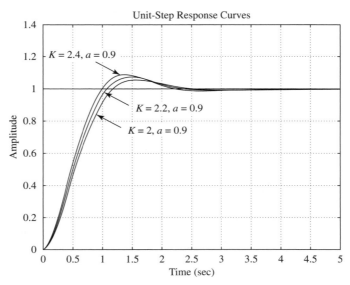

Figure 7–8
Unit-step response curves of system with $K = 2$, $a = 0.9$; $K = 2.2$, $a = 0.9$; and $K = 2.4$, $a = 0.9$.

EXAMPLE 7–5 Consider again the system shown in Figure 7–1(b). We want to find all combinations of K and a values such that the closed-loop system has a maximum overshoot of less than 15%, but more than 10%, in the unit-step response. In addition, the settling time should be less than 3 sec. In this problem, assume that the search region is

$$3 \le K \le 5 \quad \text{and} \quad 0.1 \le a \le 3$$

Determine the best choice of the parameters K and a.

MATLAB Program 7–9 gives the solution to this problem. From the sortsolution table, it looks like the first three rows are good choices. The unit-step response curves for the system corresponding to these three rows appear almost identical. Figure 7–9 shows the unit-step response curve for $K = 3.2$ and $a = 0.9$. Since this choice requires a smaller K value than most other choices, we may decide that the first row is the best choice.

MATLAB Program 7–9

```
>> t = 0:0.01:8;
>> k = 0;
>> for K = 3:0.2:5;
    for a = 0.1:0.1:3;
       num = [4*K   8*K*a   4*K*a^2];
       den = [1   6   8+4*K   4+8*K*a   4*K*a^2];
        y = step(num,den,t);
        s = 801;while y(s)>0.98 & y(s)<1.02; s = s − 1;end;
      ts = (s−1)*0.01; % ts = settling time;
      m = max(y);
      if m<1.15 & m>1.10; if ts<3.00;
         k = k+1;
         solution(k,:) = [K   a   m   ts];
        end
       end
      end
     end
>> solution

solution =

    3.0000   1.0000   1.1469   2.7700
    3.2000   0.9000   1.1065   2.8300
    3.4000   0.9000   1.1181   2.7000
    3.6000   0.9000   1.1291   2.5800
    3.8000   0.9000   1.1396   2.4700
    4.0000   0.9000   1.1497   2.3800
    4.2000   0.8000   1.1107   2.8300
    4.4000   0.8000   1.1208   2.5900
    4.6000   0.8000   1.1304   2.4300
    4.8000   0.8000   1.1396   2.3100
    5.0000   0.8000   1.1485   2.2100
```

```
>> sortsolution = sortrows(solution,3)

sortsolution =

    3.2000    0.9000    1.1065    2.8300
    4.2000    0.8000    1.1107    2.8300
    3.4000    0.9000    1.1181    2.7000
    4.4000    0.8000    1.1208    2.5900
    3.6000    0.9000    1.1291    2.5800
    4.6000    0.8000    1.1304    2.4300
    4.8000    0.8000    1.1396    2.3100
    3.8000    0.9000    1.1396    2.4700
    3.0000    1.0000    1.1469    2.7700
    5.0000    0.8000    1.1485    2.2100
    4.0000    0.9000    1.1497    2.3800

>> % Plot the response curve with the smallest overshoot shown in sortsolution table.
>>
>> K = sortsolution(1,1), a = sortsolution(1,2)

K =

    3.2000

a =

    0.9000

>> num = [4*K   8*K*a   4*K*a^2];
>> den = [1   6   8+4*K   4+8*K*a   4*K*a^2];
>> num

num =

   12.8000   23.0400   10.3680

>> den

den =

    1.0000    6.0000   20.8000   27.0400   10.3680

>> y = step(num,den,t);
>> plot(t,y)
>> grid
>> title('Unit-Step Response')
>> xlabel('t sec')
>> ylabel('Output y(t)')
```

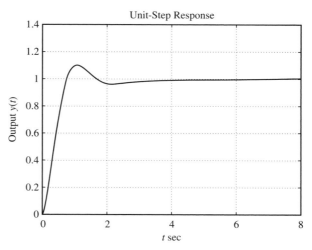

Figure 7–9
Unit-step response curve of the system with $K = 3.2$ and $a = 0.9$.

7–2 SOLVING QUADRATIC OPTIMAL CONTROL PROBLEMS WITH MATLAB

An advantage of the quadratic optimal control method over the pole-placement method is that the former provides a systematic way of computing the state feedback control gain matrix. Another advantage is that the system designed is always stable.

Quadratic optimal regulator problems. We now consider the optimal regulator problem that, given the system equation

$$\dot{\mathbf{x}} = \mathbf{A}\mathbf{x} + \mathbf{B}\mathbf{u} \tag{7-2}$$

determines the matrix **K** of the optimal control vector

$$\mathbf{u}(t) = -\mathbf{K}\mathbf{x}(t) \tag{7-3}$$

so as to minimize the performance index

$$J = \int_0^\infty (\mathbf{x}*\mathbf{Q}\mathbf{x} + \mathbf{u}*\mathbf{R}\mathbf{u})\, dt \tag{7-4}$$

where **Q** is a positive-definite (or positive-semidefinite) Hermitian or real symmetric matrix and **R** is a positive-definite Hermitian or real symmetric matrix. Note that the second term on the right-hand side of Equation (7–4) accounts for the expenditure of the energy of the control signals. The matrices **Q** and **R** respectively determine the relative importance of the error and the expenditure of this energy. We assume that the control vector $\mathbf{u}(t)$ is unconstrained.

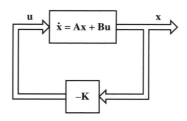

Figure 7–10
Optimal regulator system.

When the optimal control system is designed in the time domain, it is desirable to investigate the frequency-response characteristics to compensate for noise effects. The system frequency-response characteristics must be such that the system attenuates highly in the frequency range where noise and resonance of components are expected. (To compensate for noise effects, we must in some cases either modify the optimal configuration and accept suboptimal performance or modify the performance index.)

It can be shown that if the upper limit of integration in the performance index J given by Equation (7–4) is finite, then the optimal control vector is still a linear function of the state variables, but with time-varying coefficients. (Therefore, the determination of the optimal control vector involves that of optimal time-varying matrices.)

As will be seen later, the linear control law given by Equation (7–3) is the optimal control law. Therefore, if the unknown elements of the matrix $\mathbf{K}$ are determined so as to minimize the performance index, then $\mathbf{u}(t) = -\mathbf{K}\mathbf{x}(t)$ is optimal for any initial state $\mathbf{x}(0)$. The block diagram showing the optimal configuration is given in Figure 7–10.

Now let us solve the optimization problem. Substituting Equation (7–3) into Equation (7–2), we obtain

$$\dot{\mathbf{x}} = \mathbf{A}\mathbf{x} - \mathbf{B}\mathbf{K}\mathbf{x} = (\mathbf{A} - \mathbf{B}\mathbf{K})\mathbf{x}$$

In the derivations that follow, we assume that the matrix $\mathbf{A} - \mathbf{B}\mathbf{K}$ is stable or that the eigenvalues of $\mathbf{A} - \mathbf{B}\mathbf{K}$ have negative real parts.

Substituting Equation (7–3) into Equation (7–4) yields

$$J = \int_0^\infty (\mathbf{x}^*\mathbf{Q}\mathbf{x} + \mathbf{x}^*\mathbf{K}^*\mathbf{R}\mathbf{K}\mathbf{x})\, dt$$

$$= \int_0^\infty \mathbf{x}^*(\mathbf{Q} + \mathbf{K}^*\mathbf{R}\mathbf{K})\mathbf{x}\, dt$$

Let us set

$$\mathbf{x}^*(\mathbf{Q} + \mathbf{K}^*\mathbf{R}\mathbf{K})\mathbf{x} = -\frac{d}{dt}(\mathbf{x}^*\mathbf{P}\mathbf{x})$$

where **P** is a positive-definite Hermitian or real symmetric matrix. Then we obtain

$$\mathbf{x}^*(\mathbf{Q} + \mathbf{K}^*\mathbf{R}\mathbf{K})\mathbf{x} = -\dot{\mathbf{x}}^*\mathbf{P}\mathbf{x} - \mathbf{x}^*\mathbf{P}\dot{\mathbf{x}} = -\mathbf{x}^*[(\mathbf{A} - \mathbf{B}\mathbf{K})^*\mathbf{P} + \mathbf{P}(\mathbf{A} - \mathbf{B}\mathbf{K})]\mathbf{x}$$

Comparing both sides of this last equation and noting that it must hold true for any **x**, we require that

$$(\mathbf{A} - \mathbf{B}\mathbf{K})^*\mathbf{P} + \mathbf{P}(\mathbf{A} - \mathbf{B}\mathbf{K}) = -(\mathbf{Q} + \mathbf{K}^*\mathbf{R}\mathbf{K}) \tag{7-5}$$

It can be proved that if $\mathbf{A} - \mathbf{B}\mathbf{K}$ is a stable matrix, then there exists a positive-definite matrix **P** that satisfies Equation (7-5).

Hence, our procedure is to determine the elements of **P** from Equation (7-5) and see if **P** is positive definite. (Note that more than one matrix **P** may satisfy this equation. If the system is stable, there always exists one positive-definite matrix **P** that satisfies the equation. This means that, if we solve the equation and find one positive-definite matrix **P**, then the system is stable. Other **P** matrices that satisfy this equation are not positive definite and must be discarded.)

The performance index J can be evaluated as

$$J = \int_0^\infty \mathbf{x}^*(\mathbf{Q} + \mathbf{K}^*\mathbf{R}\mathbf{K})\mathbf{x}\, dt = -\mathbf{x}^*\mathbf{P}\mathbf{x}\Big|_0^\infty = -\mathbf{x}^*(\infty)\mathbf{P}\mathbf{x}(\infty) + \mathbf{x}^*(0)\mathbf{P}\mathbf{x}(0)$$

Since all eigenvalues of $\mathbf{A} - \mathbf{B}\mathbf{K}$ are assumed to have negative real parts, we have $\mathbf{x}(\infty) \to \mathbf{0}$. Therefore, we obtain

$$J = \mathbf{x}^*(0)\mathbf{P}\mathbf{x}(0) \tag{7-6}$$

Thus, the performance index J can be obtained in terms of the initial condition $\mathbf{x}(0)$ and **P**.

To obtain the solution to the quadratic optimal control problem, we proceed as follows: Since **R** has been assumed to be a positive-definite Hermitian or real symmetric matrix, we can write

$$\mathbf{R} = \mathbf{T}^*\mathbf{T}$$

where **T** is a nonsingular matrix. Then Equation (7-5) can be written as

$$(\mathbf{A}^* - \mathbf{K}^*\mathbf{B}^*)\mathbf{P} + \mathbf{P}(\mathbf{A} - \mathbf{B}\mathbf{K}) + \mathbf{Q} + \mathbf{K}^*\mathbf{T}^*\mathbf{T}\mathbf{K} = \mathbf{0}$$

which can be rewritten as

$$\mathbf{A}^*\mathbf{P} + \mathbf{P}\mathbf{A} + [\mathbf{T}\mathbf{K} - (\mathbf{T}^*)^{-1}\mathbf{B}^*\mathbf{P}]^*[\mathbf{T}\mathbf{K} - (\mathbf{T}^*)^{-1}\mathbf{B}^*\mathbf{P}] - \mathbf{P}\mathbf{B}\mathbf{R}^{-1}\mathbf{B}^*\mathbf{P} + \mathbf{Q} = \mathbf{0}$$

The minimization of J with respect to **K** requires the minimization of

$$\mathbf{x}^*[\mathbf{T}\mathbf{K} - (\mathbf{T}^*)^{-1}\mathbf{B}^*\mathbf{P}]^*[\mathbf{T}\mathbf{K} - (\mathbf{T}^*)^{-1}\mathbf{B}^*\mathbf{P}]\mathbf{x}$$

with respect to **K**. Since this last expression is nonnegative, the minimum occurs when it is zero, or when

$$\mathbf{TK} = (\mathbf{T^*})^{-1}\mathbf{B^*P}$$

Hence,

$$\mathbf{K} = \mathbf{T}^{-1}(\mathbf{T^*})^{-1}\mathbf{B^*P} = \mathbf{R}^{-1}\mathbf{B^*P} \tag{7-7}$$

Equation (7–7) gives the optimal matrix **K**. Thus, the optimal control law for the quadratic optimal control problem when the performance index is given by Equation (7–4) is linear and is given by

$$\mathbf{u}(t) = -\mathbf{Kx}(t) = -\mathbf{R}^{-1}\mathbf{B^*Px}(t)$$

The matrix **P** in Equation (7–7) must satisfy Equation (7–5) or the following reduced equation:

$$\mathbf{A^*P} + \mathbf{PA} - \mathbf{PBR}^{-1}\mathbf{B^*P} + \mathbf{Q} = 0 \tag{7-8}$$

Equation (7–8) is called the reduced-matrix Riccati equation. The design steps may be stated as follows:

1. Solve Equation (7–8), the reduced-matrix Riccati equation, for the matrix **P**. [If a positive-definite matrix **P** exists (certain systems may not have such a matrix), then the system is stable, or matrix **A** − **BK** is stable.]
2. Substitute matrix **P** into Equation (7–7). The resulting matrix **K** is the optimal matrix.

A design example based on this approach is given in Example 7–6. Note that if the matrix **A** − **BK** is stable, then the method always gives the correct result.

EXAMPLE 7–6 Consider the system shown in Figure 7–11. Assuming the control signal to be

$$u(t) = -\mathbf{Kx}(t)$$

determine the optimal feedback gain matrix **K** such that the following performance index is minimized:

$$J = \int_0^\infty (\mathbf{x}^T \mathbf{Q}\mathbf{x} + u^2)\, dt$$

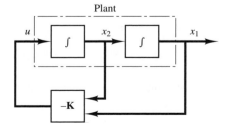

Figure 7–11
Control system.

where

$$\mathbf{Q} = \begin{bmatrix} 1 & 0 \\ 0 & \mu \end{bmatrix} \quad (\mu \geq 0)$$

From Figure 7–11, we find that the state equation for the plant is

$$\dot{\mathbf{x}} = \mathbf{A}\mathbf{x} + \mathbf{B}u$$

where

$$\mathbf{A} = \begin{bmatrix} 0 & 1 \\ 0 & 0 \end{bmatrix}, \quad \mathbf{B} = \begin{bmatrix} 0 \\ 1 \end{bmatrix}$$

We shall demonstrate the use of the reduced-matrix Riccati equation in the design of the optimal control system. Let us solve Equation (7–8), namely,

$$\mathbf{A}*\mathbf{P} + \mathbf{P}\mathbf{A} - \mathbf{P}\mathbf{B}\mathbf{R}^{-1}\mathbf{B}*\mathbf{P} + \mathbf{Q} = \mathbf{0}$$

Noting that matrix **A** is real and matrix **Q** is real symmetric, we see that matrix **P** is a real symmetric matrix. Hence, the latter equation can be written as

$$\begin{bmatrix} 0 & 0 \\ 1 & 0 \end{bmatrix} \begin{bmatrix} p_{11} & p_{12} \\ p_{12} & p_{22} \end{bmatrix} + \begin{bmatrix} p_{11} & p_{12} \\ p_{12} & p_{22} \end{bmatrix} \begin{bmatrix} 0 & 1 \\ 0 & 0 \end{bmatrix}$$
$$- \begin{bmatrix} p_{11} & p_{12} \\ p_{12} & p_{22} \end{bmatrix} \begin{bmatrix} 0 \\ 1 \end{bmatrix} [1] \begin{bmatrix} 0 & 1 \end{bmatrix} \begin{bmatrix} p_{11} & p_{12} \\ p_{12} & p_{22} \end{bmatrix} + \begin{bmatrix} 1 & 0 \\ 0 & \mu \end{bmatrix} = \begin{bmatrix} 0 & 0 \\ 0 & 0 \end{bmatrix}$$

which can be simplified to

$$\begin{bmatrix} 0 & 0 \\ p_{11} & p_{12} \end{bmatrix} + \begin{bmatrix} 0 & p_{11} \\ 0 & p_{12} \end{bmatrix} - \begin{bmatrix} p_{12}^2 & p_{12}p_{22} \\ p_{12}p_{22} & p_{22}^2 \end{bmatrix} + \begin{bmatrix} 1 & 0 \\ 0 & \mu \end{bmatrix} = \begin{bmatrix} 0 & 0 \\ 0 & 0 \end{bmatrix}$$

from which we obtain the following three equations:

$$1 - p_{12}^2 = 0$$
$$p_{11} - p_{12}p_{22} = 0$$
$$\mu + 2p_{12} - p_{22}^2 = 0$$

Solving these three simultaneous equations for p_{11}, p_{12}, and p_{22} and requiring **P** to be positive definite, we obtain

$$\mathbf{P} = \begin{bmatrix} p_{11} & p_{12} \\ p_{12} & p_{22} \end{bmatrix} = \begin{bmatrix} \sqrt{\mu + 2} & 1 \\ 1 & \sqrt{\mu + 2} \end{bmatrix}$$

From Equation (7–7), the optimal feedback gain matrix **K** is obtained as

$$\mathbf{K} = \mathbf{R}^{-1}\mathbf{B}*\mathbf{P}$$
$$= [1][0\ \ 1]\begin{bmatrix} p_{11} & p_{12} \\ p_{12} & p_{22} \end{bmatrix}$$
$$= [p_{12}\ \ p_{22}]$$
$$= [1\ \ \sqrt{\mu + 2}]$$

Thus, the optimal control signal is

$$u = -\mathbf{K}\mathbf{x} = -x_1 - \sqrt{\mu + 2}\,x_2 \tag{7–9}$$

Note that the control law given by Equation (7–9) yields an optimal result for any initial state under the given performance index. Figure 7–12 is the block diagram of this system.

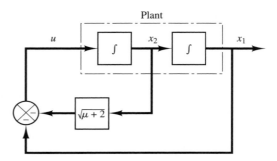

Figure 7–12
Optimal control of the plant shown in Figure 7–11.

Since the characteristic equation is

$$|s\mathbf{I} - \mathbf{A} + \mathbf{BK}| = s^2 + \sqrt{\mu + 2}\,s + 1 = 0$$

if $\mu = 1$, the two closed-loop poles are located at

$$s = -0.866 + 0.5j, \qquad s = -0.866 - 0.5j$$

These correspond to the desired closed-loop poles when $\mu = 1$.

Solving quadratic optimal regulator problems with MATLAB. In MATLAB, the command

lqr(A,B,Q,R)

solves the continuous-time, linear, quadratic regulator problem and the associated Riccati equation. This command calculates the optimal feedback gain matrix **K** such that the feedback control law

$$u = -\mathbf{K}\mathbf{x}$$

minimizes the performance index

$$J = \int_0^\infty (\mathbf{x}*\mathbf{Q}\mathbf{x} + \mathbf{u}*\mathbf{R}\mathbf{u})\, dt$$

subject to the constraint equation

$$\dot{\mathbf{x}} = \mathbf{A}\mathbf{x} + \mathbf{B}\mathbf{u}$$

Another command,

$$[K,P,E] = lqr(A,B,Q,R)$$

returns the gain matrix **K**, matrix **P**, which is the unique positive-definite solution of the associated matrix Riccati equation

$$\mathbf{PA} + \mathbf{A}*\mathbf{P} - \mathbf{PBR}^{-1}\mathbf{B}*\mathbf{P} + \mathbf{Q} = \mathbf{0}$$

and eigenvalue vector **E**.

If matrix $\mathbf{A} - \mathbf{BK}$ is a stable matrix, such a positive-definite solution **P** always exists. The eigenvalue vector **E** gives the closed-loop poles of $\mathbf{A} - \mathbf{BK}$.

It is important to note that, for certain systems, matrix $\mathbf{A} - \mathbf{BK}$ cannot be made stable, no matter what **K** is chosen. In such a case, there is no positive-definite matrix **P** for the matrix Riccati equation and the commands

$$K = lqr(A,B,Q,R)$$
$$[K,P,E] = lqr(A,B,Q,R)$$

do not give the solution. (See MATLAB Program 7–10.)

EXAMPLE 7–7 Consider the system defined by

$$\begin{bmatrix} \dot{x}_1 \\ \dot{x}_2 \end{bmatrix} = \begin{bmatrix} -1 & 1 \\ 0 & 2 \end{bmatrix} \begin{bmatrix} x_1 \\ x_2 \end{bmatrix} + \begin{bmatrix} 1 \\ 0 \end{bmatrix} u$$

Show that the system cannot be stabilized by the state-feedback control scheme

$$u = -\mathbf{K}\mathbf{x}$$

no matter what matrix **K** is chosen. (Notice that this system is not state controllable.) We define

$$\mathbf{K} = \begin{bmatrix} k_1 & k_2 \end{bmatrix}$$

Then

$$\mathbf{A} - \mathbf{BK} = \begin{bmatrix} -1 & 1 \\ 0 & 2 \end{bmatrix} - \begin{bmatrix} 1 \\ 0 \end{bmatrix} \begin{bmatrix} k_1 & k_2 \end{bmatrix}$$

$$= \begin{bmatrix} -1 - k_1 & 1 - k_2 \\ 0 & 2 \end{bmatrix}$$

Hence, the characteristic equation becomes

$$|s\mathbf{I} - \mathbf{A} + \mathbf{BK}| = \begin{vmatrix} s + 1 + k_1 & -1 + k_2 \\ 0 & s - 2 \end{vmatrix}$$
$$= (s + 1 + k_1)(s - 2) = 0$$

The closed-loop poles are located at

$$s = -1 - k_1, \quad s = 2$$

Since the pole at $s = 2$ is in the right-half s plane, the system is unstable, no matter what **K** matrix is chosen. Hence, quadratic optimal control techniques cannot be applied to this system.

Let us assume that matrices **Q** and **R** in the quadratic performance index are given by

$$\mathbf{Q} = \begin{bmatrix} 1 & 0 \\ 0 & 1 \end{bmatrix}, \quad R = [1]$$

and that we write MATLAB Program 7–10. If you are using a newer version of MATLAB, it tells you that this optimal design problem is ill posed. MATLAB tells you how you may remedy this problem. If you are using an older version of MATLAB, MATLAB solution to this problem is given by

$$K = [\text{NaN} \quad \text{NaN}]$$

(Again, NaN means "not a number.") Whenever the solution to a quadratic optimal control problem does not exist, an older version of MATLAB tells you that matrix **K** consists of NaN.

MATLAB Program 7–10

```
>> % ——— Attempt to design a quadratic optimal regulator system ———
>>
>> A = [-1   1;0   2];
>> B = [1;0];
>> Q = [1   0;0   1];
>> R = [1];
>> K = lqr(A,B,Q,R)
??? Error using ==> ss.lqr
The plant model cannot be stabilized by feedback or the optimal design
problem is ill posed.

To remedy this problem, you may
     * Make sure that all unstable poles of A are controllable through B
       (use MINREAL to check)
     * Modify the weights Q and R to make [Q   N;N'   R] positive definite
       (use EIG to check positivity)(N and N' represent cross-product terms)
```

EXAMPLE 7-8 Consider the system given by

$$\dot{\mathbf{x}} = \mathbf{A}\mathbf{x} + \mathbf{B}u$$

where

$$\mathbf{A} = \begin{bmatrix} 0 & 1 & 0 \\ 0 & 0 & 1 \\ -35 & -27 & -9 \end{bmatrix}, \quad \mathbf{B} = \begin{bmatrix} 0 \\ 0 \\ 1 \end{bmatrix}$$

The performance index J is given by

$$J = \int_0^\infty (\mathbf{x}'\mathbf{Q}\mathbf{x} + u'Ru)\, dt$$

where

$$\mathbf{Q} = \begin{bmatrix} 1 & 0 & 0 \\ 0 & 1 & 0 \\ 0 & 0 & 1 \end{bmatrix}, \quad R = [1]$$

Obtain the positive-definite solution matrix **P** of the Riccati equation, the optimal feedback gain matrix **K**, and the eigenvalues of matrix $\mathbf{A} - \mathbf{B}\mathbf{K}$.

MATLAB Program 7–11 solves this problem. (E gives the eigenvalues of matrix $\mathbf{A} - \mathbf{B}\mathbf{K}$.)

MATLAB Program 7–11

```
>> % ———— Design of quadratic optimal regulator system ————
>>
>> A = [0   1   0;0   0   1;-35   -27   -9];
>> B = [0;0;1];
>> Q = [1   0   0;0   1   0;0   0   1];
>> R = [1];
>> [K,P,E] = lqr(A,B,Q,R)

K =

    0.0143    0.1107    0.0676

P =

    4.2625    2.4957    0.0143
    2.4957    2.8150    0.1107
    0.0143    0.1107    0.0676

E =

   -5.0958
   -1.9859 + 1.7110i
   -1.9859 - 1.7110i
```

Next, let us obtain the response **x** of the regulator system to the initial condition

$$\mathbf{x}(0) = \begin{bmatrix} 1 \\ 0 \\ 0 \end{bmatrix}$$

With state feedback $u = -\mathbf{Kx}$, the state equation for the system becomes

$$\dot{\mathbf{x}} = \mathbf{Ax} + \mathbf{B}u = (\mathbf{A} - \mathbf{BK})\mathbf{x}$$

Then the system, or sys, can be given by

$$\text{sys} = \text{ss}(\mathbf{A} - \mathbf{B}*\mathbf{K}, \text{eye}(3), \text{eye}(3), \text{eye}(3))$$

MATLAB Program 7–12 produces the response to the given initial condition. The response curves are shown in Figure 7–13.

MATLAB Program 7–12

```
>> % Response to initial condition.
>>
>> A = [0   1   0;0   0   1;-35   -27   -9];
>> B = [0;0;1];
>> K = [0.0143   0.1107   0.0676];
>> sys = ss(A - B*K,eye(3),eye(3),eye(3));
>> t = 0:0.01:8;
>> x = initial(sys,[1;0;0],t);

>> x1 = [1   0   0]*x';
>> x2 = [0   1   0]*x';
>> x3 = [0   0   1]*x';
>>
>> subplot(2,2,1);plot(t,x1);grid
>> xlabel('t(sec)');ylabel('x_1')
>>
>> subplot(2,2,2);plot(t,x2);grid
>> xlabel('t (sec)');ylabel('x_2')
>>
>> subplot(2,2,3);plot(t,x3);grid
>> xlabel('t (sec)');ylabel('x_3')
```

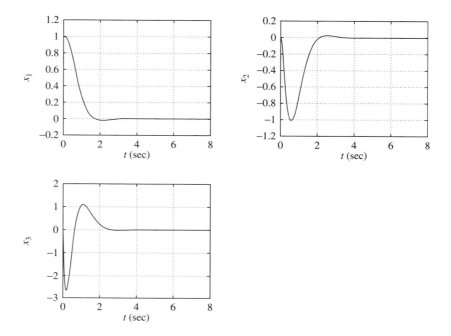

Figure 7–13
Response curves to initial condition.

EXAMPLE 7–9 Consider the system shown in Figure 7–14. The plant is defined by the state-space equations

$$\dot{\mathbf{x}} = \mathbf{A}\mathbf{x} + \mathbf{B}u$$
$$y = \mathbf{C}\mathbf{x} + Du$$

where

$$\mathbf{A} = \begin{bmatrix} 0 & 1 & 0 \\ 0 & 0 & 1 \\ 0 & -2 & -3 \end{bmatrix}, \quad \mathbf{B} = \begin{bmatrix} 0 \\ 0 \\ 1 \end{bmatrix}, \quad \mathbf{C} = \begin{bmatrix} 1 & 0 & 0 \end{bmatrix}, \quad D = [0]$$

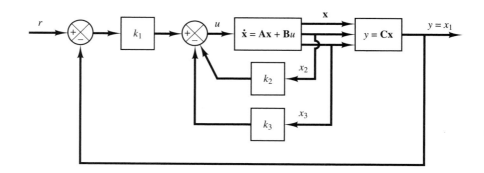

Figure 7–14
Control system.

The control signal u is given by

$$u = k_1(r - x_1) - (k_2 x_2 + k_3 x_3) = k_1 r - (k_1 x_1 + k_2 x_2 + k_3 x_3)$$

In determining an optimal control law, we assume that the input is zero, or $r = 0$.
Let us determine the state-feedback gain matrix

$$\mathbf{K} = \begin{bmatrix} k_1 & k_2 & k_3 \end{bmatrix}$$

such that the following performance index is minimized:

$$J = \int_0^\infty (\mathbf{x}'\mathbf{Q}\mathbf{x} + u'Ru)\,dt$$

where

$$\mathbf{Q} = \begin{bmatrix} q_{11} & 0 & 0 \\ 0 & q_{22} & 0 \\ 0 & 0 & q_{33} \end{bmatrix}, \quad R = r_1, \quad \mathbf{x} = \begin{bmatrix} x_1 \\ x_2 \\ x_3 \end{bmatrix} = \begin{bmatrix} y \\ \dot{y} \\ \ddot{y} \end{bmatrix}$$

To get a fast response, q_{11} must be sufficiently large compared with q_{22}, q_{33}, and r_1. In this problem, we choose

$$q_{11} = 100, \quad q_{22} = q_{33} = 1, \quad r_1 = 0.01$$

To solve the problem with MATLAB, we use the command

$$K = \text{lqr}(A,B,Q,R)$$

MATLAB Program 7–13 yields the solution to this problem.

MATLAB Program 7–13

```
>> % ------ Design of quadratic optimal control system ------
>>
>> A = [0  1  0;0  0  1;0  -2  -3];
>> B = [0;0;1];
>> Q = [100  0  0;0  1  0;0  0  1];
>> R = [0.01];
>> K = lqr(A,B,Q,R)

K =
    100.0000   53.1200   11.6711
```

Next, we shall use the matrix **K** thus determined to investigate the step-response characteristics of the designed system. The state equation for the designed system is

$$\begin{aligned}\dot{\mathbf{x}} &= \mathbf{A}\mathbf{x} + \mathbf{B}u \\ &= \mathbf{A}\mathbf{x} + \mathbf{B}(-\mathbf{K}\mathbf{x} + k_1 r) \\ &= (\mathbf{A} - \mathbf{B}\mathbf{K})\mathbf{x} + \mathbf{B}k_1 r\end{aligned}$$

and the output equation is

$$y = \mathbf{C}\mathbf{x} = \begin{bmatrix} 1 & 0 & 0 \end{bmatrix} \begin{bmatrix} x_1 \\ x_2 \\ x_3 \end{bmatrix}$$

To obtain the unit-step response, use the command

$$[y,x,t] = \text{step}(AA,BB,CC,DD)$$

where

$$AA = \mathbf{A} - \mathbf{B}\mathbf{K}, \quad BB = \mathbf{B}k_1, \quad CC = \mathbf{C}, \quad DD = D = 0$$

MATLAB Program 7–14 produces the unit-step response of the designed system. Figure 7–15 shows the response curves x_1, x_2, and x_3 versus t on one diagram.

MATLAB Program 7–14

```
>> % ——— Unit-step response of designed system ———
>>
>> A = [0   1   0;0   0   1;0   -2   -3];
>> B = [0;0;1];
>> C = [1   0   0];
>> D = [0];
>> K = [100.0000   53.1200   11.6711];
>> k1 = K(1);k2 = K(2);k3 = K(3);
>>
>> % ***** Define the state matrix, control matrix, output matrix,
>> % and direct transmission matrix of the designed systems as AA,
>> % BB, CC, and DD *****
>>
>> AA = A - B*K;
>> BB = B*k1;
>> CC = C;
>> DD = D;
>> t = 0:0.01:8;
>> [y,x,t] = step(AA,BB,CC,DD,1,t);
>> plot(t,x)
>> grid
>> title('Response Curves x_1, x_2, x_3, versus t')
>> xlabel('t Sec')
>> ylabel('x_1,x_2,x_3')
>> text(2.6,1.3,'x_1')
>> text(1.2,1.4,'x_2')
>> text(0.6,3.5,'x_3')
```

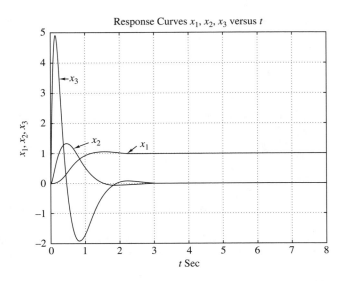

Figure 7-15
Response curves x_1 versus t, x_2 versus t, and x_3 versus t.

Appendix

This appendix discusses two subjects. Section A–1 presents a method for changing grids in root-locus diagrams or Nyquist diagrams to horizontal–vertical (x–y) grid lines. Section A–2 presents some useful results in vector–matrix analysis—in particular, the derivation of an expression for $e^{\mathbf{A}t}$ (where $\mathbf{A}$ is an $n \times n$ matrix) in terms of $\mathbf{I}, \mathbf{A}, \ldots, \mathbf{A}^n$.

A–1 CHANGING GRIDS IN ROOT-LOCUS DIAGRAMS AND NYQUIST DIAGRAMS

Consider the root-locus diagram accompanying MATLAB Program A–1.

MATLAB Program A–1

```
>> num = [1   3];
>> den = [1   5   20   16   0];
>> rlocus(num,den)
>> v = [−6   4   −6   6];axis(v)
>> grid
```

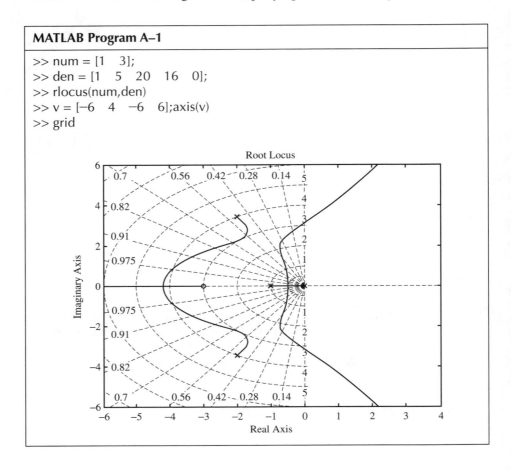

If you do not like these grids, erase them by entering grid into the program so that we have the two lines

grid
grid

The second grid command erases the grid curves already in the figure. (See the diagram accompanying MATLAB Program A–2.)

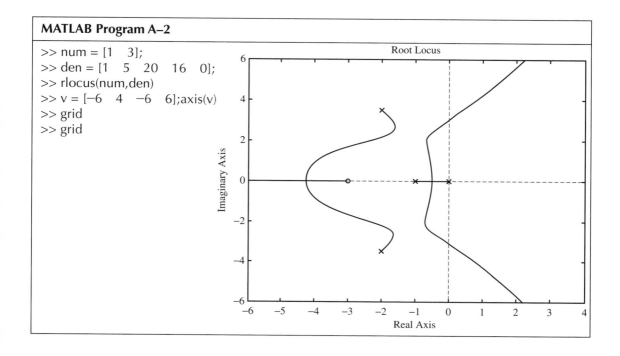

MATLAB Program A–2

```
>> num = [1   3];
>> den = [1   5   20   16   0];
>> rlocus(num,den)
>> v = [−6   4   −6   6];axis(v)
>> grid
>> grid
```

If you wish to draw x- and y-coordinate axes in any diagram (see Figure A–1), enter the following commands into the computer:

$$x1 = [xA \quad xB]; \quad (A–1)$$

$$y1 = [yA \quad yB]; \quad (A–2)$$

$$x2 = [xC \quad xD]; \quad (A–3)$$

$$y2 = [yC \quad yD]; \quad (A–4)$$

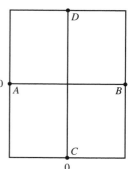

Figure A–1
Diagram showing the x-axis and the y-axis.

Here $xA, xB, xC,$ and xD are the x-coordinates of points $A, B, C,$ and D, respectively, and yA, yB, yC, yD are the y-coordinates of points $A, B, C,$ and D, respectively. Equations (A–1) and (A–2) define the x-axis, and Equations (A–3) and (A–4) define the y-axis. Hence, the command plot(x1, y1) draws the x-axis, and the command plot(x2, y2) draws the y-axis.

(See the diagram accompanying MATLAB Program A–3.) Before entering plot commands, make sure that you enter the hold command to retain the root-locus figure in the diagram. (See MATLAB Program A–3.)

MATLAB Program A–3

```
>> num = [1   3];
>> den = [1   5   20   16   0];
>> rlocus(num,den)
>> v = [-6   4   -6   6];axis(v)
>> hold
Current plot held
>> x1 = [-6   4];y1 = [0   0];
>> plot(x1,y1);  %  Draw the x-axis.
>> x2 = [0   0];y2 = [-6   6];
>> plot(x2,y2);  %  Draw the y-axis.
>> hold
Current plot released
```

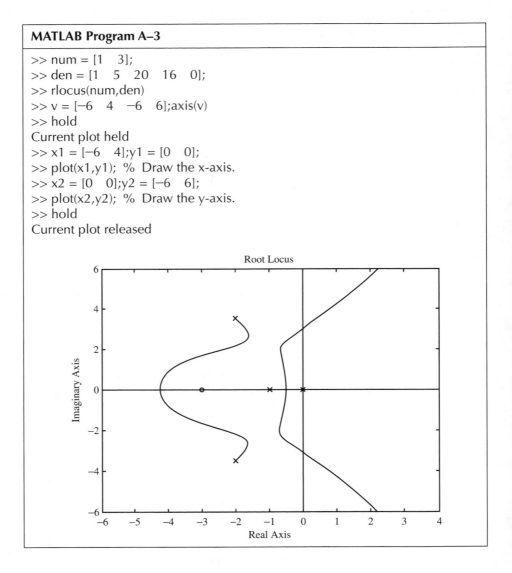

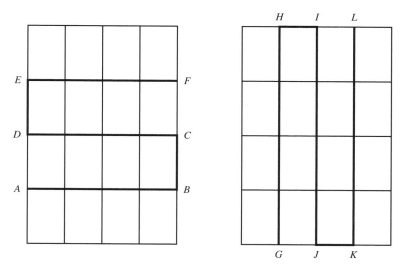

Figure A–2
Diagrams showing *x*-grids and *y*-grids.

If you wish to enter multiple *x–y* grid lines in the diagram, enter the following commands, referring to Figure A–2:

```
x1 = [xA   xB; xB   xC; xC   xD; xD   xE; xE   xF];
y1 = [yA   yB; yB   yC; yC   yD; yD   yE; yE   yF];
Plot(x1,y1)

x2 = [xG   xH; xH   xI; xI   xJ; xJ   xK; xK   xL];
y2 = [yG   yH; yH   yI; yI   yJ; yJ   yK; yK   yL];
plot(x2,y2)
```

(x1,y1) defines line *A–B–C–D–E–F* and (x2,y2) defines line *G–H–I–J–K–L*.

The diagram accompanying MATLAB Program A–4 shows x–y grids superimposed on the root-locus plot. Similar illustrative Nyquist diagrams with various grids are shown in MATLAB Programs A–5 through A–8.

MATLAB Program A–4

```
>> num = [1   3];
>> den = [1   5   20   16   0];
>> rlocus(num,den)
>> v = [-6   4   -6   6];axis(v)
>> hold
Current plot held
>> x1 = [-6   4;4   4;4   -6;-6   -6;-6   4;4   4;4   -6;-6   -6;-6   4];
>> y1 = [-4   -4;-4   -2;-2   -2;-2   0;0   0;0   2;2   2;2   4;4   4];
>> plot(x1,y1) % Draw horizontal grid lines.
>> x2 = [-4   -4;-4   -2;-2   -2;-2   0;0   0;0   2;2   2];
>> y2 = [-6   6;6   6;6   -6;-6   -6;-6   6;6   6;6   -6];
>> plot(x2,y2) % Draw vertical grid lines.
>> hold
Current plot released
```

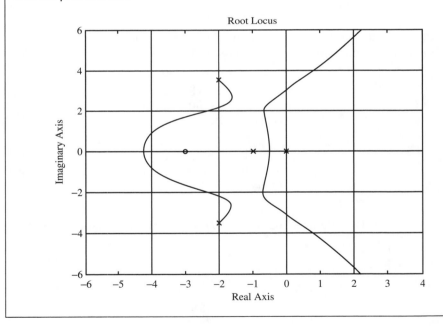

MATLAB Program A–5

```
>> num = [1];
>> den = [1   0.4   1];
>> nyquist(num,den)
>> grid
```

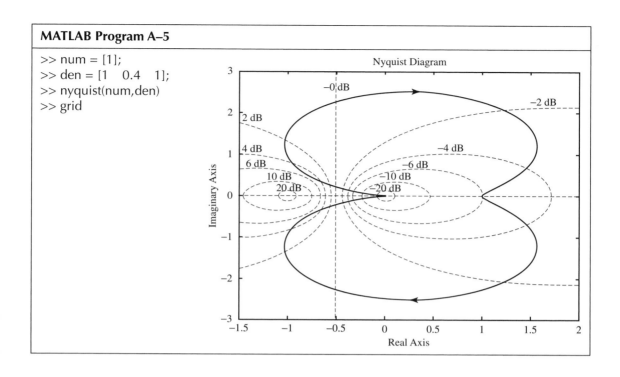

MATLAB Program A–6

```
>> num = [1];
>> den = [1   0.4   1];
>> nyquist(num,den)
>> grid
>> grid
```

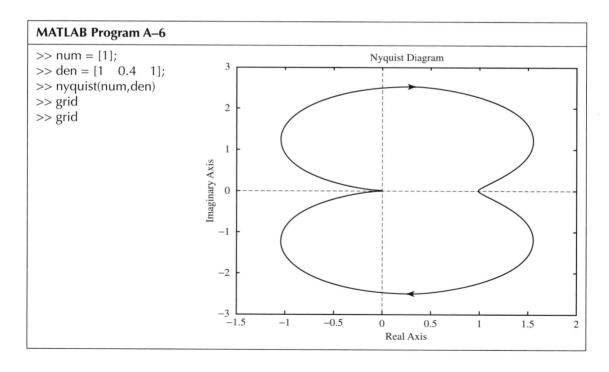

Section A–1 / Changing Grids in Root-Locus Diagrams and Nyquist Diagrams

MATLAB Program A–7

```
>> num = [1];
>> den = [1   0.4   1];
>> nyquist(num,den)
>> hold
Current plot held
>> x1 = [-1.5   2];y1 = [0   0];
>> plot(x1,y1);  % Draw the x-axis.
>> x2 = [0   0];y2 = [-3   3];
>> plot(x2,y2);  % Draw the y-axis.
>> hold
Current plot released
```

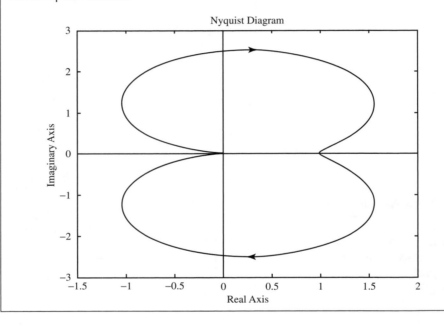

MATLAB Program A–8

```
>> num = [1];
>> den = [1   0.4   1];
>> nyquist(num,den)
>> hold
Current plot held
>> x1 = [-1.5  2;2    2;2  -1.5;-1.5  -1.5;-1.5  2;2  2;2  -1.5;-1.5  -1.5;-1.5  2];
>> y1 = [-2   -2;-2  -1;-1  -1;-1     0;0        0;0  1;1  1;1         2;2        2];
>> plot(x1,y1)  % Draw horizontal grid lines.
>> x2 = [-1   -1;-1  -0.5;-0.5  -0.5;-0.5  0;0   0;0  0.5;0.5  0.5;0.5  1;1   1;1  1.5;1.5  1.5];
>> y2 = [-3   3;3    3;3          -3;-3    -3;-3 3;3  3;3          -3;-3  -3;-3 3;3   3;3  -3];
>> plot(x2,y2);  % Draw vertical grid lines.
>> hold
Current plot released
```

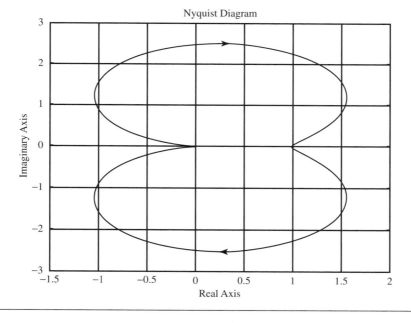

A-2 SOME USEFUL RESULTS IN VECTOR–MATRIX ANALYSIS

We next present some useful results in vector–matrix analysis that we frequently refer to in analyzing control systems in state space.

Cayley–Hamilton theorem. Consider an $n \times n$ matrix $\mathbf{A}$ and its characteristic equation,

$$|\lambda \mathbf{I} - \mathbf{A}| = \lambda^n + a_1 \lambda^{n-1} + \cdots + a_{n-1} \lambda + a_n = 0$$

The Cayley–Hamilton theorem states that the matrix $\mathbf{A}$ satisfies its own characteristic equation, or that

$$\mathbf{A}^n + a_1 \mathbf{A}^{n-1} + \cdots + a_{n-1} \mathbf{A} + a_n \mathbf{I} = \mathbf{0}$$

To prove this theorem, note that $\mathrm{adj}(\lambda \mathbf{I} - \mathbf{A})$ is a polynomial in λ of degree $n - 1$. That is,

$$\mathrm{adj}(\lambda \mathbf{I} - \mathbf{A}) = \mathbf{B}_1 \lambda^{n-1} + \mathbf{B}_2 \lambda^{n-2} + \cdots + \mathbf{B}_{n-1} \lambda + \mathbf{B}_n$$

where $\mathbf{B}_1 = \mathbf{I}$. Since

$$(\lambda \mathbf{I} - \mathbf{A}) \, \mathrm{adj}(\lambda \mathbf{I} - \mathbf{A}) = [\mathrm{adj}(\lambda \mathbf{I} - \mathbf{A})](\lambda \mathbf{I} - \mathbf{A}) = |\lambda \mathbf{I} - \mathbf{A}| \mathbf{I}$$

we obtain

$$\begin{aligned} |\lambda \mathbf{I} - \mathbf{A}| \mathbf{I} &= \mathbf{I}\lambda^n + a_1 \mathbf{I} \lambda^{n-1} + \cdots + a_{n-1} \mathbf{I} \lambda + a_n \mathbf{I} \\ &= (\lambda \mathbf{I} - \mathbf{A})(\mathbf{B}_1 \lambda^{n-1} + \mathbf{B}_2 \lambda^{n-2} + \cdots + \mathbf{B}_{n-1} \lambda + \mathbf{B}_n) \\ &= (\mathbf{B}_1 \lambda^{n-1} + \mathbf{B}_2 \lambda^{n-2} + \cdots + \mathbf{B}_{n-1} \lambda + \mathbf{B}_n)(\lambda \mathbf{I} - \mathbf{A}) \end{aligned}$$

From this equation, we see that $\mathbf{A}$ and $\mathbf{B}_i (i = 1, 2, \ldots, n)$ commute. Hence, the product of $(\lambda \mathbf{I} - \mathbf{A})$ and $\mathrm{adj}(\lambda \mathbf{I} - \mathbf{A})$ becomes zero if either of these factors is zero. If $\mathbf{A}$ is substituted for λ in this last equation, then, clearly, $\lambda \mathbf{I} - \mathbf{A}$ becomes zero. Hence, we obtain

$$\mathbf{A}^n + a_1 \mathbf{A}^{n-1} + \cdots + a_{n-1} \mathbf{A} + a_n \mathbf{I} = \mathbf{0}$$

This proves the Cayley–Hamilton theorem.

Minimal polynomial. According to the Cayley–Hamilton theorem, every $n \times n$ matrix $\mathbf{A}$ satisfies its own characteristic equation. The characteristic equation is not, however, necessarily the scalar equation of least degree that $\mathbf{A}$ satisfies. The least-degree polynomial having $\mathbf{A}$ as a root is called the *minimal polynomial*. That is, the minimal polynomial of an $n \times n$ matrix $\mathbf{A}$ is defined as the polynomial

$$\phi(\lambda) = \lambda^m + a_1 \lambda^{m-1} + \cdots + a_{m-1} \lambda + a_m, \qquad m \leq n$$

of least degree such that $\phi(\mathbf{A}) = \mathbf{0}$, or

$$\phi(\mathbf{A}) = \mathbf{A}^m + a_1 \mathbf{A}^{m-1} + \cdots + a_{m-1}\mathbf{A} + a_m \mathbf{I} = \mathbf{0}$$

The minimal polynomial plays an important role in the computation of polynomials in an $n \times n$ matrix.

Let us suppose that $d(\lambda)$, a polynomial in λ, is the greatest common divisor of all the elements of $\mathrm{adj}(\lambda \mathbf{I} - \mathbf{A})$. Then we can show that if the coefficient of the highest-degree term in λ of $d(\lambda)$ is chosen to be 1, then the minimal polynomial is given by

$$\phi(\lambda) = \frac{|\lambda \mathbf{I} - \mathbf{A}|}{d(\lambda)}$$

Note that the minimal polynomial $\phi(\lambda)$ of an $n \times n$ matrix $\mathbf{A}$ can be determined by the following procedure:

1. Form $\mathrm{adj}(\lambda \mathbf{I} - \mathbf{A})$, and write the elements of $\mathrm{adj}(\lambda \mathbf{I} - \mathbf{A})$ as factored polynomials in λ.
2. Determine $d(\lambda)$ as the greatest common divisor of all the elements of $\mathrm{adj}(\lambda \mathbf{I} - \mathbf{A})$. Choose the coefficient of the highest-degree term in λ of $d(\lambda)$ to be 1. If there is no common divisor, $d(\lambda) = 1$.
3. The minimal polynomial $\phi(\lambda)$ is then given as $|\lambda \mathbf{I} - \mathbf{A}|$ divided by $d(\lambda)$.

Lagrange's interpolation formula and Sylvester's interpolation formula. Consider the following polynomial in λ of degree $m - 1$, where we assume $\lambda_1, \lambda_2, \ldots, \lambda_m$ to be distinct:

$$p_k(\lambda) = \frac{(\lambda - \lambda_1) \cdots (\lambda - \lambda_{k-1})(\lambda - \lambda_{k+1}) \cdots (\lambda - \lambda_m)}{(\lambda_k - \lambda_1) \cdots (\lambda_k - \lambda_{k-1})(\lambda_k - \lambda_{k+1}) \cdots (\lambda_k - \lambda_m)}, \quad k = 1, 2, \ldots, m$$

Notice that

$$p_k(\lambda_i) = \begin{cases} 1, & \text{if } i = k \\ 0, & \text{if } i \neq k \end{cases}$$

Then the polynomial $f(\lambda)$ of degree $m - 1$, or

$$\begin{aligned} f(\lambda) &= \sum_{k=1}^{m} f(\lambda_k) p_k(\lambda) \\ &= \sum_{k=1}^{m} f(\lambda_k) \frac{(\lambda - \lambda_1) \cdots (\lambda - \lambda_{k-1})(\lambda - \lambda_{k+1}) \cdots (\lambda - \lambda_m)}{(\lambda_k - \lambda_1) \cdots (\lambda_k - \lambda_{k-1})(\lambda_k - \lambda_{k+1}) \cdots (\lambda_k - \lambda_m)} \end{aligned} \quad (A-5)$$

takes on the values $f(\lambda_k)$ at the points λ_k. Equation (A–5) is commonly called *Lagrange's interpolation formula*. The polynomial $f(\lambda)$ of degree $m - 1$ is determined from m independent data $f(\lambda_1), f(\lambda_2), \ldots, f(\lambda_m)$. That is, the polynomial $f(\lambda)$ passes through m points $f(\lambda_1), f(\lambda_2), \ldots, f(\lambda_m)$. Since $f(\lambda)$ is a polynomial of degree $m - 1$, it is uniquely determined. Any other representations of the polynomial of degree $m - 1$ can be reduced to the Lagrange polynomial $f(\lambda)$.

Assuming that the eigenvalues of an $n \times n$ matrix $\mathbf{A}$ are distinct, if we substitute $\mathbf{A}$ for λ in the polynomial $p_k(\lambda)$, then we get

$$p_k(\mathbf{A}) = \frac{(\mathbf{A} - \lambda_1\mathbf{I}) \cdots (\mathbf{A} - \lambda_{k-1}\mathbf{I})(\mathbf{A} - \lambda_{k+1}\mathbf{I}) \cdots (\mathbf{A} - \lambda_m\mathbf{I})}{(\lambda_k - \lambda_1) \cdots (\lambda_k - \lambda_{k-1})(\lambda_k - \lambda_{k+1}) \cdots (\lambda_k - \lambda_m)}$$

Notice that $p_k(\mathbf{A})$ is a polynomial in $\mathbf{A}$ of degree $m - 1$. Notice also that

$$p_k(\lambda_i\mathbf{I}) = \begin{cases} \mathbf{I}, & \text{if } i = k \\ \mathbf{0}, & \text{if } i \neq k \end{cases}$$

Now define

$$f(\mathbf{A}) = \sum_{k=1}^{m} f(\lambda_k) p_k(\mathbf{A})$$
$$= \sum_{k=1}^{m} f(\lambda_k) \frac{(\mathbf{A} - \lambda_1\mathbf{I}) \cdots (\mathbf{A} - \lambda_{k-1}\mathbf{I})(\mathbf{A} - \lambda_{k+1}\mathbf{I}) \cdots (\mathbf{A} - \lambda_m\mathbf{I})}{(\lambda_k - \lambda_1) \cdots (\lambda_k - \lambda_{k-1})(\lambda_k - \lambda_{k+1}) \cdots (\lambda_k - \lambda_m)} \quad \text{(A-6)}$$

Equation (A–6) is known as *Sylvester's interpolation formula* and is equivalent to the following equation:

$$\begin{vmatrix} 1 & \lambda_1 & \lambda_1^2 & \cdots & \lambda_1^{m-1} & f(\lambda_1) \\ 1 & \lambda_2 & \lambda_2^2 & \cdots & \lambda_2^{m-1} & f(\lambda_2) \\ \cdot & \cdot & \cdot & & \cdot & \cdot \\ \cdot & \cdot & \cdot & & \cdot & \cdot \\ \cdot & \cdot & \cdot & \cdots & \cdot & \cdot \\ 1 & \lambda_m & \lambda_m^2 & \cdots & \lambda_m^{m-1} & f(\lambda_m) \\ \mathbf{I} & \mathbf{A} & \mathbf{A}^2 & \cdots & \mathbf{A}^{m-1} & f(\mathbf{A}) \end{vmatrix} = \mathbf{0} \quad \text{(A-7)}$$

Equations (A–6) and (A–7) are frequently used to evaluate functions $f(\mathbf{A})$ of matrix $\mathbf{A}$—for example, $e^{\mathbf{A}t}$. [For a derivation of Equation (A–6) from Equation (A–7), see, for example, pages 811–816 of Reference 1.]

Computation of $e^{\mathbf{A}t}$. We present here the Sylvester's interpolation method. First we consider the case where the roots of the minimal polynomial $\phi(\lambda)$ of $\mathbf{A}$ are distinct. Then we deal with the case of multiple roots.

Case 1: Minimal Polynomial of **A** *Involves Only Distinct Roots.* We shall assume that the degree of the minimal polynomial of **A** is m. By using Sylvester's interpolation formula, it can be shown that $e^{\mathbf{A}t}$ can be obtained by solving the following determinant equation:

$$\begin{vmatrix} 1 & \lambda_1 & \lambda_1^2 & \cdots & \lambda_1^{m-1} & e^{\lambda_1 t} \\ 1 & \lambda_2 & \lambda_2^2 & \cdots & \lambda_2^{m-1} & e^{\lambda_2 t} \\ \cdot & \cdot & \cdot & & \cdot & \cdot \\ \cdot & \cdot & \cdot & & \cdot & \cdot \\ \cdot & \cdot & \cdot & & \cdot & \cdot \\ 1 & \lambda_m & \lambda_m^2 & \cdots & \lambda_m^{m-1} & e^{\lambda_m t} \\ \mathbf{I} & \mathbf{A} & \mathbf{A}^2 & \cdots & \mathbf{A}^{m-1} & e^{\mathbf{A}t} \end{vmatrix} = \mathbf{0} \qquad (A\text{--}8)$$

By solving Equation (A–8) for $e^{\mathbf{A}t}$, $e^{\mathbf{A}t}$ can be obtained in terms of the $\mathbf{A}^k$ ($k = 0, 1, 2, \ldots, m-1$) and the $e^{\lambda_i t}$ ($i = 1, 2, 3, \ldots, m$). [Equation (A–8) may be expanded, for example, about the last column.]

Notice that solving Equation (A–8) for $e^{\mathbf{A}t}$ is the same as writing

$$e^{\mathbf{A}t} = \alpha_0(t)\mathbf{I} + \alpha_1(t)\mathbf{A} + \alpha_2(t)\mathbf{A}^2 + \cdots + \alpha_{m-1}(t)\mathbf{A}^{m-1} \qquad (A\text{--}9)$$

and determining the $\alpha_k(t)$ ($k = 0, 1, 2, \ldots, m-1$) by solving the following set of m equations for the $\alpha_k(t)$:

$$\alpha_0(t) + \alpha_1(t)\lambda_1 + \alpha_2(t)\lambda_1^2 + \cdots + \alpha_{m-1}(t)\lambda_1^{m-1} = e^{\lambda_1 t}$$
$$\alpha_0(t) + \alpha_1(t)\lambda_2 + \alpha_2(t)\lambda_2^2 + \cdots + \alpha_{m-1}(t)\lambda_2^{m-1} = e^{\lambda_2 t}$$
$$\vdots$$
$$\alpha_0(t) + \alpha_1(t)\lambda_m + \alpha_2(t)\lambda_m^2 + \cdots + \alpha_{m-1}(t)\lambda_m^{m-1} = e^{\lambda_m t}$$

If **A** is an $n \times n$ matrix that has distinct eigenvalues, then the number of $\alpha_k(t)$'s to be determined is $m = n$. If, however, **A** involves multiple eigenvalues, but its minimal polynomial has only simple roots, then the number m of $\alpha_k(t)$'s to be determined is less than n.

Case 2: Minimal Polynomial of **A** *Involves Multiple Roots.* As an example, consider the case where the minimal polynomial of **A** involves three equal roots ($\lambda_1 = \lambda_2 = \lambda_3$) and has other roots ($\lambda_4, \lambda_5, \ldots, \lambda_m$) that are all distinct.

By applying Sylvester's interpolation formula, it can be shown that $e^{\mathbf{A}t}$ can be obtained from the following determinant equation:

$$\begin{vmatrix} 0 & 0 & 1 & 3\lambda_1 & \cdots & \dfrac{(m-1)(m-2)}{2}\lambda_1^{m-3} & \dfrac{t^2}{2}e^{\lambda_1 t} \\ 0 & 1 & 2\lambda_1 & 3\lambda_1^2 & \cdots & (m-1)\lambda_1^{m-2} & te^{\lambda_1 t} \\ 1 & \lambda_1 & \lambda_1^2 & \lambda_1^3 & \cdots & \lambda_1^{m-1} & e^{\lambda_1 t} \\ 1 & \lambda_4 & \lambda_4^2 & \lambda_4^3 & \cdots & \lambda_4^{m-1} & e^{\lambda_4 t} \\ \cdot & \cdot & \cdot & \cdot & \cdots & \cdot & \cdot \\ \cdot & \cdot & \cdot & \cdot & \cdots & \cdot & \cdot \\ \cdot & \cdot & \cdot & \cdot & \cdots & \cdot & \cdot \\ 1 & \lambda_m & \lambda_m^2 & \lambda_m^3 & \cdots & \lambda_m^{m-1} & e^{\lambda_m t} \\ \mathbf{I} & \mathbf{A} & \mathbf{A}^2 & \mathbf{A}^3 & \cdots & \mathbf{A}^{m-1} & e^{\mathbf{A}t} \end{vmatrix} = \mathbf{0} \quad (A\text{--}10)$$

Equation (A–10) can be solved for $e^{\mathbf{A}t}$ by expanding it about the last column.

Note that, just as in case 1, solving Equation (A–10) for $e^{\mathbf{A}t}$ is the same as writing

$$e^{\mathbf{A}t} = \alpha_0(t)\mathbf{I} + \alpha_1(t)\mathbf{A} + \alpha_2(t)\mathbf{A}^2 + \cdots + \alpha_{m-1}(t)\mathbf{A}^{m-1} \quad (A\text{--}11)$$

and determining the $\alpha_k(t)$'s ($k = 0, 1, 2, \ldots, m - 1$) from

$$\alpha_2(t) + 3\alpha_3(t)\lambda_1 + \cdots + \dfrac{(m-1)(m-2)}{2}\alpha_{m-1}(t)\lambda_1^{m-3} = \dfrac{t^2}{2}e^{\lambda_1 t}$$

$$\alpha_1(t) + 2\alpha_2(t)\lambda_1 + 3\alpha_3(t)\lambda_1^2 + \cdots + (m-1)\alpha_{m-1}(t)\lambda_1^{m-2} = te^{\lambda_1 t}$$

$$\alpha_0(t) + \alpha_1(t)\lambda_1 + \alpha_2(t)\lambda_1^2 + \cdots + \alpha_{m-1}(t)\lambda_1^{m-1} = e^{\lambda_1 t}$$

$$\alpha_0(t) + \alpha_1(t)\lambda_4 + \alpha_2(t)\lambda_4^2 + \cdots + \alpha_{m-1}(t)\lambda_4^{m-1} = e^{\lambda_4 t}$$

$$\cdot$$
$$\cdot$$
$$\cdot$$

$$\alpha_0(t) + \alpha_1(t)\lambda_m + \alpha_2(t)\lambda_m^2 + \cdots + \alpha_{m-1}(t)\lambda_m^{m-1} = e^{\lambda_m t}$$

The extension to other cases where, for example, there are two or more sets of multiple roots will be apparent. Note that if the minimal polynomial of $\mathbf{A}$ is not found, it is possible to substitute the characteristic polynomial for the minimal polynomial. The number of computations may, of course, be increased.

EXAMPLE A-1 Consider the matrix

$$\mathbf{A} = \begin{bmatrix} 0 & 1 \\ 0 & -2 \end{bmatrix}$$

Compute $e^{\mathbf{A}t}$, using Sylvester's interpolation formula.

From Equation (A–8), we get

$$\begin{vmatrix} 1 & \lambda_1 & e^{\lambda_1 t} \\ 1 & \lambda_2 & e^{\lambda_2 t} \\ \mathbf{I} & \mathbf{A} & e^{\mathbf{A}t} \end{vmatrix} = \mathbf{0}$$

The eigenvalues of $\mathbf{A}$ are 0 and -2. Substituting 0 for λ_1 and -2 for λ_2 in the latter equation, we obtain

$$\begin{vmatrix} 1 & 0 & 1 \\ 1 & -2 & e^{-2t} \\ \mathbf{I} & \mathbf{A} & e^{\mathbf{A}t} \end{vmatrix} = \mathbf{0}$$

Expanding the determinant yields

$$-2e^{\mathbf{A}t} + \mathbf{A} + 2\mathbf{I} - \mathbf{A}e^{-2t} = \mathbf{0}$$

or

$$e^{\mathbf{A}t} = \tfrac{1}{2}(\mathbf{A} + 2\mathbf{I} - \mathbf{A}e^{-2t})$$

$$= \frac{1}{2}\left\{ \begin{bmatrix} 0 & 1 \\ 0 & -2 \end{bmatrix} + \begin{bmatrix} 2 & 0 \\ 0 & 2 \end{bmatrix} - \begin{bmatrix} 0 & 1 \\ 0 & -2 \end{bmatrix} e^{-2t} \right\}$$

$$= \begin{bmatrix} 1 & \tfrac{1}{2}(1 - e^{-2t}) \\ 0 & e^{-2t} \end{bmatrix}$$

An alternative approach is to use Equation (A–9). We first determine $\alpha_0(t)$ and $\alpha_1(t)$ from

$$\alpha_0(t) + \alpha_1(t)\lambda_1 = e^{\lambda_1 t}$$
$$\alpha_0(t) + \alpha_1(t)\lambda_2 = e^{\lambda_2 t}$$

Since $\lambda_1 = 0$ and $\lambda_2 = -2$, the last two equations become

$$\alpha_0(t) = 1$$
$$\alpha_0(t) - 2\alpha_1(t) = e^{-2t}$$

Solving for $\alpha_0(t)$ and $\alpha_1(t)$ gives

$$\alpha_0(t) = 1, \qquad \alpha_1(t) = \frac{1}{2}(1 - e^{-2t})$$

Then $e^{\mathbf{A}t}$ can be written as

$$e^{\mathbf{A}t} = \alpha_0(t)\mathbf{I} + \alpha_1(t)\mathbf{A} = \mathbf{I} + \frac{1}{2}(1 - e^{-2t})\mathbf{A} = \begin{bmatrix} 1 & \tfrac{1}{2}(1 - e^{-2t}) \\ 0 & e^{-2t} \end{bmatrix}$$

EXAMPLE A–2 Using Sylvester's interpolation formula, compute $e^{\mathbf{A}t}$, where

$$\mathbf{A} = \begin{bmatrix} 2 & 1 & 4 \\ 0 & 2 & 0 \\ 0 & 3 & 1 \end{bmatrix}$$

The characteristic polynomial and the minimal polynomial are the same for this $\mathbf{A}$. The minimal polynomial (characteristic polynomial) is given by

$$\phi(\lambda) = (\lambda - 2)^2(\lambda - 1)$$

Note that $\lambda_1 = \lambda_2 = 2$ and $\lambda_3 = 1$. Referring to Equation (A–11), we have

$$e^{\mathbf{A}t} = \alpha_0(t)\mathbf{I} + \alpha_1(t)\mathbf{A} + \alpha_2(t)\mathbf{A}^2$$

where $\alpha_0(t)$, $\alpha_1(t)$, and $\alpha_2(t)$ are determined from the equations

$$\alpha_1(t) + 2\alpha_2(t)\lambda_1 = te^{\lambda_1 t}$$
$$\alpha_0(t) + \alpha_1(t)\lambda_1 + \alpha_2(t)\lambda_1^2 = e^{\lambda_1 t}$$
$$\alpha_0(t) + \alpha_1(t)\lambda_3 + \alpha_2(t)\lambda_3^2 = e^{\lambda_3 t}$$

Substituting $\lambda_1 = 2$, and $\lambda_3 = 1$ into these three equations gives

$$\alpha_1(t) + 4\alpha_2(t) = te^{2t}$$
$$\alpha_0(t) + 2\alpha_1(t) + 4\alpha_2(t) = e^{2t}$$
$$\alpha_0(t) + \alpha_1(t) + \alpha_2(t) = e^t$$

Solving for $\alpha_0(t)$, $\alpha_1(t)$, and $\alpha_2(t)$, we obtain

$$\alpha_0(t) = 4e^t - 3e^{2t} + 2te^{2t}$$
$$\alpha_1(t) = -4e^t + 4e^{2t} - 3te^{2t}$$
$$\alpha_2(t) = e^t - e^{2t} + te^{2t}$$

Hence,

$$e^{\mathbf{A}t} = \alpha_0(t)\mathbf{I} + \alpha_1(t)\mathbf{A} + \alpha_2(t)\mathbf{A}^2$$

$$= (4e^t - 3e^{2t} + 2te^{2t})\begin{bmatrix} 1 & 0 & 0 \\ 0 & 1 & 0 \\ 0 & 0 & 1 \end{bmatrix} + (-4e^t + 4e^{2t} - 3te^{2t})\begin{bmatrix} 2 & 1 & 4 \\ 0 & 2 & 0 \\ 0 & 3 & 1 \end{bmatrix}$$

$$+ (e^t - e^{2t} + te^{2t})\begin{bmatrix} 4 & 16 & 12 \\ 0 & 4 & 0 \\ 0 & 9 & 1 \end{bmatrix}$$

$$= \begin{bmatrix} e^{2t} & 12e^t - 12e^{2t} + 13te^{2t} & -4e^t + 4e^{2t} \\ 0 & e^{2t} & 0 \\ 0 & -3e^t + 3e^{2t} & e^t \end{bmatrix}$$

References

The books listed here are the author's previous publications and were very useful in writing this book. They present many control engineering problems solved with MATLAB.

1. Ogata, K. *Modern Control Engineering*, 4th ed. Upper Saddle River, NJ: Prentice Hall, 2002.

2. Ogata, K. *System Dynamics*, 4th ed. Upper Saddle River, NJ: Prentice Hall, 2004.

3. Ogata, K. *Solving Control Engineering Problems with MATLAB*. Upper Saddle River, NJ: Prentice Hall, 1994.

4. Ogata, K. *Designing Linear Control Systems with MATLAB*. Upper Saddle River, NJ: Prentice Hall, 1994.

5. Ogata, K. *Discrete-Time Control Systems*, 2d ed. Upper Saddle River, NJ: Prentice Hall, 1995.

Index

A

Absolute value, 24
Ackermann's formula, 315, 319
   for observer gain matrix, 329–30, 332, 340
   for state feedback gain matrix, 317
Adj $(\lambda \mathbf{I} - \mathbf{A})$, 418
Angle condition, 152
Approximation of dead time, 184
Approximation of transport lag, 184
Array division, 25
Array multiplication, 23
Available colors in MATLAB plots, 37

B

Bandwidth, 268, 275
   MATLAB approach to obtain, 269
Bode diagram, 221–33

C

Cayley-Hamilton theorem, 316, 418
Characteristic equation, 27
   for the minimum-order observer, 340
Circle, 46

Comments:
   entering in MATLAB program, 19
Complete observability, 309
Complete state controllability, 306
Complex conjugate transpose, 13
Complex matrix, 13
Complex number, 12
Conditionally stable system, 176–78, 265
Conformal mapping, 170
Conjugate transpose, 11
Constant-gain loci, 170
Constant M circles:
   a family of, 253
Constant-magnitude loci, 251
Constant N circles:
   a family of, 254–55
Constant phase angle loci, 254
Constant ω_n loci, 165, 167, 169
Constant ζ loci, 165, 167, 169
Control systems compensation:
   frequency-response approach, 271–304
   root-locus approach, 186–220
Controllability, 305–06, 312
Controllability matrix, 307
Convolution, 30

MATLAB commands discussed in this book are listed under the heading **MATLAB commands.**

D

Dead time, 180
Deconvolution, 31
Design of quadratic optimal control system, 406
Design of quadratic optimal regulator system, 403
Detectability, 310
Diagonal matrix, 16
Dual problem, 327
Dual system, 328
Duality, 328

E

$e^{\mathbf{A}t}$:
    computation of, 420–24
Eigenvalue, 28
Eigenvalue vector **E**, 401
Eigenvector, 28
Equation for the asymptotes:
    in root locus plot, 160
Error equation:
    for minimum-order observer, 339
Estimator error, 326

F

Feedback compensation, 215
for expression, 106
Full-order state observer, 325–27, 333
    equation for, 329

G

Gain crossover frequency, 262, 265, 267
Gain margin, 263–67
    negative, 264
    positive, 264
Greek letters, 37

I

Identity matrix, 16
Impulse input, 119
Impulse response, 121
Initial condition,
    response to, 404
Inner loop, 378–79
Inverted-pendulum system, 49

K

Kalman, R.E., 305

L

Lag compensation:
    frequency-response approach to, 287–98
    root-locus approach to, 194–203
Lag compensator, 194–95, 287–89

Lag-lead compensation:
    frequency-response approach to, 298–304
    root-locus approach to, 203–14
Lag-lead compensator, 204, 212, 298, 301
Lagrange polynomial, 420
Lagrange's interpolation formula, 419
Lead angle, 274
Lead compensation:
    frequency-response approach to, 271–87
    root-locus approach to, 187–93
Lead compensator, 187, 279
    Bode diagram of, 283
Line types,
    in MATLAB plots, 36
Lines of constant damping ratio ζ, 165
Log-magnitude-versus-phase plane, 256
Log-magnitude-versus-phase plot, 250–62, 290
Logarithmic plots, 35
Logical operators,
    in MATLAB, 7

M

M circles, 251
M loci,
    constant, 253
magnitude condition, 152
Mathematical model of observer, 326
MATLAB approximation of dead time, 184

MATLAB commands:
A(i,j), 8
A(i,:), 8
A(:), 8
A(:,j), 8
[A,B,C,D] = tf2ss(num,den), 66–68, 80
abs, 2
abs(A), 25
acker, 2, 322
angle, 2
angle(A), 25
ans, 2, 21
atan, 2
axis, 2, 36
axis('equal'), 36
axis('normal'), 36
axis('square'), 36
bode, 2, 221
bode(A,B,C,D), 221, 228
bode(A,B,C,D,iu), 228
bode(A,B,C,D,iu,w), 221
bode(A,B,C,D,w), 221
bode(num,den), 221–23

bode(num,den,w), 221, 224
bode(sys), 221
c2d, 2, 66
clear, 2, 18
clf, 2, 155
clock, 18
colon, 7
computer, 2
conj, 2
CONT, 310–12
conv, 2, 154
corrcoef, 2
cos, 2
cosh, 2
cov, 2
ctrb, 2
d = poly(r), 65
date, 19
deconv, 2
det, 2
diag, 2
diag(A), 16
diag(diag(A)), 16
diag(x), 16
diag(0,n), 17
diag(0:n), 17
eig, 2
eig(A), 29
eig(A − B*K), 321
end, 2, 106
exit, 2
exp, 2
expm, 2
eye, 2
eye(n), 16
feedback, 2
filter, 2
fontangle, 39
fontname, 39
fontsize, 39, 90
for, 2
for loop, 106
format long, 2, 14–15
format long e, 2, 14
format short, 2, 14–15
format short e, 3, 14
freqs, 3
freqz, 3
[G,H] = c2d(A,B,Ts), 72
GmdB = 20*log10(Gm), 268
[Gm,Pm,wcp,wcg] = margin(num,
  den), 277, 280, 284, 299, 302
[Gm,Pm,wcp,wcg] = margin(sys),
  267–68
gram, 3
grid, 3, 410

grid off, 3
grid on, 3
gtext('string'), 35
gtext('text'), 93
help, 3
hold, 3
hold off, 3
hold on, 3
i, 3
im, 237
imag, 3
impulse, 3
impulse(A,B,C,D), 118
impulse(num,den), 118
impulse(num,den,t), 118
impulse(sys), 118
impulse(sys,t), 118
inf, 3
initial, 324
initial(A,B,C,D,[initial condition],t), 142
inv, 3
inv(A), 33
iu, 228
j, 3
K = acker(A,B,J), 322–23, 343, 347,
  360, 368
K = lqr(A,B,Q,R), 401, 406
K = place(A,B,J), 322–23
K_e = acker(A',C',L)', 341, 368
K_e = place(A',C',L)', 341
[K,r] = rlocfind(A,B,C,D), 170
[K,r] = rlocfind(num,den), 170–72
[K,P,E] = lqr(A,B,Q,R), 401
legend, 3
length, 3
linspace, 3, 11
load, 3
log, 3
log10, 3
loglog, 3, 35
logm, 3
logspace, 11
logspace(d1,d2), 222
logspace(d1,d2,n), 222
lqe, 3
lqr, 3
lqr(A,B,Q,R), 400
lsim, 3
lsim(A,B,C,D,u,t), 128
lsim(num,den,u,t), 128
lsim(sys,u,t), 128
lsim(sys,u,t,x_0), 128
lsim(sys1, sys2, . . . , u,t), 128
lyap, 3
mgdB = 20*log10(mag), 221
[mag,phase,w] = bode(A,B,C,D), 222

MATLAB commands (continued):
[mag,phase,w] = bode(A,B,C,D,iu,w), 222
[mag,phase,w] = bode(A,B,C,D,w), 222
[mag,phase,w] = bode(num,den), 222
[mag,phase,w] = bode(num,den,w), 221–22, 225, 227, 246
[mag,phase,w] = bode(sys), 222
[mag,phase,w] = nichols(A,B,C,D), 257
[mag,phase,w] = nichols(A,B,C,D,iu,w), 257
[mag,phase,w] = nichols(A,B,C,D,w), 257
[mag,phase,w] = nichols(num,den), 257
[mag,phase,w] = nichols(num,den,w), 257
[mag,phase,w] = nichols(sys), 257
margin, 4
max, 4
mean, 4
median, 4
mesh, 4
mesh(t,zeta,y'), 147–50
mesh(y), 145–46
mesh(y'), 145–46
mesh(zeta,t,y), 147–50
mesh plot, 44
min, 4
minreal, 4, 83
minreal(sys), 310–11
NaN, 4, 381–83
ngrid, 4, 258–59, 261
nichols, 4
nichols(A,B,C,D), 257
nichols(A,B,C,D,iu,w), 257
nichols(A,B,C,D,w), 257
nichols(num,den), 257
nichols(num,den,w), 257
nichols(sys), 257
norm, 26–27
num2str, 4
[num,den] = pade(1,2), 185
[num,den] = residue(r,p,k), 57
[num,den] = ss2tf(A,B,C,D), 69, 158
[num,den] = ss2tf(A,B,C,D,iu), 68, 70
[NUM,den] = ss2tf(A,B,C,D,iu), 71
[num,den] = zp2tf (z,p,K), 60, 64
nyquist, 4
nyquist(A,B,C,D), 236, 241, 243
nyquist(A,B,C,D,iu), 241
nyquist(A,B,C,D,iu,w), 236, 242
nyquist(A,B,C,D,w), 236
nyquist(num,den), 236
nyquist(num,den,w), 236

nyquist(sys), 236
OBSER, 310–12
obsv, 4, 310
ode23, 4
ode45, 4
ode113, 4
ones, 4
ones(m,n), 15
ones(n), 15
ord2, 4
pade, 4
parallel, 4
%, 19
pi, 4
place, 4, 322
plot, 4
plot(re,im), 240, 245
plot(t,y), 92
plot(x,y), 34
plot(z), 35
plot3, 42
polar, 4
polar(theta,r), 273
polar(theta,rho), 35
pole, 4
poly, 4
poly(A), 27
polyfit, 4
polyval, 4
polyval(p,s), 31
polyvalm, 4
pow2(t), 54
printsys, 4
printsys(num,den), 86
printsys(num,den,'s'), 57, 86
prod, 4
pzmap, 4
quit, 4
r = abs(z), 272–73
r = rlocus(A,B,C,D,K), 159
r = rlocus(num,den), 153, 156
r = rlocus(num,den,K), 157
r = roots(d), 65
rank, 4
re, 237
[re,im,w] = nyquist(A,B,C,D), 236
[re,im,w] = nyquist(A,B,C,D,iu,w), 236
[re,im,w] = nyquist(A,B,C,D,w), 236
[re,im,w] = nyquist(num,den), 236
[re,im,w] = nyquist(num,den,w), 236, 240, 244, 246
[re,im,w] = nyquist(sys), 236
real, 4
rem, 4
residue, 4, 55–56
[r,K] = rlocus(A,B,C,D), 153

[r,K] = rlocus(A,B,C,D,K), 153
[r,K] = rlocus(num,den), 153
[r,K] = rlocus(num,den,K), 153
[r,K] = rlocus(sys), 153
rlocfind, 4, 171
rlocus, 4
rlocus(A,B,C,D), 158
rlocus(A,B,C,D,K), 153
rlocus(num,den), 152
rlocus(num,den,K), 153
rmodel, 4
roots, 4
roots(poly(A)), 28
[r,p,k] = residue(num,den), 55–56, 58, 62–63, 95
semilogx, 4, 35
semilogy, 4, 35
series, 4
shg, 4
sign, 4
sin, 4
sinh, 4
size, 4
sgrid, 165
solution, 111, 377–79, 381–86, 389, 393
sortsolution, 385–87, 389–91, 394
sphere(n), 44
sqrt, 4
sqrtm, 4
ss, 4
ss(A-BK,eye(3),eye(3),eye(3)), 324
ss2tf, 4, 66
ss2zp, 66
std, 4
step, 4
step(A,B,C,D), 87, 98–99
step(A,B,C,D,iu), 99–100
step(num,den), 86
step(num,den,t), 86
step(sys), 86, 88
step(sys,t), 86
sys = feedback(sysg,[1]), 76
sys = feedback(sysg,sysh), 76
sys = feedback(sysg,sysh,+1), 76
sys = feedback(sysg,−sysh), 76
sys = parallel(sys1,sys2), 75
sys = series(sys1,sys2), 74
sys = ss(A,B,C,D), 74
sys = tf(num,den), 74
sys_min = minreal(sys), 84
sys_tf = tf(sys_ss), 82–83
subplot, 4, 324
subplot(m,n,p), 41
sum, 4
surf, 42
switch, 4

tan, 4
tanh, 4
text, 4, 35, 93
text(X,Y,Z, 'string'), 43
tf, 4
tf2ss, 4, 66
tf2zp, 4, 66
theta = angle(z), 272–73
title, 4
trace, 4
who, 4, 18
whos, 4, 18
xlabel, 4
[X,D] = eig(A), 29–30
y = impulse(num,den), 118
y = impulse(sys), 118
y = lsim(A,B,C,D,u,t), 128
y = lsim(num,den,u,t), 128–29
y = lsim(sys,u,t), 128
y = step(num,den,t), 86
y = step(sys,t), 86, 92
[y,t] = impulse(num,den,t), 118
[y,t] = impulse(sys,t), 118
[y,t] = lsim(sys,u,t), 128
[y,t] = step(num,den,t), 86
[y,t] = step(sys,t), 86
[y,x,t] = impulse(A,B,C,D), 118
[y,x,t] = impulse(A,B,C,D,iu), 118
[y,x,t] = impulse(A,B,C,D,iu,t), 118
[y,x,t] = impulse(num,den), 118
[y,x,t] = impulse(num,den,t), 118
[y,x,t] = step(A,B,C,D,iu), 88
[y,x,t] = step(A,B,C,D,iu,t), 88
[y,x,t] = step(num,den,t), 88
ylabel, 4
z = re + i*im = re$^{i\theta}$, 272–73
zero, 6
zeros, 6
zeros(m,n), 15
zeros(n), 15
zlabel, 4
zp2ss, 66
zp2tf, 4, 66
zpk, 4
[z,p,k] = tf2zp(num,den), 59–60

End of MATLAB commands

Matrix exponential, 32
Maximum overshoot, 115–17
Maximum phase lead angle ϕ_m, 274, 281
Mechanical system, 61, 91, 137, 345
Minimum-phase system, 175–76, 263
Minimal polynomial, 418–19
Minimum-order observer, 326, 335–45
Minimum-order state observer, 326, 335–45

Minimum-order observer-based controller, 356–57
Modified Nyquist path, 234
Multiple-loop systems, 235
Multiplication:
  array, 23
  matrix, 21

N

N circles, 251, 254
Negative-feedback system, 75
  Nyquist plot for, 249
Nichols chart, 256
Nichols plot, 250, 257–61
$n \times n$ faceted sphere, 44
Non-minimum-phase system, 173–76, 265–66
Non-minimum-phase transfer function, 230
Nyquist diagram, 234
Nyquist path, 234
Nyquist plot, 234
  for positive-feedback system, 248–49
  of system defined in state space, 241
Nyquist stability analysis, 234
Nyquist stability criterion, 234

O

Observability, 305, 308, 312
  condition, 308
  matrix, 310
Observation, 325
  error, 326
Observed-state feedback control system, 333–35, 338, 341, 351
  with minimum-order observer, 340–41
Observer, 325
Observer-based controller, 349–50
  transfer function, 350
Observer-controller transfer function, 350–51
Observer error equation, 334, 339
Observer gain matrix, 326
  MATLAB approach to determine, 341
Optimal control law:
  for quadratic optimal control problem, 398
Optimal regulator system, 396
Optimal sets of parameter values, 375
Orthogonality:
  of root loci and constant-gain loci, 169
Outer loop, 378–79

P

Pade approximation of dead time, 185
Parallel compensation, 215

Parallel compensated system, 215
Partial-fraction expansion, 55
Peak time, 115–17
Performance index, 395–98, 401
Phase crossover frequency, 263, 265, 267
Phase margin, 250, 262–68
  in log-magnitude-versus-phase plot, 263
  in polar plot, 263
  negative, 264
  positive, 264
PID controller, 295, 375
PID-controlled system, 376
Plotting multiple curves, 34
Point types, 36
Polar plot, 234
Pole-assignment technique, 313
Pole placement, 313
Pole-placement problem, 315
  solving with MATLAB, 321–25
Pole-placement-with-observer approach:
  to design systems, 345–49
Pole-zero configuration:
  of lag network, 194
  of lead network, 187
Polynomial evaluation, 31
Positive-definite matrix, 397
Positive-feedback system, 75
  Nyquist plot for, 248–49

Q

Quadratic optimal control problems, 395–408
  solution to, 397–408
Quadratic optimal regulator problems, 395
  solved with MATLAB, 400–06

R

Ramp responses, 122–27
Rectangular plot, 234
Reduced-matrix Riccati equation, 398–99
Reduced-order observer, 326
Reduced-order state observer, 325
Regulator poles, 315
Regulator system, 314, 317, 351, 358
Regulator system with observer:
  design of, 358
Resonant frequency, 268–69
  MATLAB approach to obtain, 269
Resonant peak, 268–69
  MATLAB approach to obtain, 269
  frequency, 277

Response to arbitrary initial
    condition, 136
Response to arbitrary input, 128
Response to exponential input,
    130–31, 133
Response to initial condition, 138–44, 324
Riccati equation, 401
Rise time, 115–17
Robust pole placement, 321
Root-locus plots, 151

S

Second-order observer, 346
Semicolon, 7, 10
Series compensation, 215
Settling time, 115–17
Spring-mass-dashpot system, 88
Square matrix, 33
Stability analysis, 235
Stabilizability, 306, 308
Standard second-order system, 85
State feedback gain matrix, 314
State observer, 325–26
    necessary and sufficient condition
        for, 328–29
Static velocity error constant K_v, 200, 279
Step response, 56

System with observed state feedback,
    339, 345, 353
Subtraction, 19
Surf plot, 45
Sylvester's interpolation formula, 419–24

T

Tachometer feedback control
    system, 109
Three-dimensional coil, 50
Three-dimensional plot, 42, 144
Transfer function:
    of observer controller, 352–54, 357,
        360–61, 363, 370, 372
Transfer matrix, 230
Transport lag, 180, 235
Transpose, 11

U

Unit-ramp response, 122–27
Unit-step response, 86–105
Unstable controller, 362
Unstable observer controller,
    360–62, 374

V

Velocity feedback system, 216